Back to Statistics

Due to the popularity of artificial intelligence today, scientists and practitioners tend to ignore the achievements of statistical methods. This book aims to re-establish statistical methods, especially for control practitioners, for whom the task of assessing control performance is important.

The author introduces the elements of statistical theory, including basic statistical concepts and the description of distributions. Extending the most common observations toward the tails and extreme statistics, he demonstrates the robust statistics approach that deals with the tails, followed by the description of methods that can visualize and bring forward statistical properties as long as the derivative issues. By addressing statistical issues of sustainability, extreme statistics, non-Gaussianity, L-moments, tail index and index ratio diagrams, the author aims to change traditional concepts and broaden the scope of available methods. Simulation case studies and real industrial cases are also presented.

The book will be of interest to site control engineers and scholars assessing control systems and process performance.

Paweł D. Domański was born in 1967 in Warsaw, Poland. He received the M.S. degree in 1991, the Ph.D. degree in 1996 and the D.Sc. degree in 2018, all in control engineering, from the Warsaw University of Technology. Since 1991, he has been working at the Institute of Control and Computational Engineering, Warsaw University of Technology. Besides scientific research, he has participated in dozens of industrial implementations of Advanced Process Control (APC) and optimization in the power and chemical industries. He is the (co-) author of three books and more than 150 publications. His main research interest is in industrial APC applications, control performance quality assessment and optimization. Currently, his research interest is shifting toward multi-agent modeling in the supply chain management environment.

Back to Statistics

Tail-Aware Control Performance Assessment

Paweł D. Domański

CRC Press is an imprint of the
Taylor & Francis Group, an **informa** business

Designed cover image: © Renata Domańska

First edition published 2025
by CRC Press
2385 NW Executive Center Drive, Suite 320, Boca Raton FL 33431

and by CRC Press
4 Park Square, Milton Park, Abingdon, Oxon, OX14 4RN

CRC Press is an imprint of Taylor & Francis Group, LLC

ISBN: 978-1-032-67174-1 (hbk)
ISBN: 978-1-032-99537-3 (pbk)
ISBN: 978-1-003-60470-9 (ebk)

DOI: 10.1201/9781003604709

Typeset in LM Roman
by KnowledgeWorks Global Ltd.

Publisher's note: This book has been prepared from camera-ready copy provided by the authors.

Invariably, to my beloved Princess

Contents

List of Figures

List of Tables

Preface

The history of statistics is long and characterized by many twists and turns. In the last decades of the twentieth century, statistics celebrated the peak of its popularity, contributing to the success of modern minimal variance strategies, predictive and adaptive control, Kalman filtering and so forth. A wide variety of interesting solutions, algorithms and observations emerged during that period.

From the point of view of control theory, these developments focused on exploiting the assumption regarding Gaussian properties of signals. Alternative, non-Gaussian solutions developed in other fields of science, such as what happened with Extreme Value Theory in life sciences.

Now, the wheel of history has turned and statistics is losing popularity to artificial intelligence and machine learning methods. Why consider the stochastic properties indirectly, having access to a huge amount of data assuming that black boxes of machine learning themselves will find the solution hidden in the data and in a way only known to themselves? This does not require any statistical data analysis, which thus becomes obsolete.

It's hard to figure out why this is happening. Maybe statistics is treated as too complicated, maybe unnecessary or just unfashionable and unpublishable.

Certainly, there is a grain of truth in each of the above hypotheses. I, on the other hand, would like to raise another aspect, i.e. the growing gap between theoretical research and industrial reality.

Anyone who has had the opportunity to work both academically and on industrial projects recognizes the gap between the academic world and what is encountered at the industrial site. From the statistics perspective, academic solutions in the main assume Gaussian properties of signals. This also applies to the specific task of control performance assessment. However, the implementation of real projects shows that such an approach is not adequate.

This study aims to address this issue, which we observe in industrial signals in the form of their probability distribution tails. By the way, it is interesting how rarely control engineers draw a histogram. Maybe that's why the tails go unnoticed and the normal standard deviation (or variance) is still the only measure of data variation. And it is the non-Gaussian statistical methods that can practically take into account and use these properties. The book aims at bringing statistics back into the control picture.

In line with the above, this book gathers in one place a variety of statistical tools, methods and concepts that provide the engineer with adequate tools to solve the problems encountered during their control performance assessment task.

The accompanying descriptions are not meant to be fully theoretical. It is about gathering in one place and describing the potential inherent in statistical methods, which should result in a tail-aware control performance assessment.

Writing a book is a strange creative process, which occupies not only time but also the head. At this point, I would like to thank Betacom SA for allowing my head to focus and, as a result, making this book to appear.

This text, though signed only with my name, was not created in a closed, hermetic environment, without the influence of other people. It is the result of everyday conversations, reading and listening. It is the result of interactions with a huge number of people. And, in fact, it is very difficult to assess and list who, when and to what extent contributes to this work. And, in such a case, it is always easy to make an unintentional mistake and leave someone out or list them in the wrong order.

In short, I thank you all!

Warsaw, December 2024 *Paweł D. Domański*

Acknowledgments

The author acknowledges Grupa Azoty, Zakłady Azotowe "PUŁAWY" S.A. for providing the data.

I

Introduction

CHAPTER 1

The scope of the book

STATISTICS is a strange phenomenon. It exists, at least because people talk about it and teachers do lectures about it. However, we hardly believe in statistics, as we cannot see it and can hardly feel it. Nonetheless, once we meet statistics on our way, we express sincere surprise that it really exists being often hit hard by its consequences. And then we ask a naive question – why?

Actually, many legends have grown up around statistics. People's superstitions proclaim that it's difficult. Students consider it's redundant and of little use. Its glory days (60s and 70s) are long gone, and contemporary researchers are fascinated by other glistening areas, like artificial intelligence or chat-bots. Indeed, from time to time we are surprised by some unusual phenomenon or behavior. However, we try to forget about it as soon as possible and get over it, trying to live as usual. The question of why is rarely asked, and even more rarely we try to understand the answer and draw conclusions from it. This is our nature and complaining won't change anything.

We prefer to stay in our comfort zone, and any outlying experience is alien and dangerous. Control engineering, as one of the manifestations of human action, is no exception. The main success and the main interest of control engineering lies within the most common situations. We are happy with a control system, if it can keep the water level in the retention reservoir most of the time close to the demanded reference value (setpoint) and not exceeding the safety limitation (constraints). This is the main focus, the ultimate goal of the control system. Control engineer does his job to satisfy the average demand, which should work as designed for ever and a day. If there is a risk of any unforeseen or hostile situation, his work (controller tuning) tends toward *the worst*

DOI: 10.1201/9781003604709-1

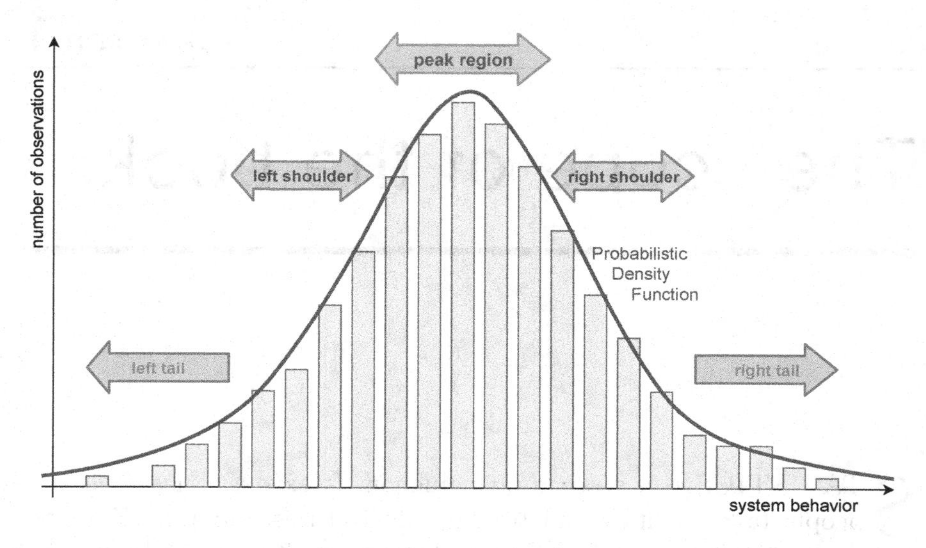

Figure 1.1: Observations: histogram, probabilistic density function and distribution regions

case. Control engineers are very rarely interested in unusual or rare situations, relegating this issue to safety and protection systems. As control engineering tasks focus on the average operation, unusual situations are very rarely taken into account.

Let us consider a simple statistical analogy. Sample control system observations, as any kind of a population data, may be visualized with a histogram; see Fig. 1.1. Once we have plotted the histogram, we may try to fit any empirical probabilistic density function (PDF) to it. Following above stipulations, control engineering focuses on the peak region, partially covering to some extent the shoulders (unconsciously and unintentionally), while the tails remain outside of his scope of interest [92].

As control engineering aims at average behavior, the methods that assess its quality do the same. Control performance assessment (CPA) by default focuses on the same – on average performance of the control system. Any anomalies or unusual behavior beyond the typical limits are generally considered erroneous, hardly noticed and not taken into account. Following this line of thinking, CPA methods focus on the peak region neglecting outlying behavior occurring in the histogram tails.

The above observations result in significant consequences, both for the control engineering and for the statistics as a mathematical apparatus that allows to reflect these unusual phenomena. Taking mental shortcuts, control engineering remains a Gaussian world whether it is explicitly stated or not. The wealth of statistical solutions and the statistical thinking remain on the fringes of mainstream automation interests. This work tries to reveal hidden or forgotten doors bringing statistics back to the control engineering.

The agenda of this narration is intended to meet the above task. The first part introduces into the subject and gives the rationale for this work. It presents in a sense personal combination of various factors originating from scientific, industrial and teaching experience. Part II describes the elements of statistical theory, which might be useful for a control engineering practitioner and a researcher. This part covers basic statistical notions, description of the distributions, which might be useful for a control engineer. The story about the most common observations is extended toward the tails and extreme statistics. The robust statistics approach that deals with the tails is followed by the description of methods that allow to visualize and bring forward statistical properties. The derivative issues (from control engineering perspective), like stationarity, homogeneity, causality, fractionality, tail indexes, etc. complete the story. Regression, though being a huge subject, is just mentioned. It isn't considered as a goal, but as a means to achieve it.

Next, Part III presents the control perspective, i.e. how to assess and how to measure performance of a control system. Statistical approach to the CPA, though historically the first, currently concerns only a small part of the whole issue and as such does not fit into the mainstream of the research interest. Statistical approaches that allow to measure a single loop performance are extended toward multi-loop and multivariate cases. The notion of the control system sustainability is then introduced as a novel research direction, which has not been considered so far. Nevertheless, in the era of industry 4.0 development, it begins to play an important role. Novel approaches that measure and visualize control system performance and its sustainability are introduced and analyzed.

The last part binds the earlier considerations together in practical case studies. The presentation of the proposed systematic statistical control performance and its sustainability assessment is followed by simulations and real industrial case studies.

The author hopes that reading this book will at least allow to notice that statistics exists, matters and is useful. It is hard to believe that statistics will suddenly become the star of the season in the multitude of much more attractive and trendy research areas. Nevertheless, the mere reminder of its existence and the collection of the statistical notions in one place makes sense and may be useful.

GLOSSARY

PDF: Probabilistic Density Function

CPA: Control Performance Assessment

CHAPTER 2

Research also owns its distribution

CONTERPORARY research is governed by its own rules and one could say that it lives its own life. It is naive to think that the science selflessly strives for the humankind future and happiness. It's not a place and time to discuss either philosophical or mundane questions of science and research. In contrary, although from the point of view of the books, it is most natural to use statistical method to see how the research behaves and whether we can discern any statistical properties in science.

One of the main products of the scientific performance is a publication. Each researcher, apart from his investigations, writes books, journal papers and submits conference speeches. And what's worse, he's obligated to do so. Since publication does not equal publication, we must try to obtain some quantifiable measure to compare them. We measure them with the number of citations, and let us do not discuss the matter of real and self-citations. We end up at measuring the number of publications, the number of citations or some combinations of them, like the Hirsch index [145].

As the citations appear systematically in time, they are available in the form of time series. If so, we can analyze them in the same as any other data. Let's do a simple exercise and try to perform the analysis of citation data. There is no single universal source of the number of citations. There are paid databases, mostly dedicated to the peer-reviewed publications, like the most common Clarivate Web of Science™ platform (`https://mjl.clarivate.com/home`) or SCOPUS (`https://www.scopus.com/`). Apart from commercially paid databases, one may get

DOI: 10.1201/9781003604709-2

access to the general open access platforms like Google Scholar (`https://scholar.google.pl/`), SCISPACE (`https://typeset.io/`) or Dimensions (`https://app.dimensions.ai/discover/publication`). It is interesting notice, what actually does not matter for our exercise, that each of the above sources delivers different numbers. We will use the SCOPUS as it allows to download data about aggregated number of publications according to some selected keywords. The database consists respective information about articles, chapters, proceedings, preprints, monographs and edited books. Collected time series include data from years 1974–2022 and they are collected according to some keywords covering common research areas of control engineering.

Fig. 2.1a shows the yearly number of publications for the selected keywords of the control engineering research: control theory, control systems, adaptive control , fuzzy control, model predictive control, PID control and process control. Actually, these are the keywords, which relate to the area of interest of this book.

Let's focus on proper data analysis, regardless of what they mean. First of all, we see that the data are trended, with the continuously increasing trend. The trend behind data might be monotonous, though the increase as such exhibits several trend breaks, which is better visible in Fig. 2.1b, which shows the same data, but in the greater scale.

Analogous data for two keywords from the currently very popular field of Artificial Intelligence (AI), i.e. Artificial Neural Networks (ANN) and Deep Learning (DL), are shown in Fig. 2.2. We see that these data look quite different. We see almost exponential growth and unprecedented popularity of AI issues in the research. Especially, the growth loos spectacular for the DL (Fig. 2.2b), which is even more stunning,

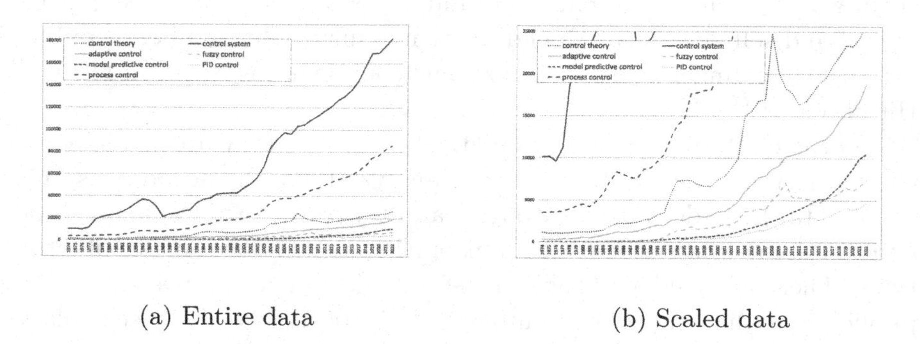

(a) Entire data

(b) Scaled data

Figure 2.1: Time series of publications in the area of control engineering

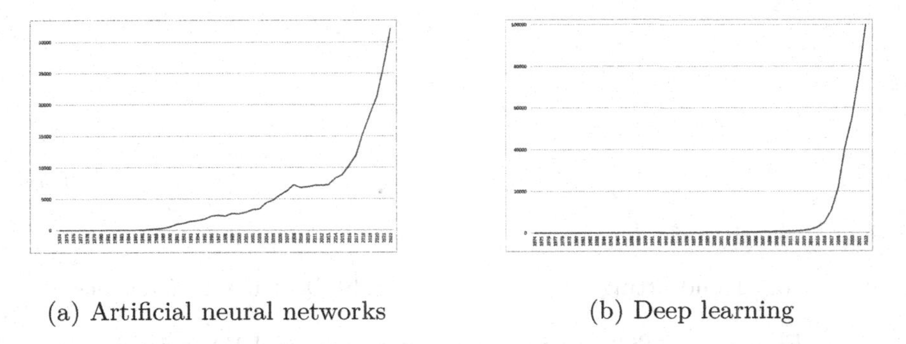

(a) Artificial neural networks (b) Deep learning

Figure 2.2: Time series for publications in the area of AI

once we notice that before mid 80s the number of publications was close to zero with only single papers. From the data analysis perspective, we notice that still data are affected by some trend with a hardly defined shape.

Finally, Fig. 2.3 shows the data related to the statistics (statistical model, robust statistics, probabilistic distribution) with added the area of control performance assessment. We also see increasing trends, however of different gains, while some research areas start to decrease.

The above summary of the publications time series observations actually constitutes the first step in any data analysis – visual data inspection. We see that the data is subject to some unknown trend. We may say that the data is trend non-stationary. Their further analysis should allow to remove observed data non-stationarity. It can be achieved in two ways. One, the most natural, though neither simple, nor evident, is to remove the trend. But the trend generally exhibits unknown

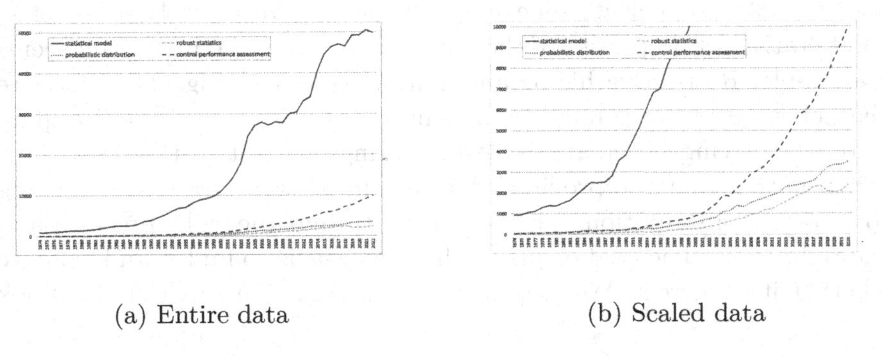

(a) Entire data (b) Scaled data

Figure 2.3: Time series for publications in statistics

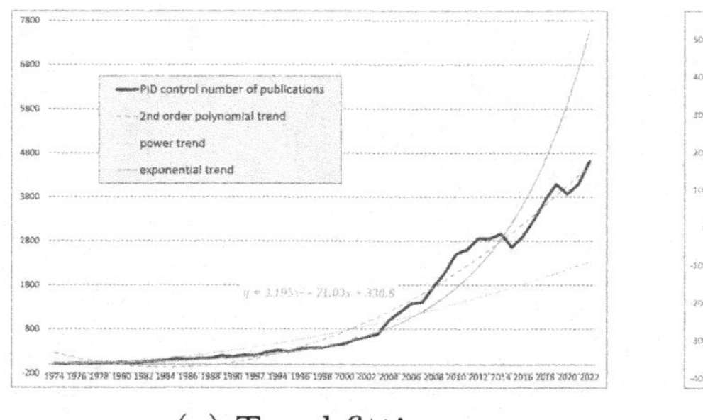

(a) Trend fitting

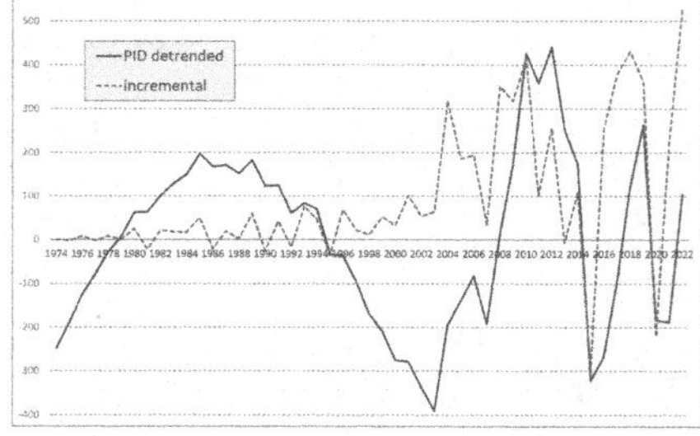

(b) Detrended time series

Figure 2.4: Time series for publications on PID control

properties and we do not know how to remove it (what should be removed). Coming back to our exercise, we may plot data for one of the variables, let's say the number of "PID control" publications. Fig. 2.4a presents possible trends that might be fitted to data: 2nd order polynomial, exponential or power trend. We see that it's hard to decide, which one better fits.

Another approach of coping with trended data is to change the original data $x(k)$ into its incremental time series $\Delta x(k) = x(k) - x(k-1)$. Once we do not exactly know what the trend to use, the incremental data approach is useful. Fig. 2.4b shows respective detrended data comparing two approaches: time series of the incremental data and their detrended version (using 2nd order polynomial). It should also be noted that the choice of detrending has a significant impact on what we receive and on the final result. Common sense and well-thought-out decisions are essential for the correctness of the analysis.

Obtained data are noisy, and it's hard to determine any visible pattern. We may only say that we do not observe any periodic oscillations and the data represent some unknown noise. Once we have no clue, what statistical properties of the data are, plotting the histogram seems reasonable. Respective histogram chart is shown in Fig. 2.5. Data are characterized by some unknown stochastic process, and we notice quite a lot of outlying observations. Anticipating the rest of the book, it is worth noting that the provided data are in no way arranged according to a normal distribution. Thus, we need to find the right function and the right tools. The goal of the further analysis is to find it and validate whether it is correct. We stop now the exercise. The goal of this book

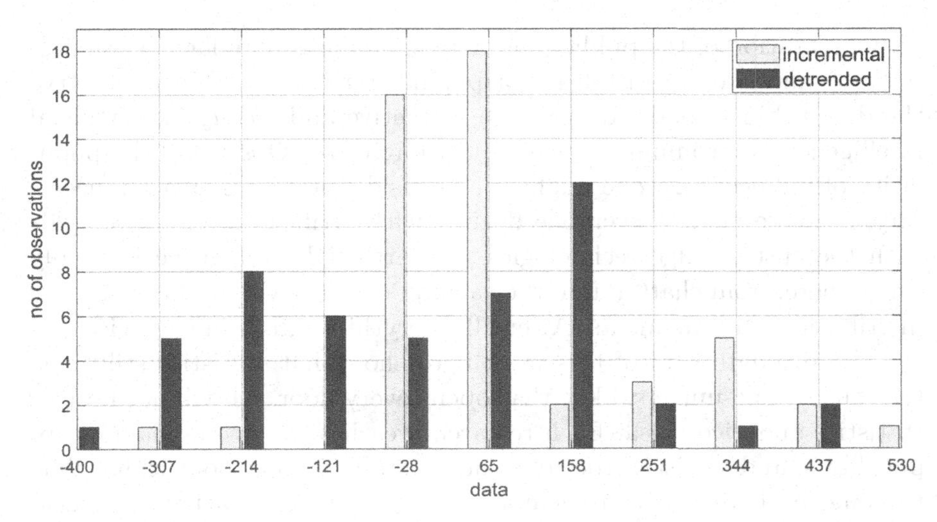

Figure 2.5: Histogram of publications on PID: incremental and detrended

is to explain how we should conduct such an analysis for any data that may come to our hands during control projects.

There are two main outcomes of this exercise. At first, we have learned the assumptions and framework of statistical analysis for a given time series. The main steps are: collecting the data, their visual inspection (and possible verification of completeness of data), detection of the trend or possible oscillatory behavior, data decomposition into the trend, oscillations and the residuum noise component, and finally the analysis of the stochastic processes behind the noise. The draft framework given above only indicates the main phases of the analysis. One of the goals of this book is to deliver the hints and the tools for conducting the analysis.

The second outcome is connected with the subject of this exercise itself. We see that even the scientific publication data are driven by some unknown, but existing stochastic process. We observe trends that an outsider might associate with fashions (why not), and we have some stochastic process that we could try to characterize using some distributions and probabilistic density function. It's not strange. Scientific research is driven by thousands or millions of humans. It is the process and we observe its realization in the form of the time series, like the number of publications or citations. The research is not an exception. It has its own distribution, as any other human activity, like the trade, the births and the deaths. This book shows how to use statistics in the time series analysis and other activities common in control engineering.

Observation of the publication data shows one more fact. Statistical research and statistics itself is not popular right now. We even observe the diminishing trends. It is not so attractive and *trendy* like artificial intelligence, data mining, machine learning or so. Observing the popularity of publications, one might even consider statistics as an unnecessary, obsolete field. If everyone dealing with artificial intelligence deals with the cutting-edge technologies, and artificial intelligence is simply everywhere, from chatGPT to the vacuum cleaner, why bother with any distributions and moments. After all, everything exists in statistics.

The ultimate goal of this work is to show that statistics still matter. There is no universal key that opens every door. AI will not replace statistics and vice versa. Each research area has its own domain of applicability, and statistics too. If we accidentally forgot about it, it's time to come back to it and remember what it can do. Statistics is back, hopefully in a new, tail-aware version.

GLOSSARY

AI: Artificial Intelligence

ANN: Artificial Neural Networks

DL: Deep Learning

DM: Data Mining

ML: Machine Learning

CHAPTER 3

Attention! Outliers!

OUTLIERS existed, exist and will exist for ever. They are a visible manifest of the beauty of the mother nature. This book is about the tail-aware perspective of the statistics. These are the outliers, which cause the tails and we cannot forget about them. If so, this study will practically constantly revolve around this concept and, for the sake of order, it should be introduced to the reader at the very beginning.

Outlier in control data is just a strange and unexpected observation, which differs from the others. For those readers who deal with real data, this is everyday life from which there is no escape. These are strange, sometimes occasional, sometimes frequent events that deviate from our expectations – from normality, however it is understood. Sometimes we don't notice them and sometimes we can't escape them. They just are. In real life we witness tornadoes, hurricane winds, extreme rainfall, droughts and floods, pandemic outbreaks; by the way, more and more often lately. Perhaps even so often that they cease to be unusual and become an everyday occurrence – normality. And although historically they occur sporadically, humanity remembers them for their serious consequences. Perhaps unusual values in the data do not have such devastating consequences, but if some abnormal behavior in the data is not the result of a simple system error but becomes the result of a cyber attack, the perspective immediately changes.

Depending on the research context and field of knowledge, it is called differently: outlier, discordant, oddity, anomaly, contamination, deviant, exception, aberration or fringelier that represents "*unusual events which occur more often than seldom*" [343].

The fact of being rare might cause them to go unnoticed. Unfortunately, there is another feature associated with the outlier. The

DOI: 10.1201/9781003604709-3

relevance of such an event might be abnormally large carrying a disproportionately high cost. Sporadic high-magnitude earthquakes and tsunamis, devastating tornadoes, damaging earthquakes or volcano eruptions, economic crises or pandemic outbreaks are extremely rare events from the historical perspective, but are well remembered by mankind for their destructive effect. Outlier is often called by other names, such as: anomaly, deviant, oddity, contamination, aberration, an exception or fringelier reflecting "*unusual events which occur more often than seldom*" [343]. Whatever the name is, they are rare and of an unknown origin.

The first literature notice of the outlier defined Dixon [74] in terms something of "*dubious in the eyes of the researcher*", while Wainer [343] named them *contaminants*. Hawkins [138] formulated the common definition denoting it as "*an observation which deviates so much from other observations as to arouse suspicions that it was generated by a different mechanism*". Other researchers have proposed different definitions, such as Johnson [177], who called them "*as an observation in a data set which appears to be inconsistent with the remainder of that set of data*", and Barnett [16] saying that "*an outlying observation, or outlier, is one that appears to deviate markedly from other members of the sample in which it occurs*".

Outliers can manifest themselves in many different forms. These may be single values in a time series that deviate at one or more moments. These may be entire pieces of data that change the properties of the series over specific time intervals. Atypical values may appear in one or many dimensions, in time and space. They have different forms.

However, the observer always faces the same dilemma. The simplest situation occurs when they are simply unnoticed for some reason. In such a situation, we are at least justified, although the fact that we cannot see something does not mean that it does not exist. However, from a reaction standpoint, there was simply nothing that could be done about the outlier itself.

The situation changes dramatically when we suddenly realize that something is wrong. In such a situation, we have two options: either pretend that nothing is happening (even though this is ignorance) and do nothing, or take some action.

It's tempting to pretend that everything is as usual, normal and safe as if nothing happens. We close our eyes and do nothing, hoping that the anomaly will just disappear itself, though it's not a solution. Our ignorance does not excuse us, but we often behave like that. People hate to leave their comfort zone.

The action itself can be of two types. The simplest solution is to simply remove any unusual observations so that they do not clutter the otherwise perfect data. This approach has a centuries-old tradition and is very common. That was Bernoulli (1777), who pointed out the practice of deleting them. In 19th century, Boscovitch proposed their removal through the *ad hoc adjustments*, in the same way as Pierce (1852), Chauvenet (1863) and Wright (1884). This practice is still alive and suggested [54]. However, it may turn out that these outliers are important and convey some information that requires action.

This book is intended to address a situation not tainted by ignorance, in which we are aware of the existence of unusual values in the data. It was Legendre (1805), who recommended not to exclude the extreme observations "*adjusted too large to be admissible*". Bessel and Baeuer (1838) claimed that their removal makes the data to become artificial.

We can identify, name and evaluate them. And finally, we can make an informed decision about what to do next. If our assessment indicates an error, then we can calmly remove such elements and our action is the result of a rational decision-making process. If we determine that the detected unusual values are significant, we take appropriate actions resulting from the current situation.

Statistical methods, which form the core of this book, are intended to support this decision-making process.

> The existence of outliers is indisputable. Nevertheless, both scientists and practitioners tend either to overlook them or just to remove them. Not noticing doesn't testify positively about us. Of course, there are many situations when their removal is justified, because they are simply errors. However, it is also common to find cases where such unusual behavior is a full-fledged observation and, more importantly, can be useful.

II

Power of statistics

CHAPTER 4

Basic notions

A STORY OF STATISTICS and an introduction into it may be written in different ways. The are many books with various approaches. This narrative is dedicated to practicing control engineers, for whom the story start with a given time series.

Let's assume that we have some discrete time set of observations, denoted as $\{x(i)\}^T$, where $i = 1, \ldots, N$ and N denotes the number of observations in the given dataset.

Statistical notions presented in the book are visualized with the same sample datasets. All time series have the same length $N = 10000$. Two of them, denoted as X_1 and X_2 are artificially generated using random number generators. The X_1 is produced as the normally distributed random numbers, while the X_2 follows heavy tailed α-stable distribution. Three other datasets, named X_3, X_4, X_5, originate from the real industrial processes. Fig. 4.1 presents the behavior of this data for first 1000 samples.

We notice various properties of the data. Simulated time series X_1 and X_2 behave according to our expectations with higher variations and rapid changes in observations of X_2 data. Simulated data look definitely different and it is difficult to classify them based only on their appearance. We may observe some noisy and periodic behavior with sudden changes in time series appearance observed for X_5 data.

Once we get time series data, we may start to talk about their basic properties. We already know the length of data denoted by N. Next, each time series is characterized by some properties. During the analyzes, which we conduct in the control engineering projects we mostly try to investigate basic statistical properties of data. Minimum and maximum

DOI: 10.1201/9781003604709-4

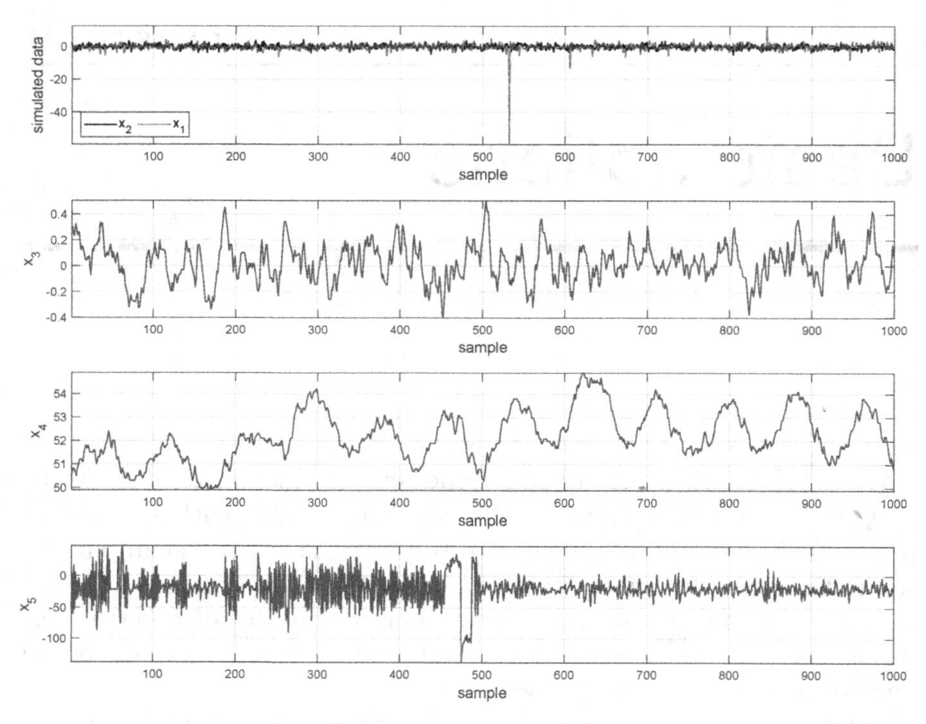

Figure 4.1: Sample time series data for $i = 1, \ldots, 1000$

values are the simplest numbers describing the data

$$\text{MIN} = min\left(x(i)\right), \tag{4.1}$$

$$\text{MAX} = max\left(x(i)\right). \tag{4.2}$$

In case of five sample datasets we obtain the following numbers, which are shown in Table 4.1. For further analysis, we have to introduce the notions statistical moments. Section 4.1 introduces basic definitions of moments, while further ones describe L-moments in Section 4.2 and other alternative estimators in Section 4.3. This chapter finishes with a discussion what to do in case of a small and very small number of data.

Table 4.1: Minimum and maximum values for sample data

		X_1	X_2	X_3	X_4	X_5
MIN	x_{min}	-5.579	-59.62	-1.111	48.90	-137.56
MAX	x_{max}	5.940	383.01	0.891	54.90	46.29

4.1 MOMENTS

In case of control engineering time series we deal with a discrete-time data and therefore the presentation is limited to the discrete versions. A random variable is described sufficiently precisely by its probability distribution. However, practical considerations determine the need to find numerical characteristics of the distribution, because these descriptions are short and enable quick comparison of distributions and data. A moment of order r $(r = 1, 2, \dots)$ of a random variable $\{X_i\}$, $i = 1, \dots N$ about an arbitrary point A is denoted as

$$\mu_r' = \frac{1}{N} \sum_{i=1}^{N} (x_i - A)^r . \tag{4.3}$$

The rth moment around the origin, i.e. in case of $A = 0$ is called the raw moment and is defined as

$$\mu_r' = \frac{1}{N} \sum_{i=1}^{N} (x_i)^r . \tag{4.4}$$

An arithmetic mean, frequently denoted as μ, constitutes the first raw moment with $r = 1$, i.e. the sum of all data x_i divided by the number of points N

$$\mu_1' = \mu = \frac{1}{N} \sum_{i=1}^{N} (x_i) . \tag{4.5}$$

Similarly, we obtain the second raw moment γ_2, assuming $r = 2$

$$\mu_2' = \frac{1}{N} \sum_{i=1}^{N} (x_i)^2 . \tag{4.6}$$

The higher order raw moments, like μ_3' and μ_4' are defined in a similar way. The moments of a random variable $\{X_i\}$ about its arithmetic mean μ are called central moments. Thus, the central moments of order $r = 1, 2, \dots$ are defined as

$$\mu_r = \frac{1}{N} \sum_{i=1}^{N} (x_i - \mu)^r . \tag{4.7}$$

The first and second central moments we obtain substituting $r = 1$ or $r = 2$ into (4.7), respectively. As it's shown below the first central moment is always equal to zero, i.e. $\mu_1 = 0$

$$\begin{aligned}\mu_1 &= \frac{1}{N}\sum_{i=1}^{N}(x_i - \mu)^1 \\ &= \frac{1}{N}\sum_{i=1}^{N} x_i - \frac{1}{N}\sum_{i=1}^{N} \cdot 1 \cdot \mu \\ &= \mu - \frac{N}{N}\cdot\mu \\ &= 0.\end{aligned} \tag{4.8}$$

The second central moment with $r = 2$ reflects the variance and is defined as

$$\begin{aligned}\mu_2 &= \frac{1}{N}\sum_{i=1}^{N}(x_i - \mu)^2 \\ &= \frac{1}{N}\sum_{i=1}^{N} x_i^2 - \mu^2 \\ &= \mu_2' - \mu^2.\end{aligned} \tag{4.9}$$

Equation (4.9) shows the relationship between the second central moment and raw moments of order 1 and 2. The second central moment is also called the variance $\mathrm{Var}(x)$, while the standard deviation denoted as σ or γ_2 is the square root of the variance

$$\sigma = \gamma_2 = \sqrt{\mu_2},\ \mu_2 = \sigma^2. \tag{4.10}$$

The first two moments are frequently supplemented by the third moment μ_3 related to skewness and the fourth one μ_4, which is directly related to the kurtosis.

$$\mu_3 = \frac{1}{N}\sum_{i=1}^{N}(x_i - \mu)^3 = \mu_3' - 3\mu_1'\mu_2' + 2(\mu_1')^2 \tag{4.11}$$

$$\mu_4 = \frac{1}{N}\sum_{i=1}^{N}(x_i - \mu)^4 = \mu_4' - 4\mu_1'\mu3' + 6(\mu_1')^2\mu_2' - 3(\mu_1')^4 \tag{4.12}$$

Formally, one would require some kind of normalization of the moments to make it scale invariant. Standardized moments of a probability

Table 4.2: First moments calculated for the reference time series

central moments		X_1	X_2	X_3	X_4	X_5
mean	$\mu = \mu'_1$	0.007	-0.007	0.005	51.79	-20.17
variance	μ_2	2.032	17.939	0.023	0.949	126.7
standard deviation	$\sigma = \sqrt{\mu_2}$	1.43	4.24	0.150	0.974	11.25
3rd moment	μ_3	0.023	73.60	-0.064	0.003	-0.407
4th moment	μ_4	2.99	6694.6	6.285	2.592	10.14

distribution offer such a possibility making the normalization by a power of the standard deviation. The first standardized moment equals to zero, the second equals to 1, and the higher ones are defined as follows:

$$\tilde{\mu}_1 = \frac{\mu_1}{\sigma^1} = 0, \tag{4.13}$$

$$\tilde{\mu}_2 = \frac{\mu_2}{\sigma^2} = 1, \tag{4.14}$$

$$\tilde{\mu}_3 = \frac{\mu_3}{\sigma^3}, \tag{4.15}$$

$$\tilde{\mu}_4 = \frac{\mu_4}{\sigma^4}. \tag{4.16}$$

The moments define properties of the probabilistic distribution of data x_i. The first moment is an example of the shift (location) measure, the second moment reflects variability, the third one data skewness, while the fourth one is concentration. Table 4.2 presents central moments evaluated for the considered benchmark time series.

The exemplary visualization of the shift and the scale impact on the distribution shape is presented in Fig. 4.2.

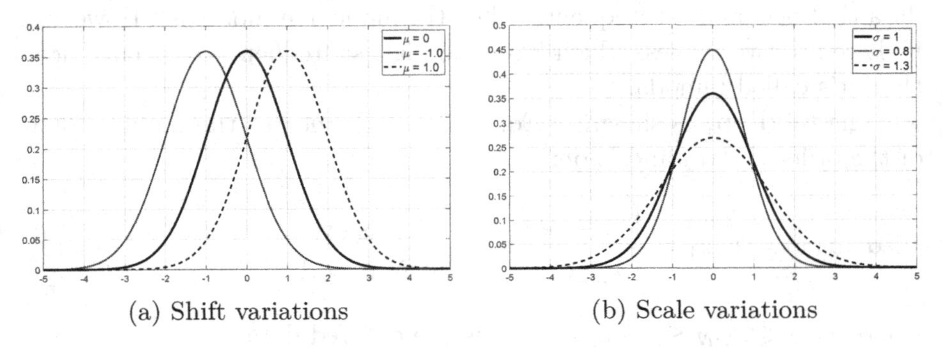

(a) Shift variations (b) Scale variations

Figure 4.2: Impact of the shift and the scale on the distribution function

Apart from the above formulations, there exist more measures and estimators of the location, variability, skewness and concentration.

4.1.1 Shift – the measure of central tendency

The shift is the measure of central tendency. It is used to represent the central value of a dataset. Arithmetic mean (4.5) can be substituted with other shift measure formulations. **Harmonic mean**, denoted as μ_{H}, which is presented in Eq. (4.17) is one of the three classical Pythagorean means and we use it in specific situations, when we need to analyze the rate of change of a given phenomenon.

$$\mu_{\mathrm{H}} = \frac{N}{\sum_{i=1}^{N} \frac{1}{x_i}}. \tag{4.17}$$

Geometric mean (4.18), denoted as μ_{G}, is the third Pythagorean mean. Due to the multiplications and the complications related to negative numbers not having real roots the use of the geometric mean is often restricted only to the positive datasets x_i.

$$\mu_{\mathrm{G}} = \sqrt[N]{\prod_{i=1}^{N} x_i}. \tag{4.18}$$

Please remember that only one average measure should be used for each case, because only one is appropriate and the others lose their meaning. Apart from above classical time series position factors we may use positional measures of location.

The **mode** $\overline{x}_{\mathrm{mo}}$ is such a value in the dataset that occurs the most frequently. The mode must be a single value. If there is no single value in data that occurs most frequently, then the mode does not exist. However, if there are two values, which have the highest frequency of occurrence, than it's called bimodal.

The **median**, as shown in Eq. (4.60), is defined as the middle point of the ordered data and is denoted as $\overline{x}_{\mathrm{med}}$.

$$\mathrm{median}\,(x) = \overline{x}_{\mathrm{med}} = \begin{cases} \frac{x_{\frac{N}{2}:N} + x_{\frac{N}{2}+1:N}}{2} & \text{if } N \text{ is even,} \\ x_{\frac{N+1}{2}:N} & \text{if } N \text{ is odd,} \end{cases} \tag{4.19}$$

where $x_{1:N} \leqslant x_{2:N} \leqslant \cdots \leqslant x_{N-1:N}$ is the ordered data.

The definition of the median is linked with the notion of quartiles. The **quartiles**, i.e. the first quartile Q_1 (25th percentile), the second Q_2

Table 4.3: Selected shift estimates for the reference time series

position estimators		X_1	X_2	X_3	X_4	X_5
arithmetic	μ	0.0071	-0.0071	0.0046	51.789	-20.173
harmonic	μ_H	0.8751	-3.2686	0.4744	51.771	-19.889
geometric	μ_G	—	—	—	51.780	—
mode	$\overline{x}_{mo}$	-5.5794	-59.6233	-1.1107	51.900	-21.670
median	$\overline{x}_{med}$	-0.0086	-0.0465	0.0040	51.800	-20.275
25th percentile	Q_1	-0.9545	-1.0065	-0.0870	51.100	-26.297
75th percentile	Q_3	0.9660	0.9448	0.0919	52.500	-14.145

(50th percentile) and the third one Q_3 (75th percentile), divide the data into four equal parts. Thus, there is approximately the same number of data in each of four sections.

Let's notice that the second quartile is exactly the median $Q_2 = \overline{x}_{med}$. Table 4.3 shows how these shift measures estimate position for the given reference data. We may notice that the geometric mean exists only for the time series X_4 because it has all positive observations. The mode value may be sometimes tricky and misleading. We may also see that one should be cautious with the harmonic mean as it may cause a hesitation

The main advantage of the mean is that it uses all the observations from the dataset, so we call it efficient in a statistical sense. The main shortcut of it is that the mean estimator is biased by the outlying observations. In contrary, the median is robust against the outliers, but it does not use all the individual observations, so it is not statistically efficient.

4.1.2 Scale – the measures of variability

Measures of variations, also called dispersion, spread or variability, describe the spread of the data. Variation and standard deviation are considered as classical dispersion measures. Another option is to use the **Average Absolute Deviation** (AAD) shown in Eq. (4.20), also called the mean absolute deviation

$$d = \text{AAD} = \frac{1}{N}\sum_{i=1}^{N} |x_i - \mu| \,. \tag{4.20}$$

The AAD uses absolute values instead of squares to circumvent the occurrence of negative differences between the observations and the mean. We may notice that there exist several variants of average absolute deviation. One option is to change the data mean estimator μ, as in this place one may use any of the presented above shift measure, i.e. not only arithmetic mean, but also harmonic, geometric, the median or the mode. In this case we obtain the measures called the mean absolute deviation around mean, around median, or around mode.

In a similar way, we may play with the average grouping measure as well. The most commonly used variants are the so-called MAD, i.e. **Median Absolute Deviation** shown in Eq. (4.21) and the **Median Absolute Deviation Around Median** (MADAM)

$$\begin{aligned} \mathrm{MAD} &= \underset{i}{\mathrm{median}} \left| x_i - \mu \right|, \\ \mathrm{MADAM} &= \underset{i}{\mathrm{median}} \left| x_i - \overline{x}_{\mathrm{med}} \right|. \end{aligned} \tag{4.21}$$

Once the mean estimator is non-zero we may define **Coefficients of Variation** (CV), which are the relative measures of dispersion related to the mean estimator. The two most common ones are

$$V_\sigma = \frac{\sigma}{\mu}, \tag{4.22}$$

$$V_{\mathrm{d}} = \frac{\mathrm{AAD}}{\mu}. \tag{4.23}$$

Following the above, we define two variation ranges

$$T_\sigma = \left]\mu - \sigma, \mu + \sigma\right[, \tag{4.24}$$

$$T_{\mathrm{d}} = \left]\mu - \mathrm{d}, \mu + \mathrm{d}\right[. \tag{4.25}$$

Actually, you can always play around with replacing the position and variability operators, but ... that would probably be an art for art's sake. Analogously to the position estimators, we may define positional measures of the variation. We may define data **range**

$$\mathrm{R} = x_{max} - x_{min}. \tag{4.26}$$

More practical and more common in practical applications is the **Interquartile Range** (IQR) . It's a value, which is calculated by subtracting the third quartile from the first, as in Eq. (4.27)

$$\mathrm{IQR} = Q_3 - Q_1. \tag{4.27}$$

Table 4.4 presents the above scale factors evaluated for the reference datasets. We may indicate two general observations. First the difference

Table 4.4: Selected scales for the reference time series

scale estimator		X_1	X_2	X_3	X_4	X_5
standard deviation	σ	1.4255	4.2354	0.1500	0.9740	11.25
variance	σ^2	2.0321	17.94	0.0225	0.949	126.7
AAD	d	1.1399	1.2619	0.1128	0.7976	8.0225
MAD		0.9599	0.9724	0.0896	0.7108	6.0855
MADAM		0.9577	0.9785	0.0893	0.7000	6.0855
coefficient	V_σ	201.5	-592.9	32.56	0.0188	-0.5579
of variation	V_d	161.1	-176.6	24.48	0.0154	-0.3977
range	R	11.52	442.6	2.00	6.00	183.85
interquartile range	IQR	1.921	1.951	0.179	1.400	12.153

between the MAD and MADAM is rather insignificant, however the AAD may differ as the average grouping estimator might be biased by outlying values. We also see that the coefficient of variation might exhibit negative values in the case of negative mean operator values. This is why in some cases people use its absolute value.

The differences between various dispersion measures are mostly connected to their vulnerability to outliers. The estimators that incorporate mean and standard deviation are biased with the outlying observations, while the ones using median, absolute value or quartiles are robust. Robustness considered as the vulnerability of the estimators to the outlying observations in time series should be always kept in mind during data analysis.

4.1.3 Skewness – the measures of asymmetry

The concept of a distribution symmetry (symmetrical probabilistic density function) is important in statistics. Thus, the task of measuring and quantifying possible deviations from symmetry is needed and we require some sizable number of asymmetry. The classical skewness factor plays that role. It uses the third central moment μ_3. We define the **skewness factor** γ_3 as the third standardized moment

$$\gamma_3 = \tilde{\mu}_3 = \frac{\mu_3}{\sigma^3}. \tag{4.28}$$

Skewness defines a measure of distortion of symmetric distribution.

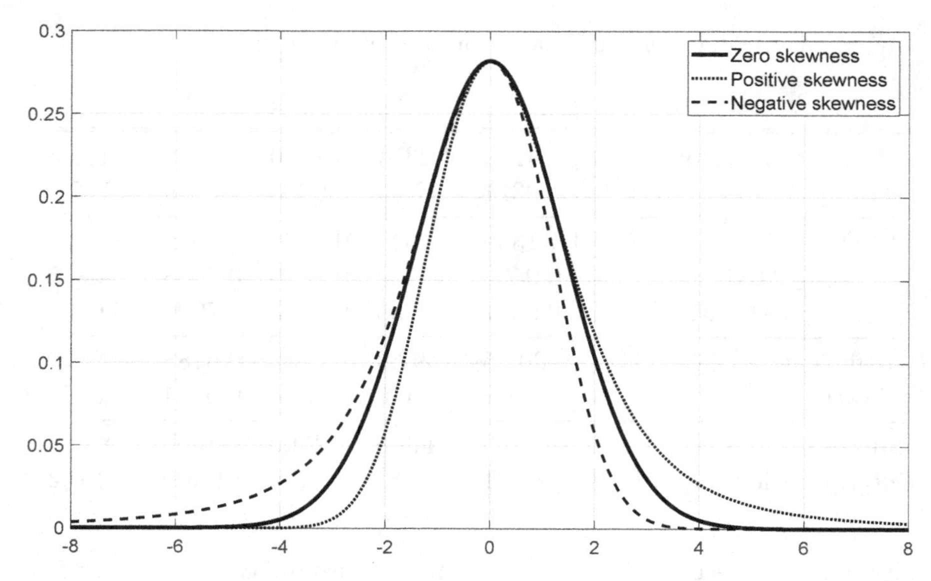

Figure 4.3: Skewness impact on distributions

It gives a number of how the given probabilistic density function deviates from a symmetric one. Zero skewness occurs when the function is symmetrical and it applies to normal data distribution. Its negative values refer to a longer or fatter left tail, while the positive values indicate the opposite situation and the tail on the right. The sign of skewness factor refers to the direction or weight of the distribution. Fig. 4.3 shows graphical representation of the skewness effect for the distribution.

Quantile-based skewness measure, which is also called as the Bowley's measure of skewness, Galton skewness or Yule's coefficient is defined as

$$S_{kq} = \frac{(Q_3 - \overline{x}_{\text{med}}) - (\overline{x}_{\text{med}} - Q_1)}{Q_3 - Q_1} = \frac{Q_3 + Q_1 - 2 \cdot \overline{x}_{\text{med}}}{Q_3 - Q_1}. \tag{4.29}$$

Classical skewness factor γ_3 is also referred to as Pearson's moment coefficient of skewness or simply the moment coefficient of skewness, however it might not be confused with Pearson's other skewness coefficients. We define **Pearson's first coefficient of skewness** or the Peaerson mode skewness in the relation to the standard deviation or

AAD

$$Sk1_\sigma = \frac{\mu - \overline{x}_{\text{mo}}}{\sigma}, \tag{4.30}$$

$$Sk1_\text{d} = \frac{\mu - \overline{x}_{\text{mo}}}{d}. \tag{4.31}$$

Pearson's second skewness coefficient or the Peaerson median skewness related to the standard deviation or AAD is defined as

$$Sk2_\sigma = \frac{\mu - \overline{x}_{\text{med}}}{\sigma}, \tag{4.32}$$

$$Sk2_\text{d} = \frac{\mu - \overline{x}_{\text{med}}}{d}. \tag{4.33}$$

Table 4.5 presents the above skewness factors evaluated for the reference datasets. We notice that the caution is demanded about the skewness sign, as different estimators might result in opposite signs. This observation is caused by the different vulnerability and sensitivity of estimators to the outliers.

Skewness gives the quantified measure of data symmetry, however specific measures are vulnerable to data properties, especially to the distribution tails and outliers. The user must pay special attention when choosing an estimator and always take into account the nature of data. Even more attention should be paid to the interpretation of obtained results.

Table 4.5: Selected skewness estimators for the reference time series

estimator	X_1	X_2	X_3	X_4	X_5
γ_3	0.0232	73.5976	-0.0637	0.0034	-0.4072
S_{kq}	0.0150	0.0160	-0.0178	0.0000	0.0090
$Sk1_\sigma$	3.919	14.076	7.436	-0.114	0.133
$Sk1_\text{d}$	4.901	47.245	9.891	-0.139	0.187
$Sk2_\sigma$	0.0110	0.0093	0.0038	-0.0111	0.0091
$Sk2_\text{d}$	0.0137	0.0312	0.0050	-0.0135	0.0127

4.1.4 Kurtosis – the measures of concentration

The classical formulation of the concentration factor called **kurtosis** uses the notion of the fourth central moment μ_4 and is

$$\gamma_4 = \frac{\mu_4}{(\sigma^2)^2}. \quad (4.34)$$

The kurtosis is limited below be the value equal to the squared skewness plus 1, as

$$\gamma_4 \geq \gamma_3 + 1, \quad (4.35)$$

however it is not bounded above (the upper limit is infinite).

The number $K = \gamma_4 - 3$ is called the excess coefficient or the excess kurtosis. Unfortunately, the excess coefficient is sometimes also called kurtosis, which may lead to confusion. The excess kurtosis is a term which we use to measure the extent to which a distribution deviates from the normal PDF in terms of the peak and tails.

It is assumed that a statistical population has normal distribution, when $\gamma_4 = 3$. Such a distribution is called the mesokurtic or medium-tailed. We say that the distribution of some feature in a population is leptokurtic (slender) or heavy-tailed when $\gamma_4 > 3$. In contrary, when $\gamma_4 < 3$, the distribution is said to be platykurtic (flattened) or thin tailed. Fig. 4.4 shows a graphical representation of the kurtosis effect on the distribution.

The concept of tails starts to be more visible with the introduction of the kurtosis, however its full definition and discussion will be presented in detail later in the discussion. However, it is already clear at this point that awareness of the existence of tails is already manifested in the early stage of basic definitions of statistical moments. We call the **percentile concentration coefficient** a number

$$\gamma_{4,P} = \frac{\frac{1}{2}(Q_3 - Q_1)}{P_{90} - P_{10}}, \quad (4.36)$$

where P_{90} is the 90th percentile (90% of points lie below this value) and P_{10} is the 10th percentile (10% of points lie below this value). Once the distribution is normal, the percentile concentration factor $\gamma_{4,P} = 0.263$.

Table 4.6 combines the kurtosis estimators for the reference datasets. We see a big different between the classical estimator and the percentile one. The interpretation of the data is straightforward with the classical kurtosis γ_4. We may notice that the data X_1 are relatively close to being

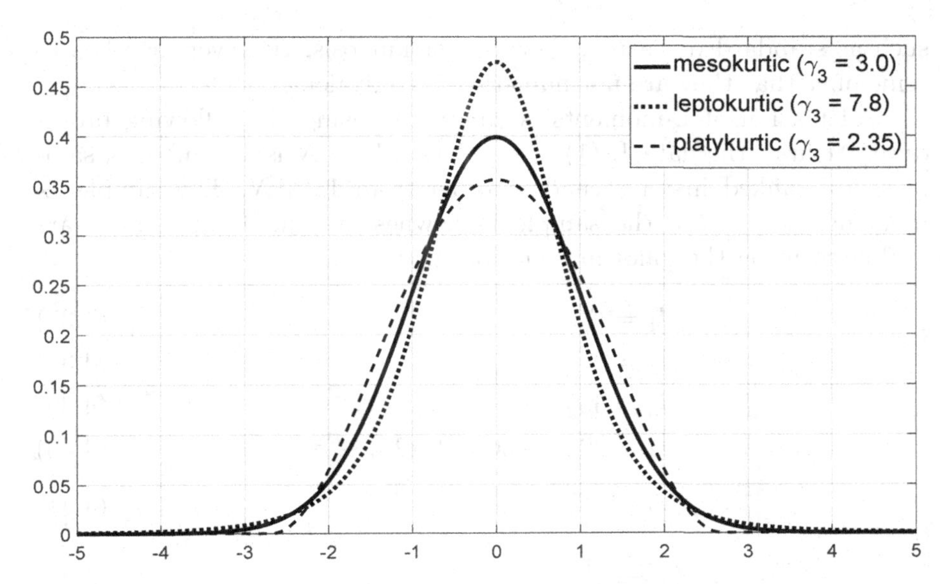

Figure 4.4: Kurtosis impact on distributions

normally distributed. Dataset X_4 is platykurtic, while the three others are clearly leptokurtic.

> Kurtosis is the measure of data concentration and the simplest approach to the information about the data tailness and potential existence of outlying observations.

4.2 L-MOMENTS

L-moments are introduced as linear combinations of order statistics [149]. The research of order statistics is related to the statistics of ordered random variables and samples. The well-known and already presented minimum, maximum, and median are likely the most popular order statistics. L-moments are direct analogs to the presented above product moments,

Table 4.6: Selected kurtosis estimators for the reference time series

estimator	X_1	X_2	X_3	X_4	X_5
γ_4	2.985	6694.6	6.285	2.592	10.136
$\gamma_{4,P}$	0.2640	0.2606	0.2565	0.2800	0.2480

such as standard deviation, skewness or kurtosis. However, one has to remember that they are not numerically equivalent.

Calculation of L-moments is conducted using the following procedure. At first, the data $\{x(1), \ldots, x(N)\}$, where N is a number of samples, are ranked in an ascending order from 1 to N. The sample L-moments $(l_1, \ldots, l_4)$, the sample L-skewness τ_3 and L-kurtosis τ_4 are evaluated using the following relationships

$$l_1 = \beta_0, \tag{4.37}$$

$$l_2 = 2\beta_1 - \beta_0, \tag{4.38}$$

$$l_3 = 6\beta_2 - 6\beta_1 + \beta_0, \tag{4.39}$$

$$l_4 = 20\beta_3 - 30\beta_2 + 12\beta_1 - \beta_0, \tag{4.40}$$

$$\tau_3 = \frac{l_3}{l_2}, \tag{4.41}$$

$$\tau_4 = \frac{l_4}{l_2}, \tag{4.42}$$

where

$$\beta_j = \frac{1}{N} \sum_{i=j+1}^{N} x_i \frac{(i-1)(i-2)\cdots(i-j)}{(N-1)(N-2)\cdots(N-j)}. \tag{4.43}$$

L-moments are analog to conventional moment estimates, and the first L-moment denoted l_1 is just the arithmetic mean. The second L-moment, called the L-scale and denotes as l_2, measures a random variable's variation. Apart from the second L-moment, we may use the dimensionless function of L-moments, being an analogy the CV – coefficient of variation [149]. It is denoted as L-Cv and derived by the following equation

$$\text{L-Cv} = \tau_2 = \frac{l_2}{l_1}. \tag{4.44}$$

Higher order L-moments, i.e. l_3 and l_4 are considered as measures of symmetry and concentration, respectively [315, 149]. The rth order L-moment ratios [152] are defined as

$$\tau_r = \frac{l_r}{l_w}, r = 3, 4, 5, \ldots \tag{4.45}$$

The ratio τ_3 is called the sample L-skewness and the τ_4 is known as the L-kurtosis. Due to their definition they can achieve the full range of values of the population ratios in contrary to the traditional definitions of skewness and kurtosis. These ratios satisfy the condition $|\tau_r < 1|$ and allow to measure independently of scale the shape of the distribution. Skewness introduces an alternative bound on the L-kurtosis existing for any distribution function with finite expected value

$$\frac{1}{4}\left(5\tau_3^2 - 1\right) \leq \tau_4 < 1 \qquad (4.46)$$
$$\frac{1}{4} \leq \tau_4 < 1$$

It is proven [149] that if the mean value of a given random variable exists, then all the higher L-moments exist as well. The set of factors $\{l_1, l_2, \tau_3, \tau_4\}$ can be used to summarize any univariate distribution. It must be noted that L-moments do not exist for probability density functions that do not have finite means, like the Cauchy distribution, or General Pareto (GPO) and Generalized Extreme Value (GEV) for some shape parameters values [316].

The use of L-moments significantly improves conventional methodology. They introduce new characterization of the shape of distributions and allow to estimate the distribution parameters. It is done through fitting of the sample empirical L-moments to the exact theoretical L-moments of the distribution. L-moments τ_3 and τ_4 play the role of the goodness-of-fit measure. L-moments can be theoretically evaluated for various PDFs. Theoretical relationships for different univariate PDFs can be found in [149, 150, 191].

Unlike product moments, L-moments allow for robust and almost unbiased L-moments estimation, even for very small samples. Furthermore, they are less sensitive to the distribution's tails [259]. Above properties suit them almost perfectly and allow them to describe data commonly characterized by a moderate or high skewness. Above advantages make the quite attractive for a distributional analysis of various function: Gaussian and non-Gaussian; thin and heavy tailed; symmetrical and asymmetrical. Moreover, they might be used for other tasks, like the regional frequency analysis [121], homogeneity testing [153, 316], discordancy testing [153, 246, 98].

Despite above characteristics, they did not gain global understanding. L-moments are widely used in life sciences with a multiple successful applications to the extreme events analysis in climatology[293], hydrology [186, 225], astronomy [267] or medicine [221]. Only recently have the first works appeared showing the possibilities of L-moments for the purpose of assessing the performance of control or the sustainability of control systems [98].

Table 4.7: L-moments calculated for the reference time series

L-moments	X_1	X_2	X_3	X_4	X_5
l_1	0.0071	-0.0071	0.0046	51.79	-20.17
l_2	0.805	0.921	0.081	0.554	5.885
τ_2	113.7	-128.9	17.67	0.011	-0.292
τ_3	0.0072	0.0334	0.0080	-0.0046	0.0018
τ_4	0.1203	0.2138	0.1719	0.0910	0.2115

The literature shows several extensions and generalizations of the L-moments, like trimmed moments (termed TL-moments) [31], LQ-moments [241], which replace the expectations by estimators like the median, Gastwirth estimator and the trimmed mean, LH moments [346], which use linear combinations of higher-order statistics, LL-moments [22] giving higher weights to small observations or PL-moments (partial L-moments) [347], which are introduced for characterizing the upper distribution part. It's worth to note that the TL- and LQ-moments are more robust than L-moments and always exist. The trimming, which is used by the TL-moments is the simplest extension of the L-moments and as such is just straightforward to be used.

Table 4.7 shows the estimated values of the consecutive L-moments evaluated for the reference dataset. We notice that the L coefficient of variance may exhibit negative values, which is due to the negative value of the L-shift l_1. That's why in some applications researchers use its absolute value, i.e. L-Cv $= |\tau_2|$.

> L-moments are a reasonable alternative to the traditional formulation of statistical moments. They provide estimates that are robust to outliers and the fat tails they generate, and enable correct calculations for short samples, which is a serious but poorly recognized challenge in research. Moreover, they allow the analysis of probability distributions in a much broader scope, not only in relation to a single time series, but also provide space for comparing and grouping several datasets.

4.3 ROBUST MOMENTS ESTIMATORS

Above stipulations present various estimators of statistical moments. Their origins vary depending on the selected approach, though some difference between them was already demonstrated. Some of them are said to be sensitive to outliers, while the others not being robust. The fact that outliers exist have an incalculable impact on the data analysis. They increase estimated data variance and simultaneously decrease the power of statistical tests [252]. They deteriorate statistical properties of the data, as for instance data normality, at the same time contribute to the distributional tails [325] and finally bias moments estimators and the regression analysis [287].

Following the discussion on outliers from Chapter 3 the introduction of the robust statistics confirms that we notice outliers. We know that they exist in data, and that spoil the analysis. Thus, we can consciously and intentionally propose an approach to moment estimation that will minimize or even remove the negative impact of outliers by treating the data as if they did not exist, i.e. ignoring their significance.

The above assumptions pose an analytical challenge. Robust statistics and robust estimators of the statistical factors, like the shift, the scale or the shape can be found in the research for a long time [138] however, it was only Huber's works [157] that attracted the attention of a wide scientific audience and gave a new impetus for further research and practical applications. Proposed variety of robust estimators achieve high performance and robustness for data exhibiting different distributions, especially for Gaussian PDF.

Once we talk about the robustness we need to be able to measure it. Apparently, the breakdown point is a measure of testing the robustness of a statistical procedure. It was introduced by Hampel in 1968 [132]. We simply define it as a fraction of the data that are introduced through contamination without making the estimator biased. The highest possible value of the breakdown point is 50%, which means that half of data are contaminated. Higher values cannot appear as then the contamination starts to be normality and vice versa.

The traditional arithmetic mean estimator has a zero breakdown point, while the median reaches 50%, which is the highest achievable. In case of the scale estimators, standard deviation exhibits 0% breakdown point, while MAD has the highest possible value of 50%. Interested readers may find more on the breakdown point in [118].

The arithmetic average value, called the mean μ, given by Eq. (4.5) is the most popular and known estimator of a true (expected) value of a random variable. It is the shift (location or position) estimator of the considered data. Unfortunately, the estimator loses its accuracy in the presence of outlying observations, so we call it a non-robust estimator. In contrary, its robust analog properly describes the data properties whatever their content is. Apparently, the median value $\overline{x}_{\text{med}}$ described by Eq. (4.60) is a reasonable and still simple choice, as it sustains robustness. Median value indicates, where the middle of a dataset is.

Let's follow the Huber's attention and reformulate the problem to the functional estimation approach. In the location-scale model we generate the univariate observations $x_1, \ldots, x_n$ as

$$x_i = \mu + \sigma\epsilon_i \tag{4.47}$$

where ϵ_i are independent and identically distributed random variables with the reference distribution function F_0 for $i = 1, \ldots, n$. The model distribution for the observations is described as $F_{\mu,\sigma}$. The parameters μ and σ denote the location and the scale parameters, respectively. If F_0 is normal, then μ is the mean and σ the standard deviation.

Actually, there is an infinite number of possible position and scale estimators. Let's denote T_n the estimate of position and S_n of scale. We demand that these estimators are affine equivariant, i.e.

$$T_n\left(ax_1 + b, \ldots, ax_n + b\right) = aT_n\left(ax_1, \ldots, ax_n\right) + b, \tag{4.48}$$

$$S_n\left(ax_1 + b, \ldots, ax_n + b\right) = |a|\, S_n\left(ax_1, \ldots, ax_n\right), \tag{4.49}$$

for any a and b. Moreover, the estimator should be precise according to its statistical efficiency.

We already know that median is simple location and MADAM a well known scale estimator. Robust statistics introduces the family of so called M-estimators being a solution to the estimating equation. They are highly efficient and robust, but they are not described by any explicit expression.

The class of the introduced M-estimators includes the maximum likelihood estimator (MLE). The log-likelihood formulation of the assumed model distribution $F_{\mu,\sigma}$ is

$$\sum_{i=1}^{N}\left\{\log f_0\left(\frac{x_i - \mu}{\sigma} - \log\sigma\right)\right\}, \tag{4.50}$$

The M-location estimator denoted as $\hat{\mu}$ is defined as a solution to the equation

$$\frac{1}{n}\sum_{i=1}^{n}\psi\left(\frac{x_i - \hat{\mu}}{\sigma_0}\right) = 0, \tag{4.51}$$

where $\psi(.)$ is any non-decreasing odd function (called the influence function) the $\hat{\mu}$ is a location estimator and σ_0 is a preliminary assumed scale estimator, like MADAM. In a similar way, we define M-estimator for the scale $\hat{\sigma}$ as solution to:

$$\frac{1}{n}\sum_{i=1}^{n}\rho\left(\frac{x_i-\mu_0}{\hat{\sigma}}\right)=1, \tag{4.52}$$

where $\rho(.)$ is a loss function, σ is a location estimator and μ_0 is a preliminary location estimator, like median. The loss function $\rho(.)$ should meet the following properties:

- being continuous, though it's not need, but makes numerical calculations easier,
- being symmetrical ($\rho(\xi)=\rho(\xi)$), although in real cases errors does not have to be equally distributed around the mean,
- being strictly positive ($\rho(\xi)\geq 0$) and integrable,
- being an increasing monotonous function of x_i ($\rho(\xi_i)\geq\rho(\xi_j)$ for $|\xi_i|>|\xi_j|$),
- $\rho(0)=0$,
- preferentially being convex as it guarantees the solution uniqueness for problems, which are described by linear models.

Please note, that the influence function is a differential of the loss function, i.e. $\rho(.)$.

$$\psi(\xi)=\frac{\partial\rho(\xi)}{\partial\xi} \tag{4.53}$$

Depending on the selected influence/loss function, we obtain various M-estimators. Literature shows quite a large variety of such function [224, 71]. Here we will present only two representative examples: Huber and logistic functions. They can be confronted with the arithmetic mean , which is represented by the least squares (LS) estimator, also known as the ℓ_2

$$\rho(\xi)=\frac{1}{2}\xi^2 \longrightarrow \psi(\xi)=\xi \tag{4.54}$$

and the least absolute value (LAV), also known as the ℓ_1 estimator

$$\rho(\xi)=|\xi| \longrightarrow \psi(\xi)=\begin{cases}-1 & \text{for } \xi<0\\ +1 & \text{for } \xi>0\end{cases}. \tag{4.55}$$

The ℓ_2 estimator is not robust, as its influence function $\psi(\xi)$ is linear, what causes that the influence of outlying numbers on the estimation is unlimited and increases proportionally to the increasing magnitude of the outlier.

Huber loss $\rho_{\mathrm{H}}(\xi)$ and influence $\psi_{\mathrm{H}}(\xi)$ functions are given as:

$$\rho_{\mathrm{H}}(\xi) = \begin{cases} \frac{1}{2}\xi^2 & \text{for } |\xi| \leq k_{\mathrm{H}} \\ k_{\mathrm{H}}|\xi| - \frac{1}{2}k_{\mathrm{H}}^2 & \text{for } |\xi| > k_{\mathrm{H}} \end{cases}, \tag{4.56}$$

$$\psi_{\mathrm{H}}(\xi) = \begin{cases} \xi & \text{for } |\xi| \leq k_{\mathrm{H}} \\ k_{\mathrm{H}} \cdot \operatorname{sgn}(\xi) & \text{for } |\xi| > k_{\mathrm{H}} \end{cases}. \tag{4.57}$$

The parameter k allows to tune the function. We may relate it in respect to the Normal distribution efficiency. Literature suggests the following selection $k_{\mathrm{H}} = 1.345$ [276].

The logistic functions $\rho_{\mathrm{L}}(\xi)$ and $\psi_{\mathrm{L}}(\xi)$ are given by

$$\rho_{\mathrm{L}}(\xi) = k_{\mathrm{L}}^2 \ln \left[\cosh\left(\frac{\xi}{k_{\mathrm{L}}}\right)\right], \tag{4.58}$$

$$\psi_{\mathrm{L}}(\xi) = k_{\mathrm{L}} \tanh\left(\frac{\xi}{k_{\mathrm{L}}}\right). \tag{4.59}$$

The work [147] suggests to use the following coefficient value $k_{\mathrm{L}} = 1.205$. Fig. 4.5 presents graphical representation of the respective ℓ_2, ℓ_1, Huber and logistic estimators.

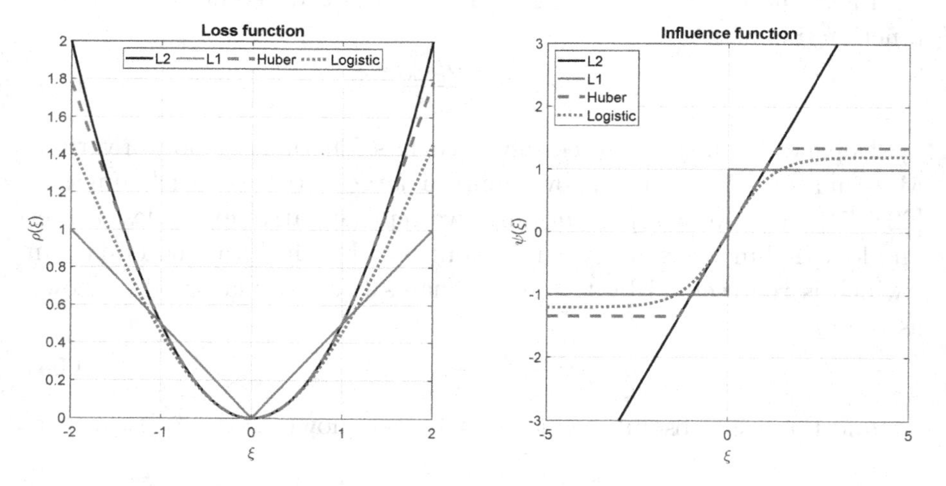

Figure 4.5: Selected loss and influence functions

Robust methods have been developed to estimate location, scale, and regression parameters for time series affected by outliers and this feature is the most interesting in our case. The fact that the robust approach is not sufficiently used in control is probably connected with the fact that they have appeared and are mostly investigated in the research context of the chemical engineering. It is clearly reflected by the ratio between the number of publications in chemistry and control engineering.

Finally, Table 4.8 presents summary of the robust estimation of M-estimators for the position and the scale. As the reference, the mean and standard deviation are added. We may notice some interesting observations. Time series X_1 seems to be not far away from being normally distributed. It is due to the fact that robust scale estimator is quite close to standard deviation. Dataset X_2 exhibits tails as the robust estimator is significantly lower than the traditional scale. Datasets X_3 and X_5 are probably also tailed, but to lower extent. Time series X_4 presents behavior that is seemingly difficult to interpret, as standard deviation is smaller than the M-scale. The reason might lie in the non-stationary behavior, oscillatory character and/or the fact that the signal is subject to a trend. Verification of the time data shown in Fig. 4.1 initially confirms above observations, however further validation is required using histograms, PDF fitting or statistical tests.

Robust statistics offers consecutive estimators, which are based on data normality assumption combined with a conscious desire to minimize the impact of outliers. Comparison of the traditional and robust estimators provides the observer with the initial information about time series properties, indicates further analytical steps and formulates first hypotheses about data properties.

Table 4.8: Robust M-estimators calculated for the reference time series

M-estimators		X_1	X_2	X_3	X_4	X_5
position	$\hat{\mu}$	0.007	-0.007	0.005	51.79	-20.17
	M-Huber	0.005	-0.037	0.004	51.79	-20.17
	M-logist	0.005	-0.038	0.004	51.79	-20.18
scale	$\hat{\sigma}$	1.426	4.235	0.150	0.974	11.25
	M-logist	1.432	1.455	0.134	1.037	9.112

4.4 SMALL SAMPLE STATISTICS

The subject of the small sample statistics evaluation is very important, though frequently it does not meet enough attention and is neglected. Its impact on the analysis is visualized with the following example. Let's assume that we have some normally distributed noise $\mathcal{N}(0, 1)$, i.e. with the mean equal to zero $\mu = 0$ and standard deviation equal to one $\sigma = 1$ with $n = 1000$ observations.

We divide the time series into smaller sequences of length equal to $\{1000; 500; 200; 100; 50; 20; 10; 5\}$, respectively and for each subset we calculate its mean and standard deviation. Fig. 4.6 presents graphically evaluated values. We definitely see that the results obtained are far from the theoretical values for the fundamental population. The smaller the sample size is, the higher scattering of the results might be. This effect is well-known and addressed in statistics.

The **weak law of large numbers** says if the X_i is an infinite sequence of independent and identically distributed i.i.d. random variables with the expected value $\mathbb{E}(X_i) = \mu$, than the sample average $\overline{X}_n$ converges to the expected value as $n \to \infty$. As we see in our example it is difficult to be achieved and we do not know what should be the sample size n to be sure that we have converged enough.

It is also worth mentioning here the **central limit theorem** as it describes the size and the distribution of variations around the mean

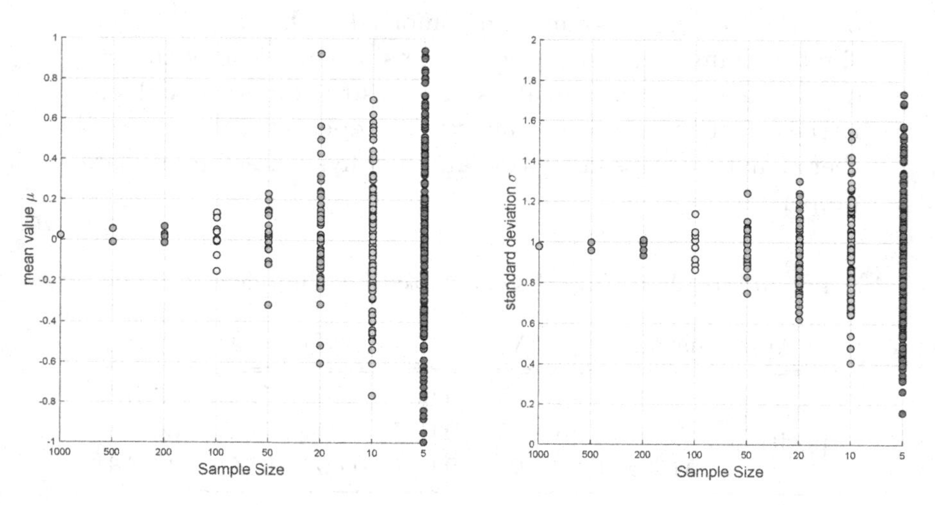

Figure 4.6: Sample size impact on estimates of mean and standard deviation

during this convergence. It states (Lindeberg-Lévy version) that if a sequence of independent and identically distributed i.i.d. random variables with mean $\mu < \infty$ and variance $\sigma^2 < \infty$, and X_n being the sample of size n, then as $n \to \infty$ the random variables $\sqrt{n}\left(\overline{X}_n - \mu\right)$ converge in distribution to the Gaussian $\mathcal{N}\left(0, \sigma^2\right)$.

Statistics says a lot about asymptotic behavior as $n = \infty$, however we are not quite sure what happens if n is not very large [325]. Especially, we cannot say, whether the sample size is enough for proper estimation of the mean and further higher moments as well, like standard deviation. Thus, one needs approaches what to do in case of small samples.

The notion of the small size is also undefined. In some reports, it is considered as between 3 and 8 [288] or lower than 25 [222]. Some authors define other limits or distinguish between very small samples and moderately small samples. The exact definition will be found in the description of the appropriate methods.

4.4.1 Small sample mean estimation

The first common approach is just to use the average value, i.e. the arithmetic mean μ estimator as in Eq. (4.5). The second approach is to use the sample median $\overline{x}_{\text{med}}$ – Eq. (4.60). Hodges–Lehmann estimator (HL) [133] is another option:

$$\overline{x}_{\text{HL}} = \text{median}\left\{\left(\frac{x_i + x_j}{2}\right); 1 \leq i < j \leq n\right\}. \tag{4.60}$$

The k-trimmed average can be considered as the next option, which is the mean value of the data except the k smallest and k largest observations [288]

$$\mu_{\text{k-trim}} = \text{mean}\left\{x_{k+1:n}, \ldots, x_{n-k:n}\right\}. \tag{4.61}$$

Hozo estimator [154] proposes the explicit formulation originating from the quartiles approach. For the sample size n we get:

$$\overline{x}_{\text{Hozo}} = \begin{cases} \frac{\text{MIN}+2\cdot\overline{x}_{\text{med}}+\text{MAX}}{4} & \text{for } n \leq 25 \\ \overline{x}_{\text{med}} & \text{for } n > 25 \end{cases}. \tag{4.62}$$

We may use such a simple expression even if we don't know the real sample size

$$\overline{x}_{\text{Hozo}} = \begin{cases} \frac{\min+2\cdot\overline{x}_{\text{med}}+\max}{4} & \text{for } n \leq 25 \\ \overline{x}_{\text{med}} & \text{for } n > 25 \end{cases}. \tag{4.63}$$

Wan [344] proposes another quartile approach defining

$$\overline{x}_{\mathrm{Wan}} = \frac{Q_1 + \overline{x}_{\mathrm{med}} + Q_3}{3}, \tag{4.64}$$

while Bland's method [33] suggests to use the following formulation

$$\overline{x}_{\mathrm{Bland}} = \frac{\mathrm{MIN} + 2 \cdot Q_1 + 2 \cdot \overline{x}_{\mathrm{med}} + 2 \cdot Q_3 + \mathrm{MAX}}{8} \tag{4.65}$$

The three above estimators can be improved by weighting between the estimator and the median [222], however it introduces another degree of freedom associated with the weight value and makes its use uncertain. One should remember that some authors [288] propose to use in-explicit formulations with M-estimators, however their evaluation might be an over-complication in relation to the task simplicity.

> Small sample size approaches are rather less formal with a lower formalities than the asymptotic estimators used for normal/large sample size. generally, the small sample cases are more difficult in the analysis, with less formal approaches being subject to assumptions and interpretations.

4.4.2 Small sample scale estimation

The tradition suggests to use classical standard deviation σ formulation as in Eq. (4.10). Similarly to the mean small sample size estimators it is also proposed to use AAD (4.20), MAD or MADAM estimator (4.21). Other authors [288] propose to use modified MADAM estimator

$$\mathrm{MADAM_r} = b_n \cdot 1.4826 \cdot \underset{i}{\mathrm{median}} \left| x_i - \overline{x}_{\mathrm{med}} \right| \tag{4.66}$$

or the Q estimator [285]

$$\sigma_{\mathrm{Q}} = c_n \cdot 2.2219 \left\{ |x_i - x_j| \, ; i < j \right\}_{(l)}, \tag{4.67}$$

where the last part is the lth ordered value among this set of $\binom{n}{2}$ values and $l = \binom{h}{2} \approx \binom{n}{2}/4$ with $h = \lfloor n/2 \rfloor + 1$ ($\lfloor t \rfloor$ is the largest integer $\leq t$). The constants b_n and c_n are small-sample correction factors which make these estimators unbiased.

We may also use the trimming, like the k-trimmed range

$$\mathrm{range_{k-trim}} = \left| x_{n-k:n} - x_{k+1:n} \right| \tag{4.68}$$

or k-trimmed standard deviation

$$\sigma_{\text{k-trim}} = \sigma \left\{ x_{k+1:n}, \ldots, x_{n-k:n} \right\}. \tag{4.69}$$

The data range can be also used in some basic way [37], i.e.

$$\text{range}_C = \frac{\max - \min}{C}, \tag{4.70}$$

where $C \in [3; 4]$ varying according to the sample size [154].

The mean of successive differences (MSSD) is another approach [239]. It is calculated by taking the sum of squared differences between consecutive observations, then taking the mean of that sum divided by two

$$\sigma_{\text{MSSD}} = \sqrt{\frac{\sum \left(x_{i+1} - x_i\right)^2}{2\left(n-1\right)}}. \tag{4.71}$$

Finally, data minimum, maximum, and median can be used to evaluate an estimate of standard deviation for small sample sizes using the so-called Hozo estimate [154]

$$\sigma_{\text{Hozo}} = \sqrt{\frac{1}{12}\left(\frac{(\min - 2 \cdot \overline{x}_{\text{med}} + \max)}{4} + (\max - \min)^2\right)}. \tag{4.72}$$

As sample data, which are exemplifying presented statistical notions, do not represent small sample size data, the respective moments are not evaluated for them. Concluding, the scientist may find some simple and suitable approximation of data shift and scale.

> Small samples appear much more frequently than we think. Though we should remember that it matters which estimator we chose, and once we suspect that the number of data is not large enough, it's reasonable to consider small sample size approach. Simple k-trimming might be enough, to save us from making serious mistakes.

4.5 HISTOGRAMS

A histogram is the basic type of diagram discovering the statistical properties of any time series. At the same time, it is a tool that should be used at first during the data analysis process. This applies to both statisticians and control engineers. It is traditionally utilized to investigate data distribution and to help in the discovering of the underlying PDF function shape for a given data. The histogram allows to analyze time series basic statistical properties, like its position, variability and the most important its shape. The histogram shape delivers the first, visual assessment of the potential type of the probabilistic density function that might characterize the generation mechanism behind the considered dataset. One may try to determine in a qualitative day the data position, symmetry, concentration or the tails.

We start building the histogram by splitting data into intervals, which are called bins. As the result, each bin consists of some number of observations with the values lying within the limits of the bin. Once we have the bins and the number of data within each bin we draw a respective bar plot. Sample histograms are sketched in Fig. 4.7, which presents diagrams for five reference time series.

The construction of the histogram is simple, however we have to define one parameter, i.e. the number of bins related to the data range. It's very important. Fig. 4.8 shows two histograms prepared with the same, i.e. time series X_1. The diagram to the left shows the case with the low number of bins (large bins width). In such a case it is extremely difficult to distinguish the real, underlying histogram shape (PDF function). In contrary, the right plot shows data divided into the very large of narrow bins. Now, it is hard to reveal the frequency distribution (underlying pattern) of the time series.

The specific situation with the bin number selection is visible with the histogram of the dataset X_2. It uses intentionally increased number of bins (400), as the data exhibit observations that are positioned significantly far away from the majority. These are outliers and we see how they disturb the histogram clarity. Fig. 4.9 allows to visualize this issue.

To the left we have the histogram with its bins placed within the full range of data. As the outliers seriously increase the range we need to have large number of bins to have enough of them near the main part of histogram. Those single outliers that stretch the histogram are more clearly visible in the lower left diagram, which is scaled in OY axis to the level of five observations. To get any shapes for such wide ranges we require a lot of bins; in this case 400. The large histogram on the right is limited only to the narrow range $\langle -6; +6 \rangle$ around the mean of data and covers the main bulk of data. Now, to reflect the distribution shape we need only 50 of them, and the outlying observations are included in the first (the most left) and in the last (the most right).

We should remember that the area of the bar matters as it represents the frequency of observations for each bin. The bar height doesn't have to

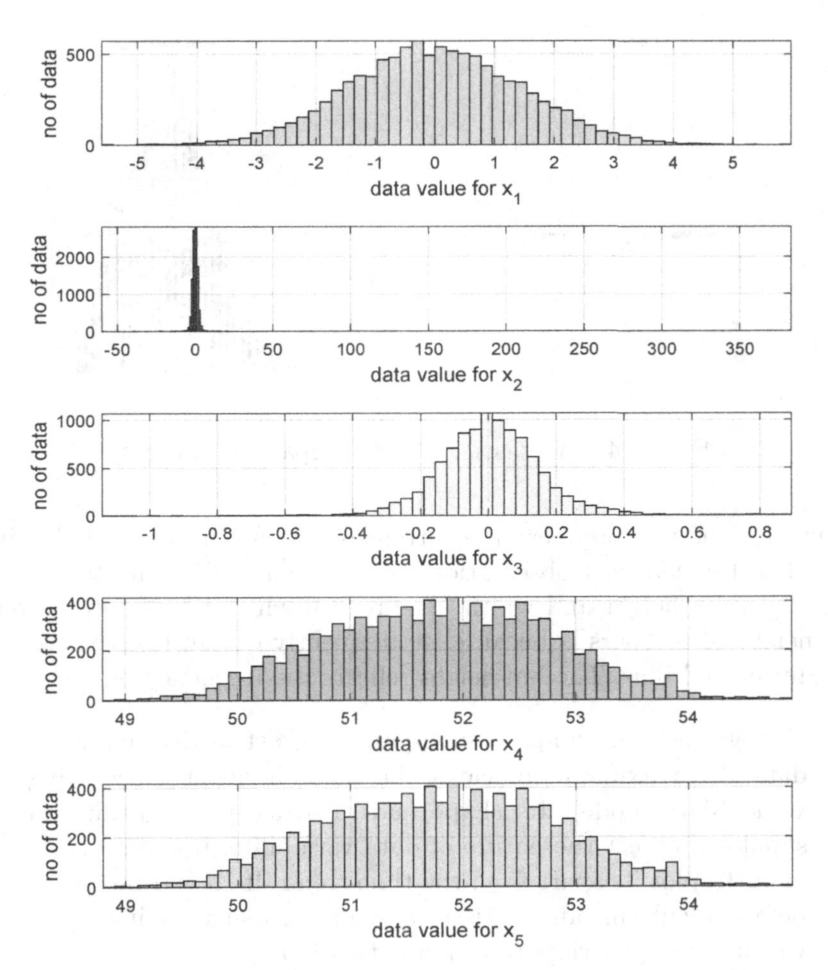

Figure 4.7: Histograms for reference data (X_2 has 400 bins, others 60)

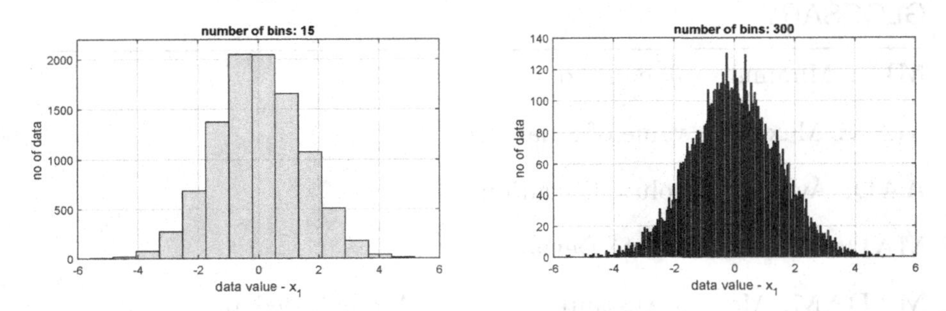

Figure 4.8: Impact of bins on the histogram shape

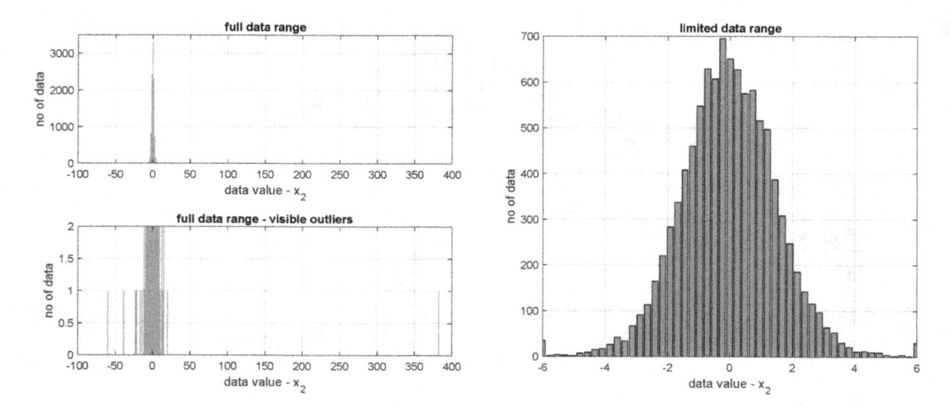

Figure 4.9: Dataset X_2 – the impact of outliers

point out, the real number of occurrences that fits into each individual one. The frequency of observations for a given bin is reflected by the product of its height and width. The reason why we take into account the height of the bars is because we frequently obtain histograms with equally spaced bins. The bin height reflects the frequency only then.

> Histogram is the simplest, non-parametric statistical model of data. Its maximum indicates the most frequently occurring value (data mode). Its shape and tails decay, the peak and shoulders reflect the nature of data variability. It's almost obvious to plot it, apart from the time trend. It should be drawn before calculating any statistical factors, as its nature indicates, which are appropriate to describe the data.

GLOSSARY

MIN: Minimum value of a dataset

MAX: Maximum value of a dataset

AAD: Average Absolute Deviation

MAD: Median Absolute Deviation

MADAM: Median Absolute Deviation Around Median

CV: Coefficient of Variation

IQR: InterQuartile Range

GPO: Generalized Pareto

GEV: Generalized Extreme Value

MLE: Maximum Likelihood Estimator

LAV: Least Absolute Value

LS: Least Squares

MSSD: Mean of Successive Differences

CHAPTER 5

Normal distribution – the peak

NORMAL DISTRIBUTION, also named the Gaussian, constitutes the focal point in the majority of statistical stipulations. Someone might even get the impression that without the normal distribution and Gauss's work, statistics would not exist.

Normal probabilistic density function is described by a specifically formulated function of a given data X_i and is characterized by two statistical factors: the mean μ and the standard deviation σ (5.1). The Gaussian PDF function is symmetrical around the mean, i.e. data position and its standard deviation measure the broadness (variability). The function, commonly named the bell-shaped, is formulated as follows:

$$\mathcal{F}^{\text{norm}}(x)_{\mu,\sigma} = \frac{1}{\sqrt{2\pi\sigma^2}} e^{-\frac{(x-\mu)^2}{2\sigma^2}}. \tag{5.1}$$

Its coefficients (statistical factors): μ and σ always exist and can be analytically obtained. They are equivalent to the data average value Eq. (4.5) and (4.10) present the formulation in the discrete time. Exemplary shapes of the normal PDF function are presented in Fig. 5.1 showing how their values change the PDF shape.

Statistical factors of the Gaussian function are frequently used as common descriptors of the data. Some users exchange the standard deviation with the variance $Var(x) = \sigma^2$. Both indexes are used alternately.

These properties may be validated formally with specific normal distribution hypothesis tests. One of them is Kolmogorov-Smirnov normality test (5.2). In statistics, the Kolmogorov-Smirnov test (KS) is a non-parametric test for the equality of continuous, one-dimensional

DOI: 10.1201/9781003604709-5

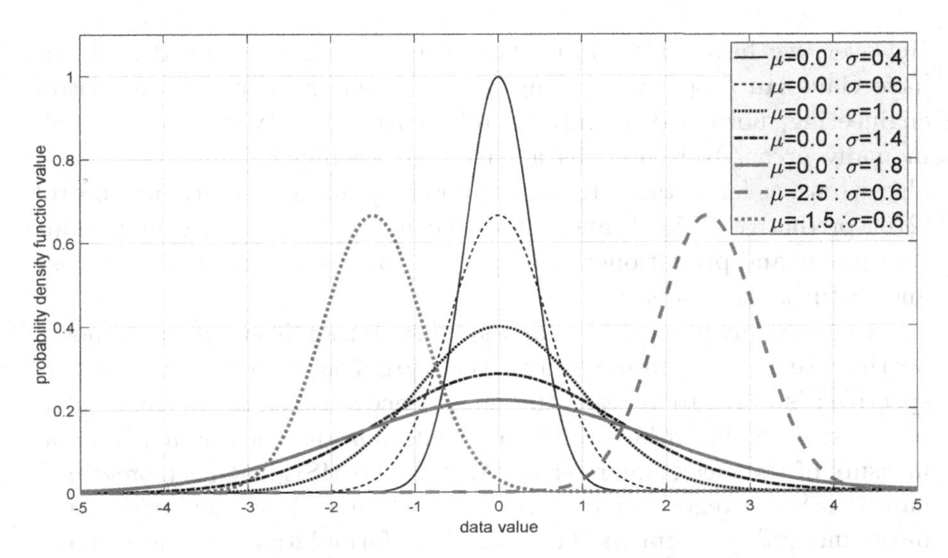

Figure 5.1: Effect of factors on normal probabilistic density function

probability distributions that can be used to compare a sample with a reference PDF. The Kolmogorov-Smirnov statistic quantifies a distance between the empirical distribution function of the sample and the cumulative distribution function of the reference distribution. The Kolmogorov-Smirnov serves as a goodness of fit test. In the special case of testing for normality of the distribution or fitting with other probabilistic density functions

$$KS = \max_x \left| \hat{F}_1(x) - \hat{F}_2(x) \right| . \tag{5.2}$$

Another approach is to check normality hypothesis through skewness or α stability. Skewness test is based on the difference between the data's skewness and value of zero, with some selected p-value (value $p = 0.05$ is used in the paper). In a very similar way, the α stability hypothesis is tested, when distance from value of 2 of α stability index is checked with some selected p-value ($p = 0.05$). Besides, there are also many other tests, like Smirnov-Cramer-von Mises [58], Anderson-Darling [8], Shapiro-Wilk [306], Lilliefors [213]. Experience shows that visual inspection and the histogram analysis enable better assessment, however it is more time-consuming than simple normal hypothesis validation with some of the above tests.

Importance of Gaussian measures and their common acceptance is unquestionable. Majority of researchers and control engineers use them.

However, we have to be aware that they are valid, while the signals have Gaussian properties. In engineering sciences, especially in control engineering, normal distribution probabilistic density function is used in many areas. It is used as a backbone assumption for many basic algorithms, as for instance Kalman filter [179], minimum variance control [270], predictive [277, 61] and adaptive control [352, 123]. In simulations researchers and practitioners incorporate Gaussian function as the noise and disturbance model.

Among many other fields Gaussian distribution is utilized as a model for the estimation residuum during the regression modeling. In that perspective it's worth to mention the equivalence between the normal distribution scale coefficient of standard deviation σ and the popular integral measure of the mean square error (MSE). The MSE metrics (sometimes named ISE – integral square error or the ℓ_2-norm) is calculated as the mean integral (or sum in the discretized formulation) of the squared residua over some time period $k = 1, \dots, N$

$$\mathrm{MSE} = \frac{1}{N} \sum_{k=1}^{N} \left(y(k) - \hat{y}(k) \right)^2, \tag{5.3}$$

where N denotes the number of data, y the real observation and $\hat{y}$ its estimated value.

It was shown that for large sets of industrial control error histograms only minority (percentage share $\approx 6\%$ of the loops) has normal distribution [77]. Majority of them exhibits fat tails and therefore, the α-stable distribution seems to be the most suitable one (more than 60%). Above observation leads to the conclusions that other probabilistic density functions and the associated measures should be discussed, which appears in the following paragraphs.

The considered exemplary datasets are used for the visualization of normal distribution fitting using classical factors and their robust equivalents. Fig. 5.2 presents the respective visualization for datasets X_1 and X_3. We see that the first time series x_1 is close to be normal, at least from the draft visual perspective. In contrary, the other dataset is more peaked than normal.

The following two datasets (X_4 and X_5) that are sketched in Fig. 5.3 exhibit different features. The left histogram, dedicated to the X_4 dataset might be interpreted as normal, however we observe nonuniform and jagged dependency, which even visually does not allow to assign a given dependency to a normally distributed function. The other histogram for

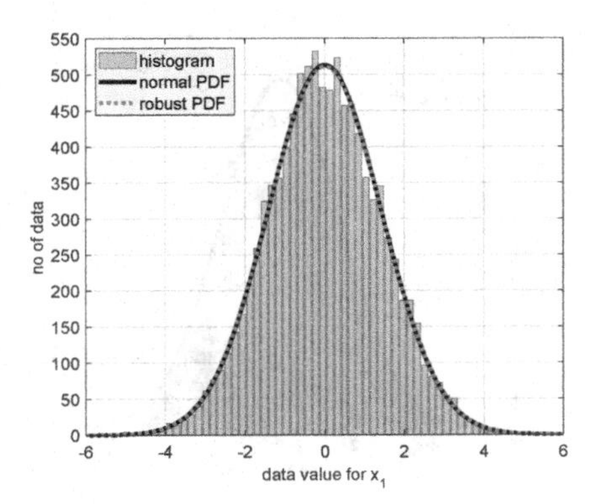

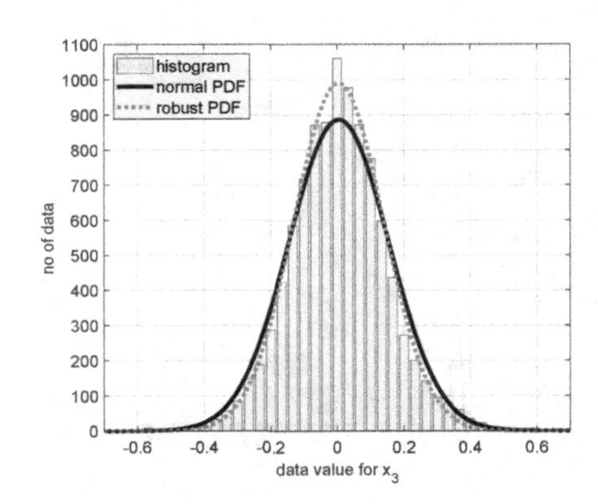

Figure 5.2: Histogram and normal distribution fitting for sample datasets: X_1 and X_3

X_5 time series presents clearly fat-tailed distribution, as the standard deviation is seriously overestimated, comparing to its robust counterpart.

Fig. 5.4 addresses the most evident case for the X_2 dataset with extremely peaked distribution that exhibits very long tails, which are caused by the very distant outlying observations. Interpretation in such cases is sensitive to the decision of an analyst. The left histogram is plotted using all the data, even those outlying one. Once we discard them (subjective assumption), we get the right plot, which shows much lower bias generated by the outlying observations. Though this discussion focuses only on the normal distribution fitting, we observe that the

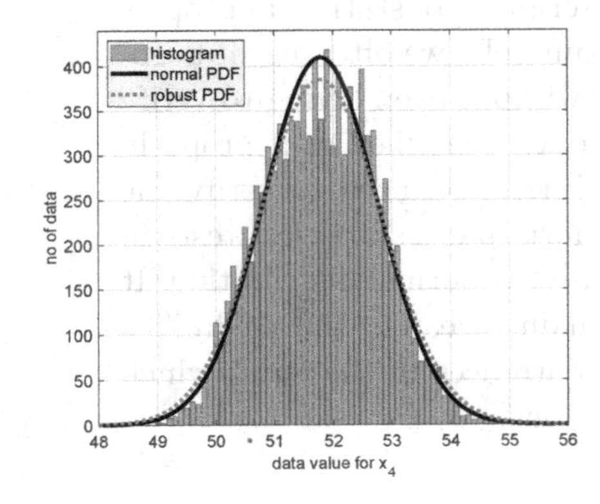

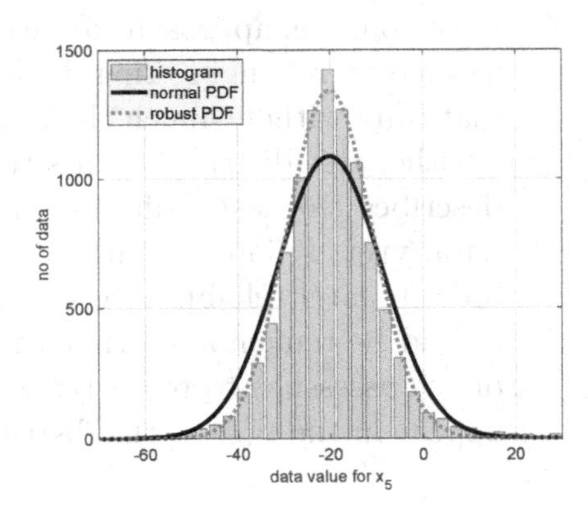

Figure 5.3: Histogram and normal distribution fitting for sample datasets: X_4 and X_5

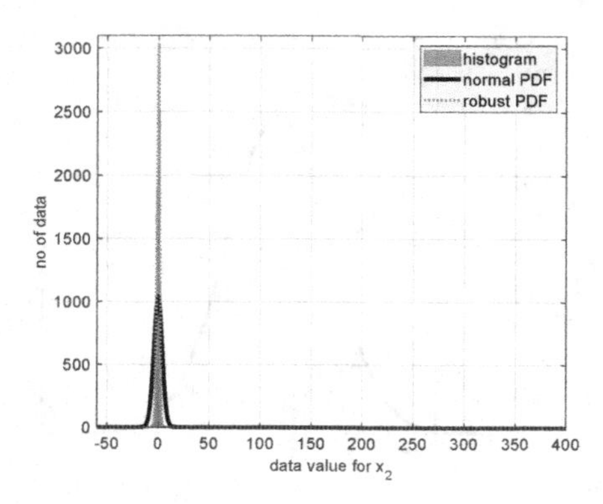

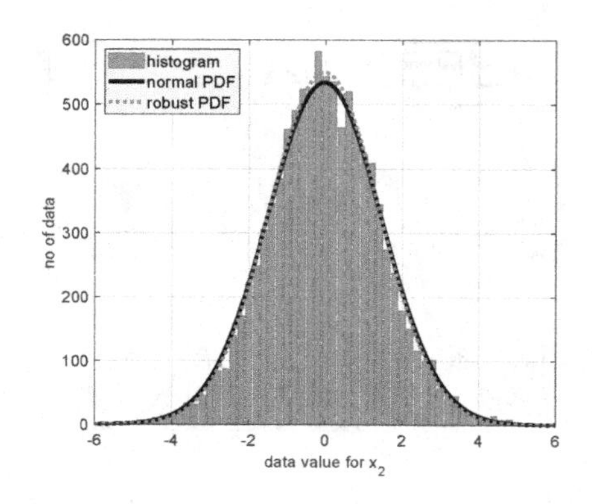

Figure 5.4: Histogram and normal distribution fitting for dataset X_2

ambivalent decision on the removal of outliers may have consequences on further analysis and the interpretation of phenomena behind the data.

This example shows how the PDF fitting works and how the histograms can be used in the analysis. We see that even relatively simple plots allow to discover quite fundamental properties of data and allow to make initial decisions, which define the directions for further steps. The comparison of the normal density function with the histogram focuses on the most frequent observations and describes how the distribution peak is shaped. Further analysis and the use of other, non-Gaussian functions, allows to observe what happens in other parts of the histogram – in tails.

> Normal Gaussian distribution constitutes fundamental and the most common approach to the description of statistical properties in control engineering. Unfortunately, we often use it automatically without much thought, without even trying to verify whether this distribution fits the data and whether it properly describes stochastic phenomena. First, you need to verify the data, visually interpret it and perform basic statistical tests in order to later reliably assess and use the normal distribution. It should be remembered that the main threat to the credibility of the assessment are outlying occurrences in the data, which appear in the tails of the distributions.

GLOSSARY

MSE: Mean Square Error

ISE: Integral Square Error

CHAPTER 6

Non-Gaussian distributions – the tails

TAILS are not the main focus in control engineering. Control theory and practice have never, or maybe very rarely, tried to focus on unusual situations and very infrequent events. If they were observed, they were either omitted or simply removed. The control system must always, i.e. under normal conditions, function properly. In unusual situations, the system is either switched to manual operation mode or emergency or protection systems are activated. Taking into account unusual situations, e.g. disturbances or uncertainties, is achieved by means of the robust control, design safety margins or the controller tuning for the worst case, of course the known one.

Actually, the opposite situation appears in other areas of science, like for instance the life sciences. In meteorology, the common and the most frequent phenomena are of the least interest. Who cares about average river discharge or average rainfall? In contrary, everybody is interested in the high flooding or the drought. Thus, the extreme events meet the highest interest.

Therefore, these sciences use a lot of statistics and normal distribution is not the main focus. As the analysis of extreme events is used, the highest attention is put to outliers and statistical models that incorporate them. Actually, the extreme statistics deals mostly with the one sided, asymmetric distributions (rainfall cannot be negative) and will be presented in the following Chapter 6. However, there exist symmetrical tailed distributions that will be presented in this chapter. Three types of the probabilistic density functions are addressed: Laplace (also called double exponential), t-Student and the family of α-stable distributions.

DOI: 10.1201/9781003604709-6

The tails have many names and there exists probably a hesitation about their interpretation. We may read about thin, heavy, fat or long tails. We need to clarify and explain these notions before we go beyond the normal distribution. We have introduced basic properties of the histogram in Fig. 1.1 and the tails appear there. The name of heavy tails is used to address different phenomena, sometimes it means that the distribution follows the power law, sometimes it means that it is scale-free. In other cases it refers to the stable distribution, or subexponential, or that it exhibits an infinite variance. Moreover, the name heavy tails is relative. Thus, let us introduce a threshold point.

We use the exponential distribution as that threshold. The distribution is said to be heavy-tailed if its tail is heavier than any exponential distribution [245]. The formal definition is defined in terms of the cumulative distribution function and can be found in the literature. Though this threshold between heavy and light tails could be considered arbitrary it is useful and let us keep it.

Generally, we may speak about the one-sided distributions (described in Chapter 7), when we mostly talk about the right-side tail. The left-side tail we just define for the right tail with $-X$. In case of two-sided distribution, like Gauss or Laplace, we get two tails. Once the distribution is symmetric (Gauss, Laplace, Cauchy) the left tail equals to the right one. The situation is different for the asymmetric distributions. According to definition, normal (Gaussian) function is light-tailed. The visualization of normal distribution light tail behavior is sketched in Fig. 6.1.

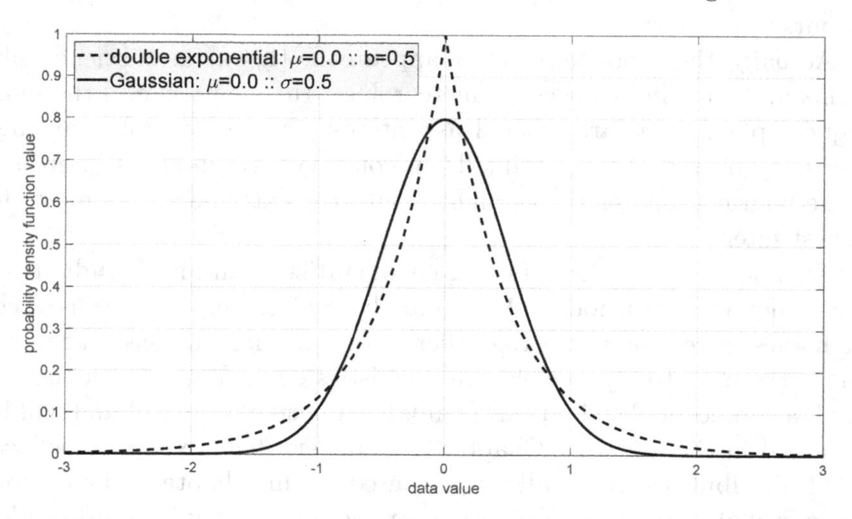

Figure 6.1: Two-sided Gaussian light tails versus exponential distribution

There is one more issue that requires to be explained, i.e. various adjectives used to name the tails. We already know what are light and heavy tails. Thin tail is just another name of the light tail. The long tails are considered to be a sub-type of the heavy tails. For such distributions there is a large number of observations far from peak part of the distribution. More detailed discussions on that subject can be found in [113]. The term of a fat tail is often synonymous to the heavy tail, while sometimes it is also considered as a sub-type of heavy tails. Generally, such a difference is arbitrary and more of a mathematical concept, therefore let assume that both fat and heavy tails apply to the same feature [53].

The phenomena of tails and their existence is one issue. The reason for them is another one. Generally, they are caused by the observations, which appear in too large numbers than usual, far from the expected value. As we already know, these are outlying observations. Thus, the outliers are considered to generate the heavy tails. The source of outliers is just another issue, which will be considered later.

6.1 LAPLACE FUNCTION

The Laplace distribution is also called the double exponential probabilistic density function . It is built as a function of differences between two independent variables with identical exponential distributions. It is named after Pierre-Simon Laplace (1749–1827), who discovered it in 1774. Its PDF is given by the following equations

$$\mathcal{F}^{\text{Lap}}_{\mu,b}(x) = \frac{1}{2b} e^{-\frac{|x-\mu|}{b}}, \tag{6.1}$$

where $\mu \in \mathbb{R}$ is a shift factor, while $b > 0$ represents the scale.

Its shape decays exponentially with a rate depending on the parameter b. This function is symmetric and does not have any shape parameters. Exemplary shapes of Laplace PDF are presented in Fig. 6.2. Following the heavy-tailedness definition, Laplace PDF plays the role of a discriminant between the light- and heavy-tailed functions.

Estimation of Laplace function factors can be done using maximum likelihood [176] or Bayesian approaches [210]. We use the MLE approach implemented as a standard Matlab function. Recently, the Laplace distribution increases its popularity due to its robustness properties. Laplace PDF is popular in engineering sciences [195], particularly in signal processing, fracture problems and navigation.

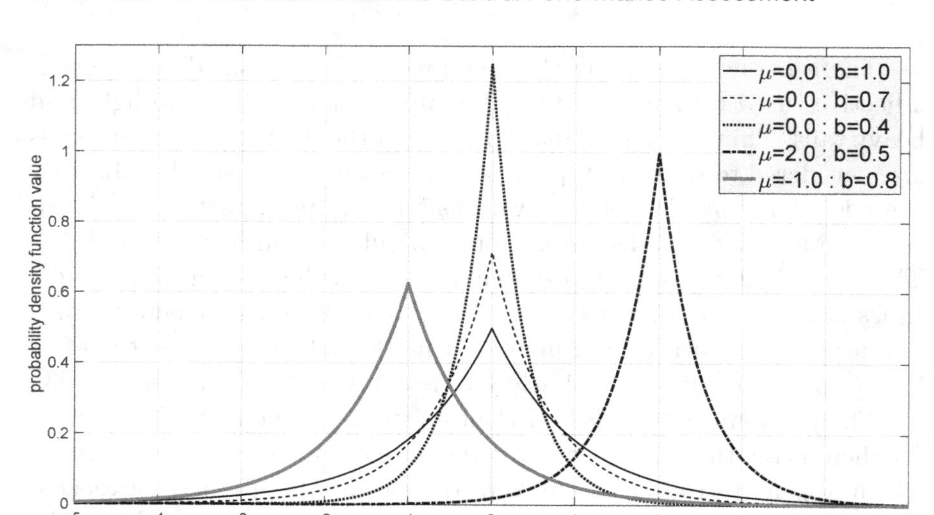

Figure 6.2: Effect of factors on Laplace probabilistic density function

With the analogy to Gaussian distribution is also used as a model for the estimation residuum during the regression modeling. In that perspective it's worth to mention the equivalence between the double-exponential distribution, $\overline{x}_{\text{med}}$ and scale measure MAD – Eq. (4.21). It appears that the MLE estimator of μ is the sample median

$$\hat{\mu} = \text{median}\,(x)\,, \tag{6.2}$$

and the MLE estimator of scale b is the MAD

$$\hat{b} = \frac{1}{N}\sum_{k=1}^{N} |y(k) - \hat{\mu}|\,. \tag{6.3}$$

Following the above statistical inference there is an equivalence of scale coefficient b and the popular integral measure of the mean absolute error (MAE). We calculate this metrics (sometimes named IAE – integral absolute error or the ℓ_1-norm) as the mean integral (or sum in the discrete form) of the absolute residual values over some time period $k = 1, \ldots, N$

$$\text{MAE} = \frac{1}{N}\sum_{k=1}^{N} |y(k) - \hat{y}(k)|\,, \tag{6.4}$$

where N denotes the number of data, y the real observation and $\hat{y}$ its estimated value.

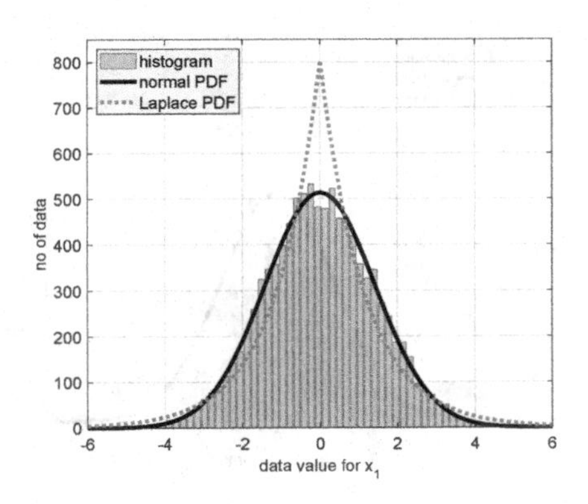

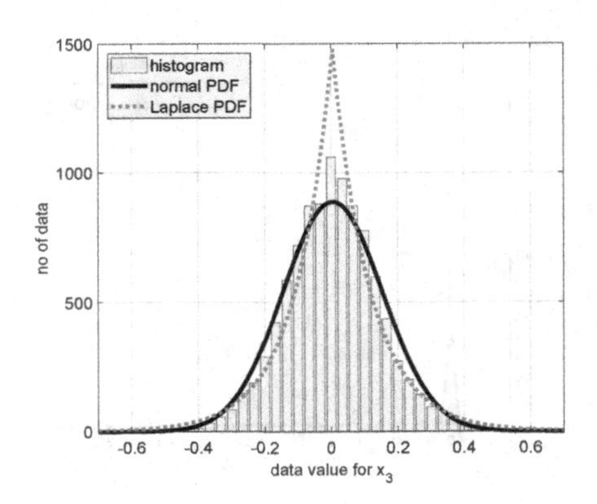

Figure 6.3: Laplace distribution fitting for sample datasets: X_1 and X_3

Finally let's fit Laplace distribution to the sample data used in the book. Fig. 6.3 presents datasets X_1 and X_3. As we have already noticed in the previous chapter, the first time series x_1 is close to the Gaussian PDF and Laplace distribution is not a good choice in that case. Dataset X_3 is neither normally distributed nor with Laplace function.

The other two datasets (X_4 and X_5) shown in Fig. 6.4 have different properties. The X_4 dataset is clearly closer normal than to the Laplace, while the X_5 is not as clear and Laplace function seems to be more appropriate, mostly in the region of the shoulders and tails.

Finally, the Fig. 6.5 refers to the X_2 dataset with extremely peaked distribution with long and heavy tails. The left histogram is plotted using all the data, even those outlying one. In that case Laplace function is a lot better. Once we discard those distance occurrences (subjective

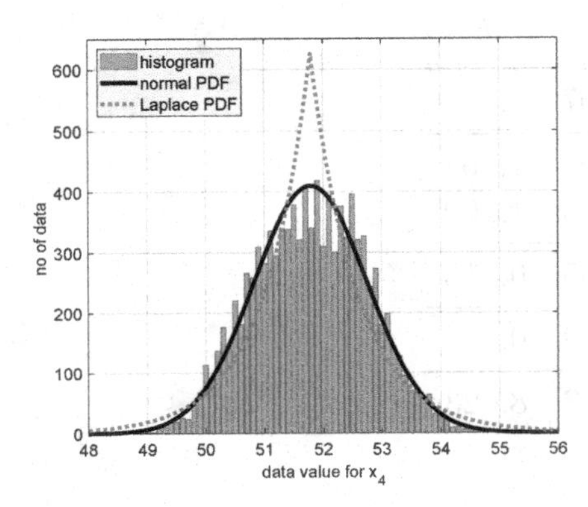

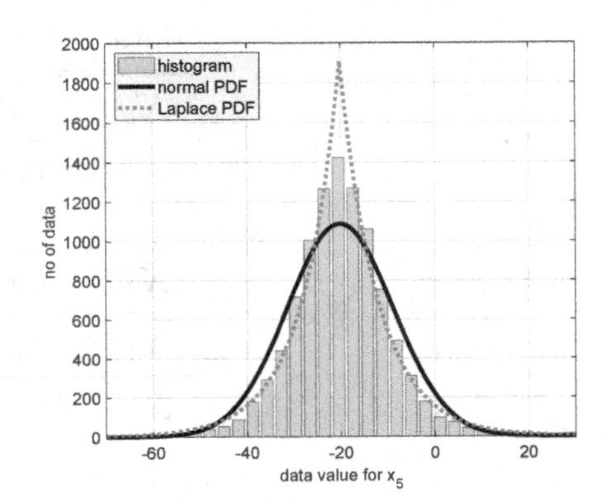

Figure 6.4: Laplace distribution fitting for sample datasets: X_4 and X_5

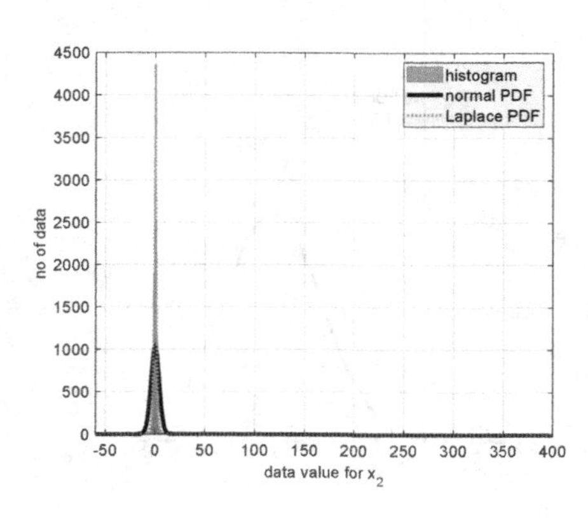

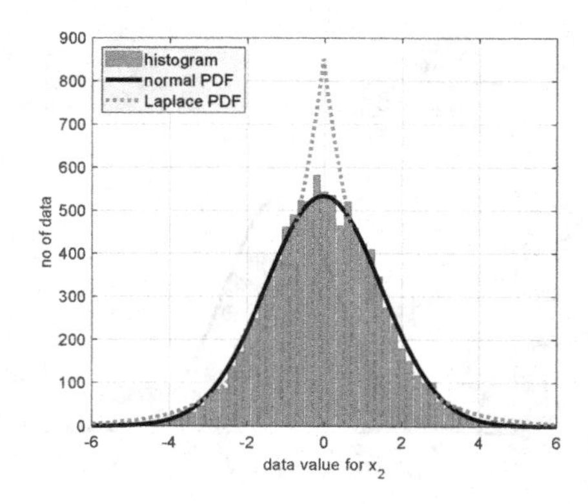

Figure 6.5: Laplace distribution fitting for dataset X_2

assumption), we get the right plot, which is less biased by outliers. As we see that case starts to be closer to Gaussian. As previously, we observe that the arbitrary decision on the outliers removal introduces consequences on the results interpretation. Table 6.1 summarizes for further comparison fitted parameters for all datasets.

This example continues discussion on the PDF properties and differences between them. It confirms that relatively simple plots allow to discover quite fundamental properties of data and allow to make initial decisions, which define the directions for further steps. It also shows that though Laplace distribution is a discriminant between light and heavy tails it is not so frequent in control engineering practice.

Table 6.1: Laplace distribution fitted parameters

data	μ	b
X_1	0.007	1.140
X_2	-0.007	1.262
	-0.033	1.169
X_3	0.005	0.113
X_4	51.789	0.798
X_5	-20.173	8.023

6.2 THE T-STUDENT FUNCTION

The t-distribution, which also called the Student's t-distribution, is a type of probability density function that generalizes normal function. The function shape is similar to the bell of Gaussian distribution, but holds heavier tails. The t-Student PDF was discovered in the second part of the 19th century with works of Helmert and Lűroth followed by the formalization as the Pearson type IV distribution introduced by Karl Pearson [257]. Its shape is described by the following function

$$\mathcal{F}_{\nu}^{\mathrm{t}}(x) = \frac{\Gamma\left(\frac{\nu+1}{2}\right)}{\sqrt{\nu\pi}\Gamma\left(\frac{\nu}{2}\right)}\left(1+\frac{x^2}{\nu}\right)^{-\frac{\nu+1}{2}}, \tag{6.5}$$

where $\nu > 0$ is a degree of freedom and $\Gamma(\alpha)$ a gamma function

$$\Gamma(\alpha) = \int_0^{\infty} x^{\alpha-1}e^{-1}dx. \tag{6.6}$$

The t-Student function gives simple forms in case of specific ν values:

- for $\nu = 1$ we get the Cauchy distribution,
- for $\nu = \infty$ we get the normal distribution.

Sample shapes of t-distribution are sketched in Fig. 6.6. We often use the t-Student function to estimate population parameters for small sample sizes or in case of the unknown variances. The t-distribution, due to heavier tails, better suits to capture extreme values than normal distributions. t-Student function is symmetric. Its shape is similar to the bell shape of a normally distributed variable $\mathrm{N}(0,1)$, being a little bit lower and wider. The functions converge to the Gaussian distribution $\mathrm{N}(0,1)$, as ν increases. Thus, ν is also named as the normality factor.

There exists the natural extension of the t-Student's distribution, which introduces to the function the shift μ and the scale $\sigma > 0$. In such case we obtain a generalized Student's t-distribution function (also named as a t-Location-Scale Distribution)

$$\mathcal{F}_{\mu,\sigma,\nu}^{\mathrm{t}}(x) = \frac{\Gamma\left(\frac{\nu+1}{2}\right)}{\sqrt{\nu\pi}\Gamma\left(\frac{\nu}{2}\right)\sigma}\left(1+\frac{1}{\nu}\left[\frac{x-\mu}{\sigma}\right]^2\right)^{-\frac{\nu+1}{2}}. \tag{6.7}$$

The estimation of the generalized Student's t-distribution function factors can be obtained using different methods, like the Expectation/Conditional Maximization Either (ECME) algorithm [216] or with

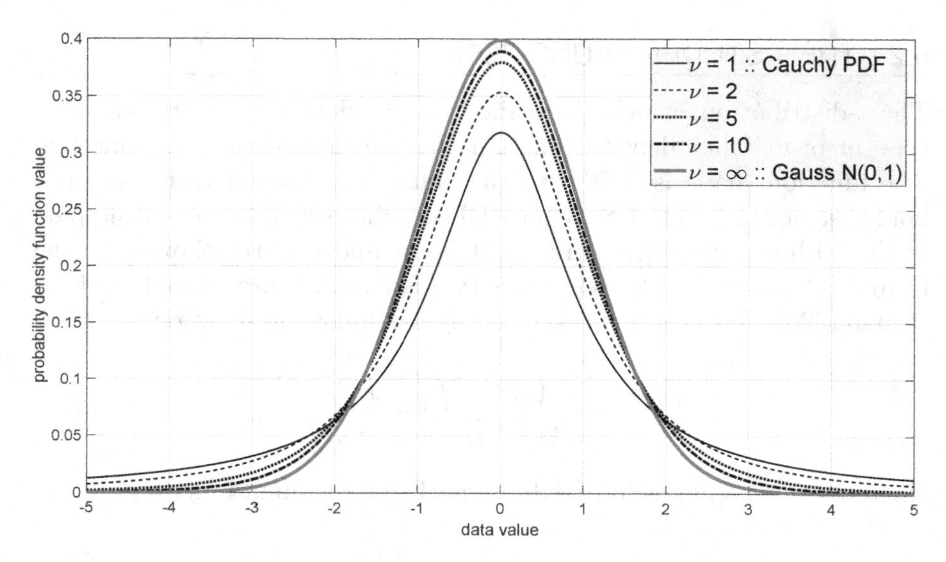

Figure 6.6: Examples of Student's t probabilistic density functions

the constrained nonlinear Nelder-Mead minimization [203]. We use the ECME method in the existing research. The t-Student distribution is mostly used in the hypothesis testing, while the fact that it captures heavy tails is rather not existing in control engineering yet.

Fig. 6.7 presents the visualization for datasets X_1 and X_3. As previously, the x_1 time series is close to the being Gaussian and t-distribution works fine. Dataset X_3 is much better fitted with the t-distribution.

The other two datasets (X_4 and X_5) shown in Fig. 6.8 are also satisfactory captured with the generalized Student's t-distribution.

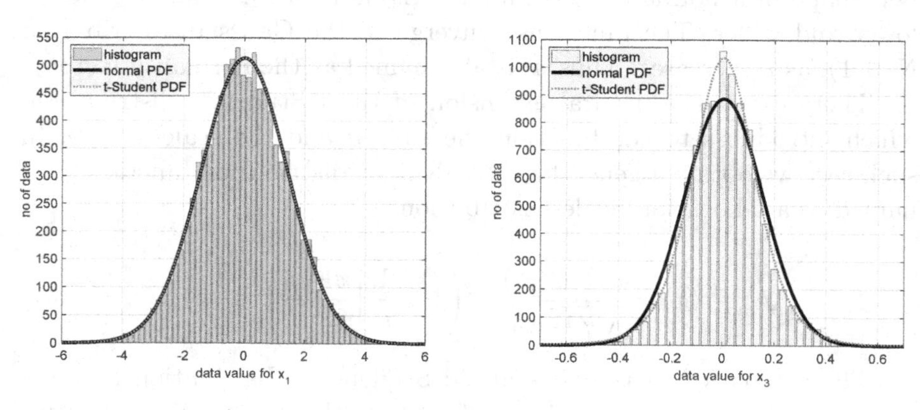

Figure 6.7: Generalized Student's t-distribution fitting for X_1 and X_3

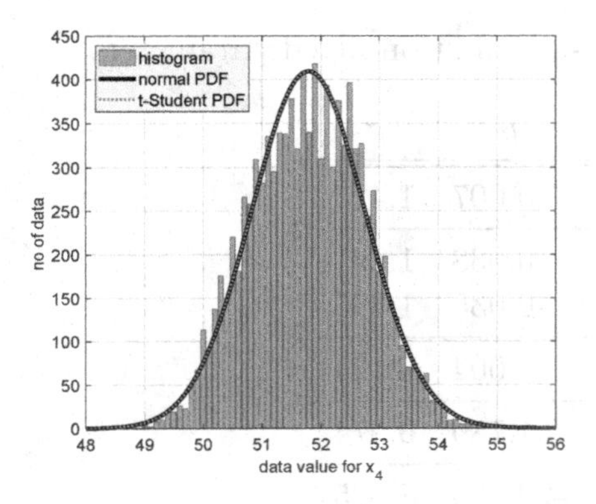

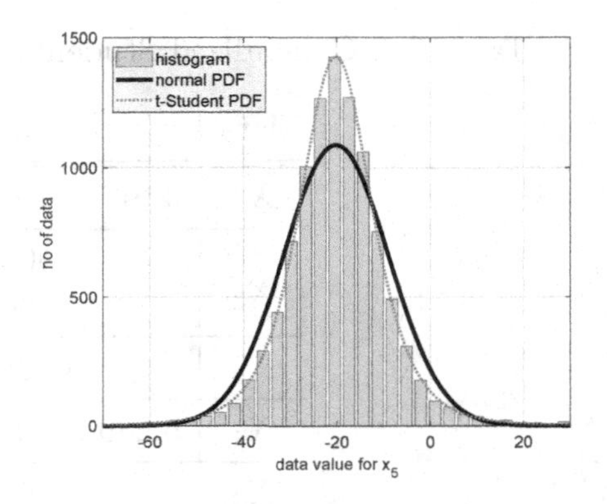

Figure 6.8: Generalized Student's t-distribution fitting for X_4 and X_5

Finally, the Fig. 6.9 refers to the X_2 dataset with extremely peaked distribution with long and heavy tails. The left histogram is plotted using all the data, even those outlying one. In that case generalized Student's t-distribution underestimates the scale factor making very slim peak function. Once we discard distance occurrences (subjective assumption), we get the right plot, which is less biased by outliers. In that case the t-distribution fits better. Table 6.2 summarizes the parameters.

This example confirms previous observations that relatively simple histogram and PDF fitting plots allow to discover fundamental properties of data. They simplify initial decisions. Moreover, the use of the generalized Student's t-distribution improves the hypotheses aboiut data properties through the introduction of heavy tails into the comparison.

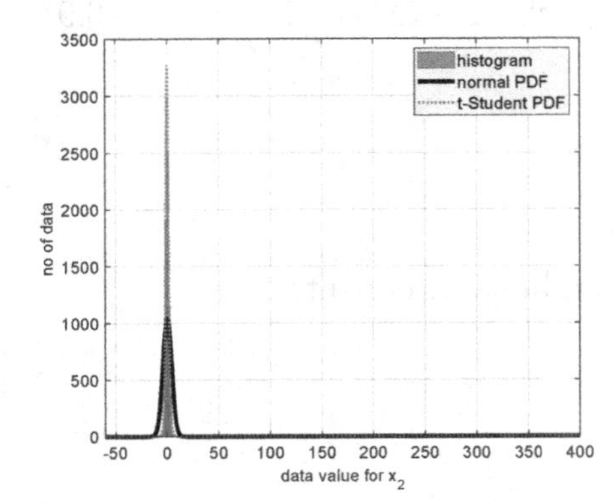

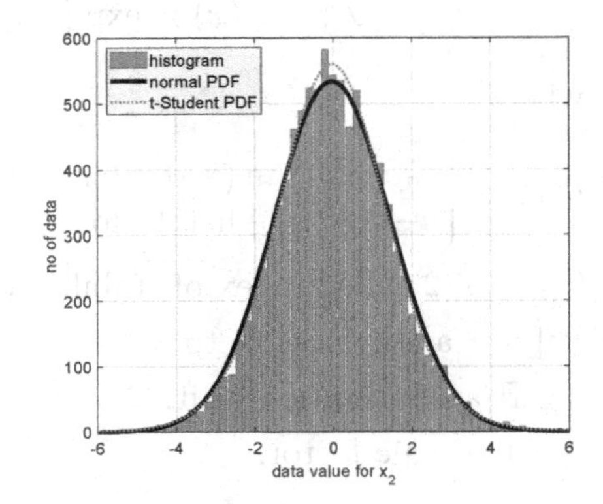

Figure 6.9: Generalized Student's t-distribution fitting for X_2

Table 6.2: Generalized Student's t-distribution fitted parameters

data	ν	μ	σ
X_1	4.917×10^5	0.007	1.425
X_2	5.503	-0.038	1.291
	16.900	-0.035	1.395
X_3	6.438	0.004	0.124
X_4	6.229×10^6	51.789	0.974
X_5	4.153	-20.199	8.075

6.3 THE α-STABLE FAMILY

The α-stable distributions family constitutes an interesting alternative to the previously introduced functions. Given independent and identically distributed random variables $X_1, X_2, \dots, X_n$ and X, the dataset X is said to follow an α-stable distribution, if there exists such a positive constant C_n and a real number D_n that the following relation is satisfied

$$X_1 + X_2 + \cdots + X_n \stackrel{d}{=} C_n \cdot X + D_n, \tag{6.8}$$

where $\stackrel{d}{=}$ denotes equality in distribution. The α-stable distribution cannot be described in a closed PDF form and is expressed through the following characteristics equation

$$\mathcal{F}^{\text{stab}}_{\alpha,\beta,\delta,\gamma}(x) = \exp\left\{i\delta x - |\gamma x|^{\alpha}\left(1 - i\beta l(x)\right)\right\}, \tag{6.9}$$

where

$$l(x) = \begin{cases} sgn(x)\tan\left(\frac{\pi\alpha}{2}\right) & \text{for } \alpha \neq 1 \\ -sgn(x)\frac{2}{\pi}\ln|x| & \text{for } \alpha = 1 \end{cases},$$

$0 < \alpha \leq 2$ called index of stability or stability exponent,

$|\beta| \leq 1$ a skewness factor,

$\delta \in \mathbb{R}$ a distribution shift,

$\gamma > 0$ a scale factor.

It must be noted that for $\alpha < 2$ the 2nd and higher moments do not exist. The characteristics function is shaped by four coefficients. There are special cases with a closed form of the PDF:

- $\alpha = 2$ denotes independent realizations, when for $\alpha = 2$, $\beta = 0$ we obtain exact normal distribution $\mathrm{N}\left(\delta, \sqrt{2}\gamma\right)$, i.e. $\gamma^2 = 2\gamma^2$,
- $\alpha = 1$ and $\beta = 0$ reflects the Cauchy case ,
- $\alpha = 0.5$ and $\beta = 1$ denotes the Lévy case,
- $\alpha = 1$ and $\beta = 1$ combinations address the Landau PDF, while
- $\alpha = 3/2$ and $\beta = 0$ a Holtsmark distribution.

The last three cases are not considered in this work and the narration will address specifically the Cauchy function. It should be noted that in case of skewness factor $\beta = 0$ we obtain the so-called symmetric α-stable distribution, denoted as SαS. The examples of the α-stable functions are shown in Fig. 6.10.

The α-stable distribution has more degrees of freedom than previously described functions, as it is parameterized by four parameters: shift, scale and two shape coefficients. Shift coefficient δ keeps information about function position, but it should not be confused with the mean. Scaling factor γ represents the distribution broadness and measures fluctuations of the respective variable. Parameter β informs about distribution skewness. This factor has analogous meaning to the com-

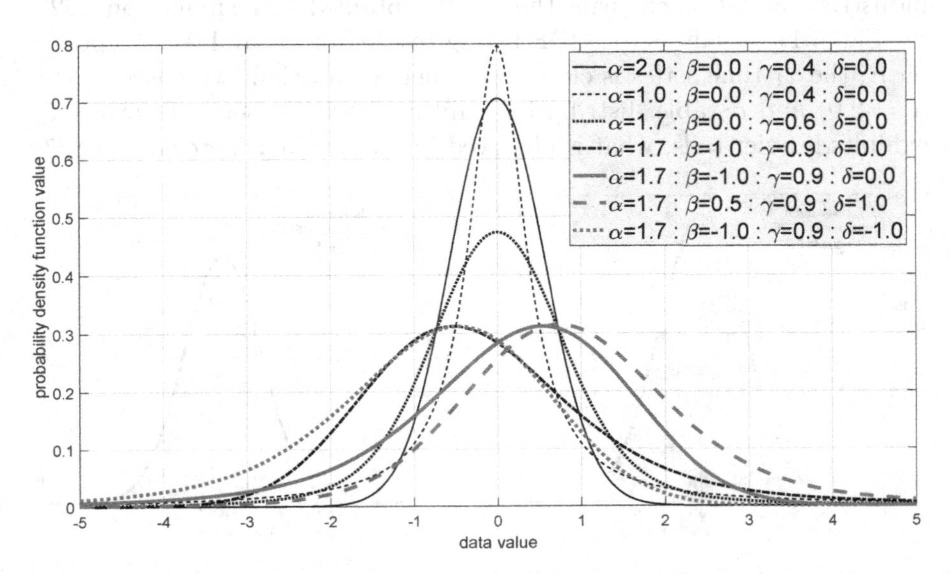

Figure 6.10: Examples of the α-stable probabilistic density functions

mon skewness factor. The second shape factor, i.e. the stability index α is responsible for the heaviness of tails.

Though stable distribution may exhibit infinite variance, it is not anything wrong. We shouldn't eliminate stable functions from the control engineering analyzes. A lot of statistical processes exhibit the "infinite variance", however may have finite, and often very well-behaved, mean deviations [325].

We may estimate the α-stable PDF coefficients in many ways [290]: like using the most common iterative Koutrouvelis approach based on the characteristics function estimation [197], fast but not very accurate quantiles algorithm [235], logarithmic moments method [201] or the maximum likelihood (MLE) approach [36], which achieves the highest accuracy but at the highest calculation cost. The Koutrouvelis approach is used in this book as a well-balanced compromise between the accuracy and the calculation time.

Industrial data analysis shows that it well suites to describe tailedness in control engineering problems. The research on α-stable functions has shown that in several cases they can be even better than the Gaussian "predecessors". Analysis of the loops controlled by the PID [81] and GPC predictive controller [99] has shown their robustness to various noises and disturbances having different statistical properties. Their industrial validation confirms the results obtained with simulations [82].

Fig. 6.11 visualizes the PDF fitting results for X_1 and X_3. As previously, the x_1 time series is close to the being Gaussian (we observe that $\alpha = 2.0$) and αstable distribution captures that property. Dataset X_3 exhibits heavier tails, what is observed in the stability factor $\alpha = 1.867$.

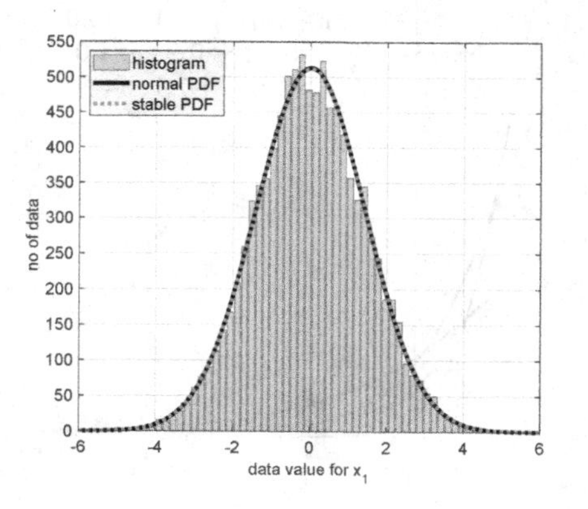

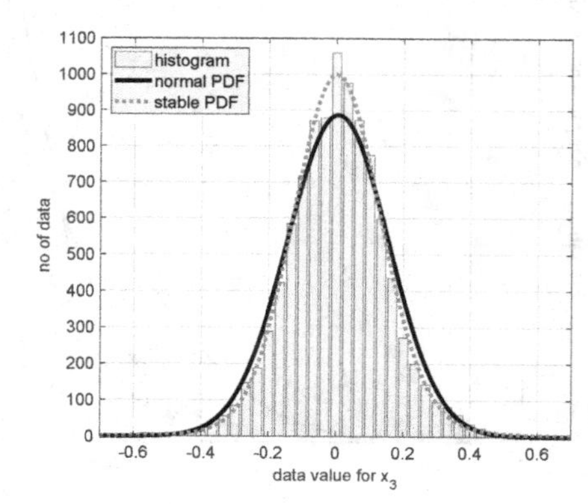

Figure 6.11: The α-stable distribution fitting for X_1 and X_3

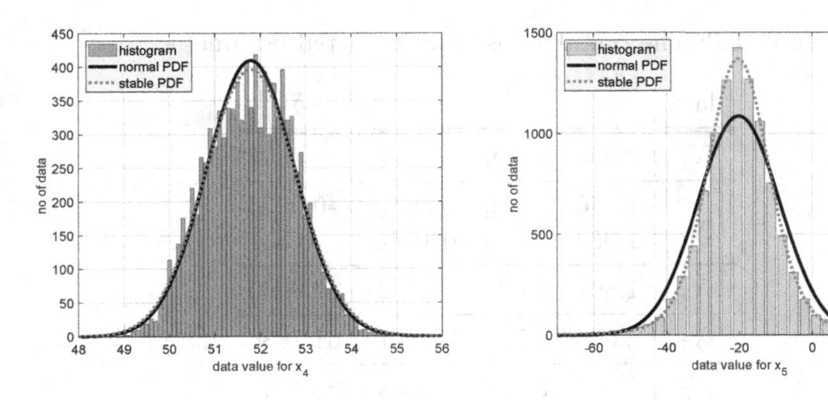

Figure 6.12: The α-stable distribution fitting for X_4 and X_5

The next datasets (X_4 and X_5) shown in Fig. 6.12 are also properly reflected by the α-stable distribution. Data X_4 exhibits independent realizations (Gaussian property) with $\alpha = 2.0$, while time series X_5 has heavier tails $\alpha = 1.759$. It appears that this data are characterized with the heaviest tails measured with the index of stability α, which confirms visual observations.

Finally, the Fig. 6.13 refers to the X_2 dataset with extremely peaked distribution with long and heavy tails. Its stability factor equals to $\alpha = 1.876$. The left histogram is plotted using all the data, even those outlying one. Once we discard outliers (subjective assumption), we get the right plot, which is less biased. This visual observation is confirmed by the increases stability index $\alpha = 1.963$ In that case the t-distribution fits better. Table 6.3 summarizes all the parameters.

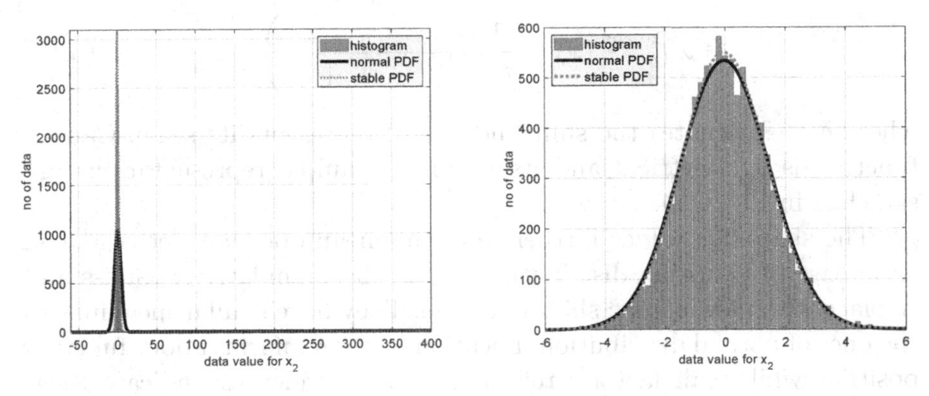

Figure 6.13: The α-stable distribution fitting for X_2

Table 6.3: The α-stable distribution fitted parameters

data	α	β	γ	δ
X_1	2.000	1.010	-1	0.01
X_2	1.876	1.016	0.0657	-0.0294
	1.963	1.020	0.1671	-0.031
X_3	1.867	0.094	0.2592	0.0065
X_4	2.000	0.712	0.0381	51.7886
X_5	1.759	6.346	0.0638	-20.1243

This example confirms previous observations. We may also see that due to the fact that the α-stable distribution has the highest number of coefficients and its distribution function can be shaped with the highest degrees of freedom we obtain the best fitting. Moreover, interpretation of the phenomena behind data is the easiest as each statistical factor $\alpha, \beta, \gamma, \delta$ provides another piece of the puzzle to build an overall picture.

6.3.1 Cauchy probabilistic density function

Cauchy PDF is a member of the family of stable functions and a representative example of the heavy-tailed distribution. It is known under different names, such as Lorentz, Cauchy-Lorentz or a Breit-Wigner distribution, or Lorentz(-ian) function . The Cauchy distribution is described by the following function

$$\mathcal{F}_{\delta,\gamma}^{\text{Cauchy}}(x) = \frac{1}{\pi\gamma}\left(\frac{\gamma^2}{(x-\delta)^2+\gamma^2}\right), \tag{6.10}$$

where $\delta \in \mathbb{R}$ denotes the shift and $\gamma > 0$ represents its scale. Cauchy function is symmetrical and its graphical sample representations are sketched in Fig. 6.14.

The shape for values further from the mean does not decay as fast as in case of Gaussian distribution. It also has, similarly to Gauss and Laplace, two parameters: shift and scale. They have similar meanings to the ones of normal distribution. Location factor δ informs about function position, while scale factor γ reflects function slenderness indicating signal variability. Similarly to normal standard deviation optimal value of γ is unknown and we have to more rely on its relative changes. Though

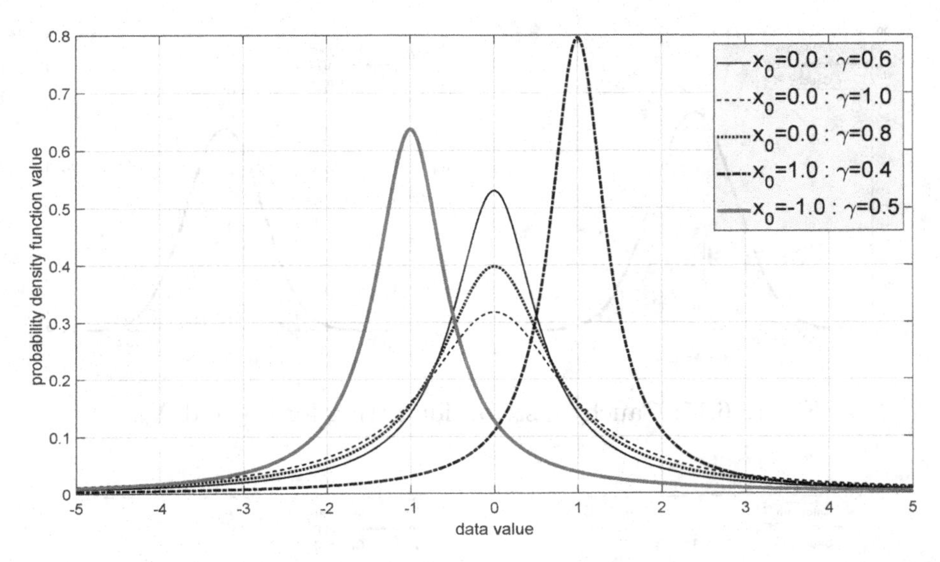

Figure 6.14: Examples of Cauchy probabilistic density functions

this function has the greatest popularity in physics, its feasibility has been also shown in control engineering research [81, 99, 82].

There are several methods dedicated toward finding the proper Cauchy distribution for the certain time series [301]. The fitting of the PDF shape to the histogram with the maximum likelihood estimation is one of the most popular and robust ones [112]. The method using the successive quadratic programming solver [11] is used in this work.

Figs. 6.15...6.17 show the fitting results for all considered datasets. It is done in the same way as previously. Table 6.4 summarizes estimated factors of the Cauchy functions.

Table 6.4: Cauchy distribution fitted parameters

data	μ	γ
X_1	-0.018	0.8761
X_2	-0.049	0.8929
	-0.049	0.8846
X_3	0.004	0.0823
X_4	51.807	0.6357
XX_5	-20.241	5.5936

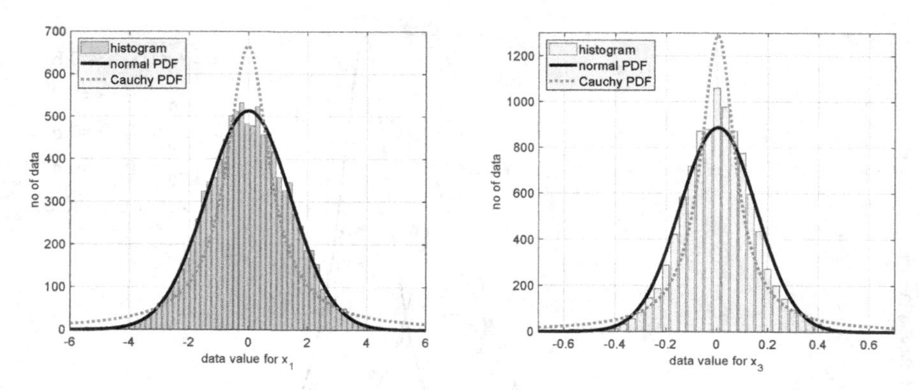

Figure 6.15: Cauchy distribution fitting for X_1 and X_3

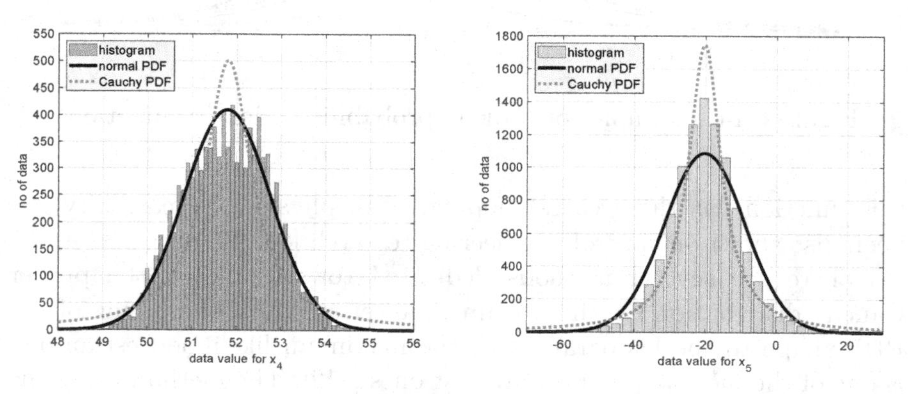

Figure 6.16: Cauchy distribution fitting for X_4 and X_5

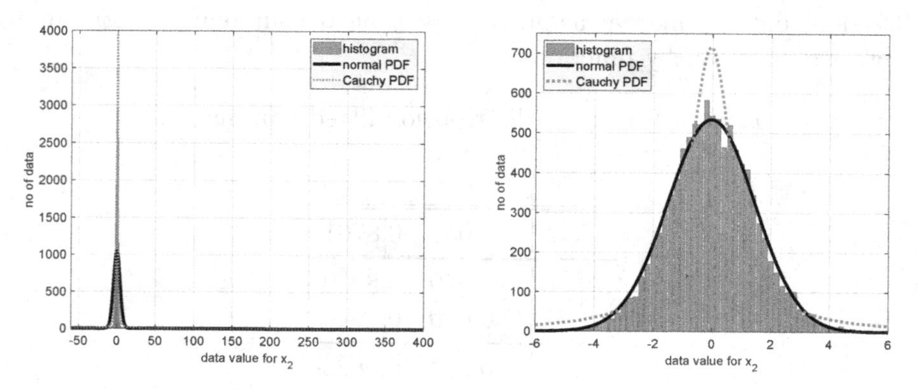

Figure 6.17: Cauchy distribution fitting for X_2

We can notice that in none of the considered cases Cauchy function shows proper fit. It is clearly too narrow and the heavy tails it takes into account are not reflected in the data. Only in the case of the X_5 dataset does its fitting begin to approach the data histogram, but the lowest stability index α occurs for this dataset.

The fact that this function does not fit in the cases under consideration does not discriminate against it and is worth remembering, although the theoretically stable α-distribution includes it as a special case.

6.4 CONCLUDING REMARKS

This chapter introduces four tailed functions that might be considered as an alternative to the normal distribution in case of data biased by tails. We show that different data files exhibit various statistical properties, starting from the dataset X_1, which is considered Gaussian, up to X_5 which exhibits heaviest tails.

The discussions included in the last two chapters will be nicely concluded by the comparison of all the functions in one picture. Dataset X_5 seems to be the most interesting as its tails are the most significant. Fig. 6.18 presents its histogram and six fitted functions: classical Gaussian and its robust version, Laplace, generalized t-Student, Cauchy and α-stable.

Generally, the visual comparison is only qualitative and might be confusing. Therefore, we may calculate the MAE indexes of the fitting residuum between the probabilistic density function of a given distribution and the histogram. The evaluated indexes are presented in Table 6.5 ordered from the worst fitted to the best one.

We clearly see that two limiting distributions, i.e. Gaussian ($\alpha = 2.0$) and Cauchy ($\alpha = 1.0$) are the worst and the least suitable. Two functions with the highest number of shaping degrees of freedom ensure the best fit. The α-stable function, which is characterized by four coefficients (shift, scale and two shape factors), assures the closest realization of the histogram. Similarly, the generalized Student's t-distribution function, which possesses three coefficients (shift, scale and shape factor) allows

Table 6.5: The MAE residuum indexes for PDF fitting to the X_5

	Cauchy	Gauss	Laplace	robust	t-Student	α-stable
MAE	35.42	27.39	19.32	10.97	5.43	5.35

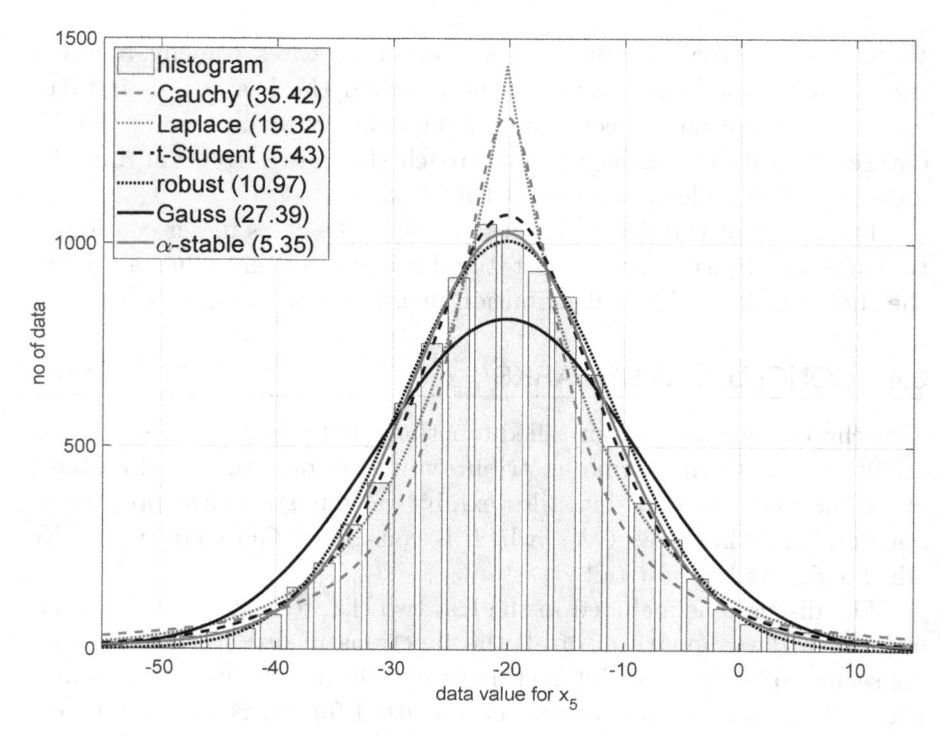

Figure 6.18: Histogram and various PDFs fitting efficiency for X_5

almost similar fitting. We assume that once the histogram starts to be asymmetric the superiority of the stable function would be more visible.

These two functions take into account the tails as the significant element of the represented shape and its properties are resembled in their parameters. They are **tail-aware**, in simple words. They seriously take into account all the outlying observations assuming that they carry important information.

The fitting with the robust functions presents an opposite tail sensitivity – the **tail-ignorant** approach. The methodology that uses robust statistics neglects tails. It focuses on the peak and shoulder parts of the histogram with no attention to the tails. The approach leads to the removal of all outliers, no matter what they are and what information they contain. This established control engineering approach, which is hardly agreeable, ignores and loses (consciously or not, it doesn't matter) a part of knowledge about the data. It should be used only consciously and only in justified cases.

The selection of proper distribution function in the time series analysis matters and mustn't be done haphazardly, without any reflection.

GLOSSARY

MAE: Mean Absolute Error

IAE: Integral Absolute Error

ECME: Expectation/Conditional Maximization Either

SαS: Symmetric α-Stable distribution

MLE: Maximum Likelihood Estimation

CHAPTER 7

Extreme statistics – the power of tails

EXTREME STATISTICS are rare as extremes occur infrequently. Though they appear occasionally, the consequences of their appearance can be very serious. Extreme statistics or the extreme value theory is also named as the extreme value analysis (EVA). This is the part of the statistics that addresses the extreme deviations of some random variable. The use of the extreme statistics has developed during last 60 years. The applications of EVA are common in life sciences (meteorology, hydrology, oceanology) or economics, because in those areas extreme events, which exceed typical behavior have significant relevance.

The extreme value distributions are the limiting functions of the variable extremes (minima or maxima) of a set of random variables. The most popular extreme statistics distribution is a Gumbel distribution, which occurs to be the limiting function for the Gaussian and exponential distributions [9]. The above definition of the problem entails a natural consequence regarding the requirements for the probability distributions that would be of interest. Such a PDF function should only take values of one sign (mostly positive), such as we encounter with rainfalls, river discharges, or financial risk levels.

Direct definition of the extreme theory as such might have a very limited interest in engineering, especially in the domain of control. One could imagine the impact of the meteorological phenomena on technical installations, like for instance river dam reservoir level and flood control systems. However, there are quite a lot of situations, where variables which are under control take only positive values with a tendency to be close to zero and rare exceeding observations. Emissions, like carbon

DOI: 10.1201/9781003604709-7

monoxide or particulate matter levels at the combustion stack are good examples. Communication traffic data form another origin of extreme-like datasets, where the problems are quite analogous to hydrology or emissions.

The above examples of technical issues share one common element. Our interest no longer focuses on typical (average) values. Extreme (abnormal, outlying) observations become more interesting, as they are the ones that, although rare, determine the safety of the system. Extremely high precipitation can cause flooding, anomalies in high or low temperatures may cause system shutdown, abnormally high traffic on a telecommunications network can cause it to be blocked, and emissions that exceed environmental limits will result in high penalties or even the need to shut down the entire installation. Therefore, it is worth to introduce statistic models and methods used in the extreme theory to the wider control engineering community.

The presentation of selected distributions starts from the simplest ones, i.e. those that exhibit only scale factor followed by functions with one and two additional shape factors. These distributions apply to one sided asymmetrical data, which generally has only positive (or negative) values and their expected, required or the most frequent, value is zero. Data used previously to exemplify statistical issues are improper. This story is visualized with separate datasets. Four time series are used: $Y_1, \ldots, Y_4$. They have the same length $N = 10080$ and originate from the industry. Fig. 7.1 presents data time trends.

Following previous discussions one may evaluate basic statistics for them. Despite the fact that nothing prevents their determination, already a simple look at these data shows that they differ from previous data (compare Fig. 4.1). This may mean that the classical calculation of the mean, standard deviation, etc. is perhaps not adequate. The above observation is key. You can't evaluate some statistical indicators without a second thought just because everyone is calculating them, or just because in previous analyzes we used them. Decisions should be made consciously and relevant to the situation. In such a situation it is suggested to extend visual data inspection. Fig. 7.2 presents evaluated histograms for all five datasets.

Completely different nature of data is apparent at a first glance. They take only positive values, while the shape of the histogram is decidedly asymmetrical and one-sided (compare previously used data in Fig. 4.7). The peak zone is narrow, while on the positive side a very long tail stretches. It represents rare, but distant from the peak zone observations.

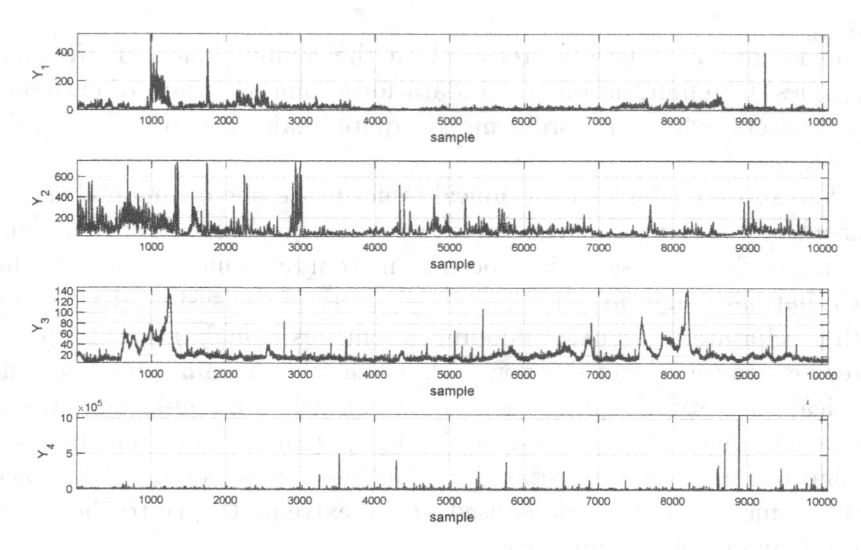

Figure 7.1: Sample $Y_1, \ldots, Y_4$ data for $i = 1, \ldots, 2500$

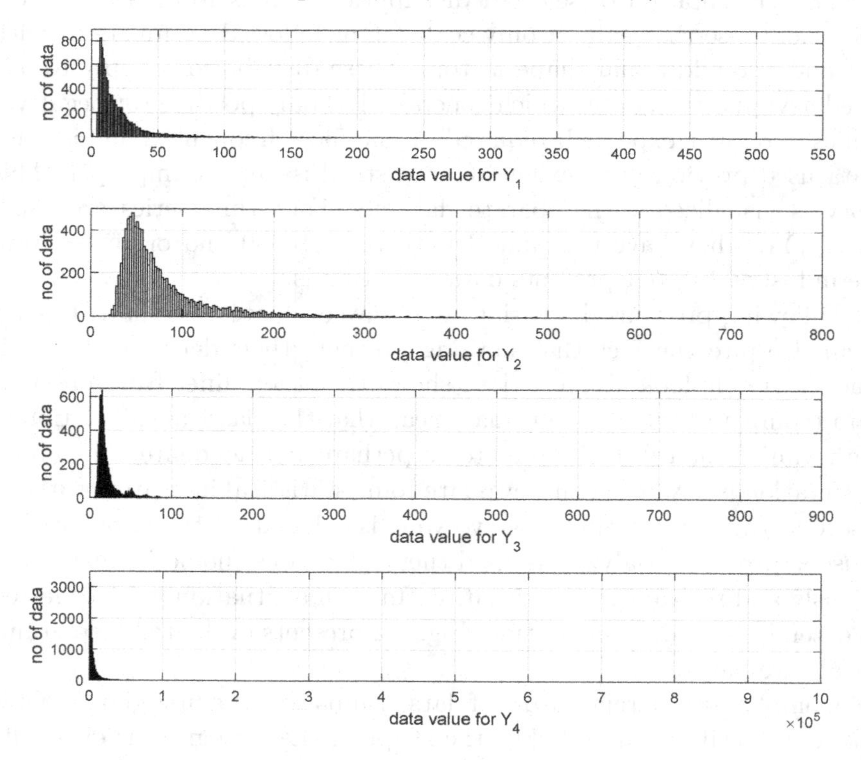

Figure 7.2: Histogram plots for $Y_1, \ldots, Y_4$ data

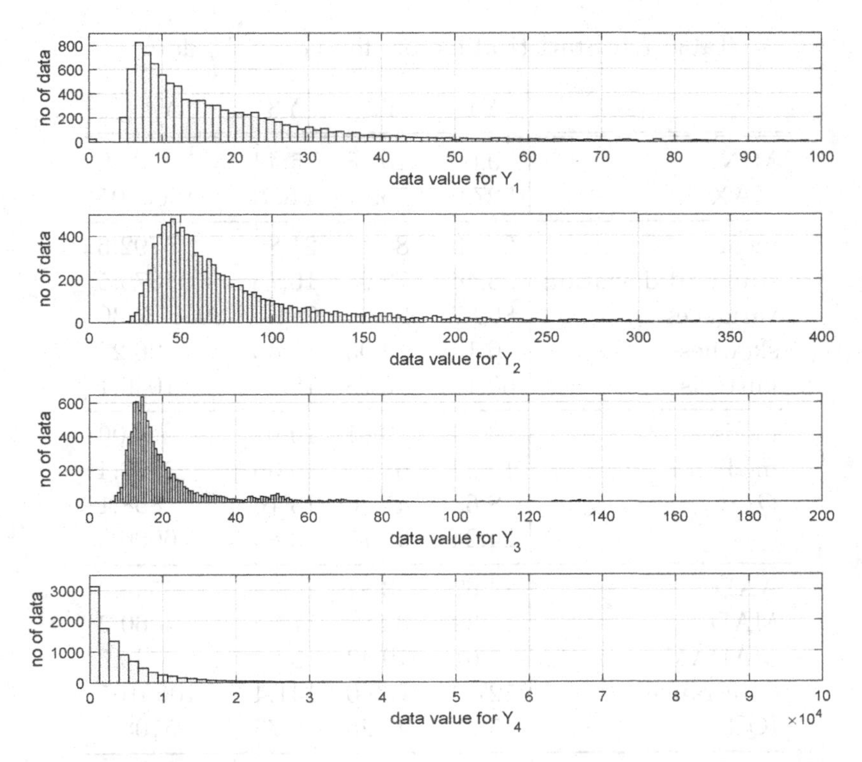

Figure 7.3: Histograms for $Y_1, \ldots, Y_4$ data – the peak area

Such an observation must immediately turn on a red light for us that the standard approach should be changed. The observed right tails decisively affect any statistical coefficient estimates. And from the point of view of further analysis, the shape of the peak zone is important. The next diagrams included in Fig. 7.3 show on an enlarged scale this part of the histogram graph. We see that despite the existing long tail its decay shape and simultaneously the peak zone differ, what means that the data exhibit various statistical properties.

As in case of improper PDF assumption the estimators might be misleading; the evaluation of statistical factors should be performed with caution. Table 7.1 presents the most popular statistical factors potentially applicable during data analysis.

The use of MIN and MAX values is straightforward, obvious and without any doubts. Following that the data range is also proper. However, the others are not so obvious. Fig. 7.4 shows the reason. Once we calculate classical estimator of mean and standard deviation we obtain

Table 7.1: Statistical factors for $Y_1, \ldots, Y_4$ data

	Y1	Y2	Y3	Y4
MIN	0.00	20.08	5.44	0.00
MAX	527.9	759.1	146.9	1060105
mean	21.90	84.17	21.85	5792.6
standard deviation	28.46	72.68	16.78	19879.5
variance	809.7	5283.0	281.6	395195530
skewness	6.13	4.03	3.46	26.22
kurtosis	60.14	26.78	18.77	1040.1
mode	6.59	49.61	10.01	0.00
median	14.22	61.26	16.36	2777.4
Q_1	8.63	45.50	13.16	898.1
Q_3	24.24	92.75	22.86	6006.3
AAD	14.67	43.87	10.30	5545.2
MAD	11.32	33.88	7.55	4060.2
MADAM	6.61	19.43	3.99	2248.9
data range	527.9	739.0	141.4	1060105
IQR	15.61	47.25	9.70	5108.3
l_1	21.90	84.17	21.85	5792.6
l_2	10.06	29.80	6.94	3793.0
τ_2u2	0.46	0.35	0.32	0.65
L-skewness	0.51	0.47	0.49	0.58
L-kurtosis	0.35	0.31	0.33	0.43
M-logist shift	17.29	71.23	18.41	3813.2
M-logist scale	10.05	30.03	6.20	3252.4

some distribution function, which by no means fits the data. Therefore, these shift and scale estimators are improper. Similar observation applies to their robust variants. We may use quartiles, like median, the first Q_1 or the third Q_3 quartile as they are independent on any distribution assumption. Also L-moments might be used. Unfortunately, other estimators are biased and they might lead to falsified decisions.

Above short discussion aims at showing that the statistical factors should not be used without prior data inspection that can be easily done using the time series plotting and histogram review. This step will protect us from unknowingly making basic mistakes that could affect

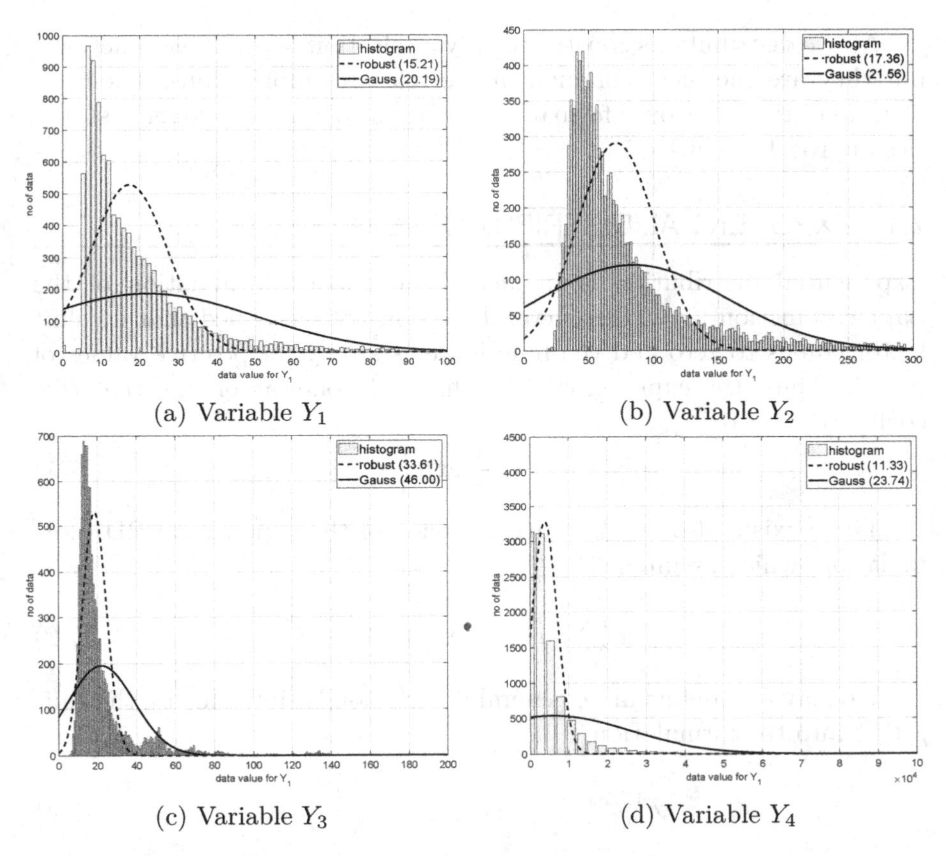

(a) Variable Y_1

(b) Variable Y_2

(c) Variable Y_3

(d) Variable Y_4

Figure 7.4: Histogram and normal PDFs fitting for extreme sample data

further analysis. Knowing that we can introduce the probabilistic density functions, which are appropriate for the analysis of one-sided data:

- Exponential (EXP),
- Gamma (GAM),
- Lognormal (LGN),
- Weibull (WEI),
- Generalized Extreme Value (GEV),
- General Pareto (GPA),
- Generalized Logistic (GLO),
- Four-parameter Kappa family (K4P).

Above distributions constitute only a selection of available functions, but they are the most common in the relevant applications. Their description brings not only formal issues, but also explains extreme statistics approach as such.

7.1 EXPONENTIAL DISTRIBUTION (EXP)

Exponential distribution is in fact the one-sided formulation of the Laplace function [283]. In general definition, it is assumed that its shift factor equals to zero and the function exists only for positive values of $x \geq 0$. Thus, the exponential PDF has only one factor, i.e. the scale coefficient $\beta > 0$

$$\mathcal{F}_{\beta}^{\text{EXP}}(x) = \frac{1}{\beta} \exp\left(-\frac{x}{\beta}\right). \tag{7.1}$$

There exists also an alternative version of the exponential PDF formulation, which assumes $\lambda = \frac{1}{\beta}$

$$\mathcal{F}_{\lambda}^{\text{EXP}}(x) = \lambda \exp\left(-\lambda x\right). \tag{7.2}$$

We may also define more general description by introducing the shift $\mu \in \mathbb{R}$ into the formulation

$$\mathcal{F}_{\mu,\beta}^{\text{EXP}}(x) = \frac{1}{\beta} \exp\left(-\frac{x-\mu}{\beta}\right). \tag{7.3}$$

Exemplary shapes of the exponential distribution function are sketched in Fig. 7.5. Following the heavy-tailedness definition and by the analogy to Laplace function, exponential distribution shape plays the role of a discriminant between the light- and heavy-tailed functions.

Moments of Exponential distribution exist and are given by:

$$\text{mean} = \beta = \frac{1}{\lambda}, \tag{7.4}$$

$$\text{median} = \beta \ln 2 = \frac{1}{\lambda} \ln 2, \tag{7.5}$$

$$Q_1 = \beta \ln \frac{4}{3} = \frac{1}{\lambda} \ln \frac{4}{3}, \tag{7.6}$$

$$Q_3 = \beta \ln 4 = \frac{1}{\lambda} \ln 4, \tag{7.7}$$

$$\text{mode} = \mu, \tag{7.8}$$

$$\text{range} = \left(\mu, +\infty\right), \tag{7.9}$$

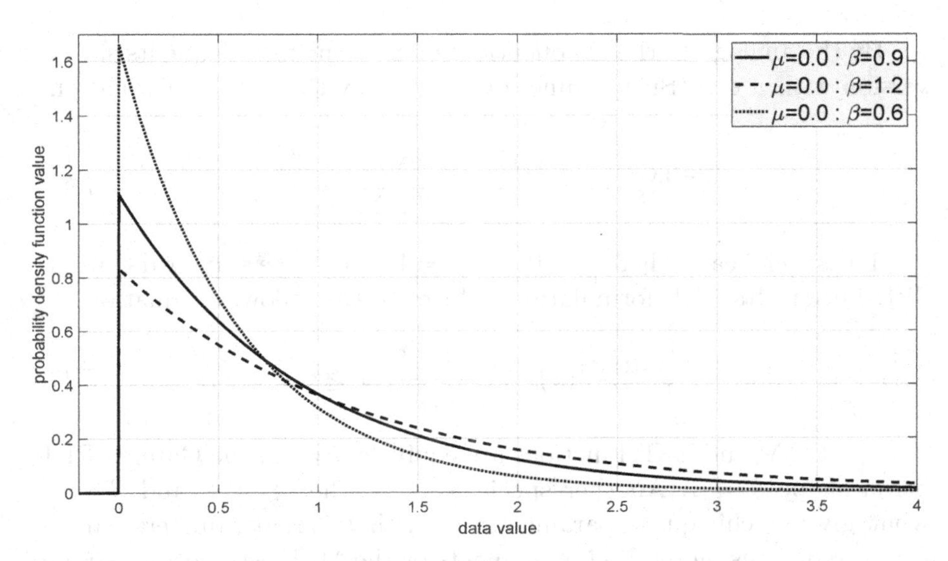

Figure 7.5: Examples of Exponential probabilistic density functions

$$\text{standard deviation} = \beta = \frac{1}{\lambda}, \tag{7.10}$$

$$\text{coefficient of variation} = 1, \tag{7.11}$$

$$\text{skewness} = 2, \tag{7.12}$$

$$\text{kurtosis} = 9. \tag{7.13}$$

The exponential distribution is primarily used in reliability analyzes to model the data with a constant failure rate [23], which is reflected in the hazard plot, which is simply equal to a constant value.

7.2 GAMMA FUNCTION (GAM)

The origins of the Gamma distribution can be traced back to Laplace considerations, who defined it as a *precision constant* distribution. Similarly to the exponential function Gamma the PDF is defined assuming zero shift value $x \geq 0$, i.e. $\mu = 0$ [283]. It's defined as follows

$$\mathcal{F}_{\beta,\xi}^{\text{GAM}}(x) = \frac{\left(\frac{x}{\beta}\right)^{\xi-1} \exp\left(-\frac{x}{\beta}\right)}{\beta\Gamma(\xi)}, \tag{7.14}$$

where $\beta > 0$ denotes *scale*, $\xi > 0$ *shape* and $\Gamma(\cdot)$ is a Gamma function

$$\Gamma(x) = \int_0^\infty t^{x-1} e^{-t} dt. \tag{7.15}$$

By the analogy to the exponential function one may define its shifted version with $\mu \in \mathbb{R}$. Such a function is given by the following equation

$$\mathcal{F}^{\text{GAM}}_{\mu,\beta,\xi}(x) = \frac{\left(\frac{x-\mu}{\beta}\right)^{\xi-1} \exp\left(-\frac{x-\mu}{\beta}\right)}{\beta\Gamma(\xi)}. \tag{7.16}$$

The special case where $\mu = 0$ and $\beta = 1$ we call the standard Gamma distribution. Its PDF formulation reduces to the following equation

$$\mathcal{F}^{\text{stGAM}}_{\xi}(x) = \frac{x^{\xi-1}e^{-x}}{\Gamma(\xi)}, \; x \geq 0. \tag{7.17}$$

The GAM and EXP functions are equivalent when the Gamma PDF has a shape $\xi = 1$. Another special case is when $\xi = \frac{\nu}{2}$ and $\beta = \frac{1}{2}$, what gives a chi-square parameterized with ν. Its parameters can be fitted to data using method of moments or the MLE estimation. Sample Gamma distribution shapes are shown in Fig. 7.6. Main statistics of Gamma distribution are given explicitly, however the median doesn't have a closed-form formulation:

$$\text{mean} = \xi, \tag{7.18}$$

$$\text{mode} = \begin{cases} \xi - 1, & \text{for } \xi \geq 1, \\ 0, & \text{for } \xi < 1, \end{cases} \tag{7.19}$$

$$\text{range} = \langle 0, +\infty), \tag{7.20}$$

$$\text{standard deviation} = \sqrt{\xi}, \tag{7.21}$$

$$\text{coefficient of variation} = \frac{1}{\sqrt{\xi}}, \tag{7.22}$$

$$\text{skewness} = \frac{2}{\sqrt{\xi}}, \tag{7.23}$$

$$\text{kurtosis} = 6 + \frac{6}{\xi}. \tag{7.24}$$

Gamma function has been successfully used in various applications, such as economy (insurance models) [2], telecommunication [162] (wireless communication), meteorology [7] (rainfall modeling), or medicine [59] (cancer frequency or neurology signals).

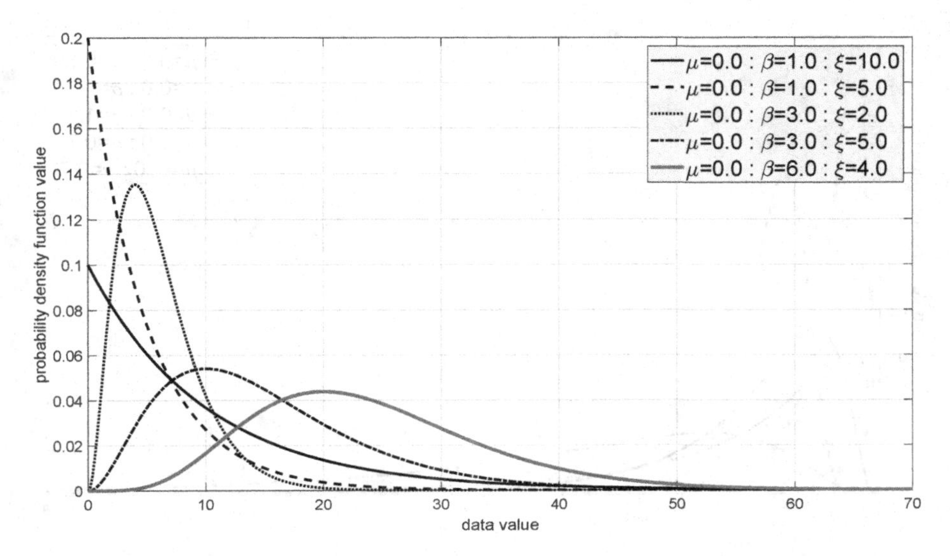

Figure 7.6: Examples of Gamma probabilistic density functions

7.3 LOGNORMAL FUNCTION (LGN)

A given variable is said to be lognormally distributed if some $Y = \ln(X)$ is normally distributed and "ln" denotes the natural logarithm [283]. General formulation of the lognormal distribution is given by function

$$\mathcal{F}^{\mathrm{LGN}}_{\mu,\sigma}(x) = \frac{1}{(x-\mu)\,\sigma\sqrt{2\pi}} \exp\left(\frac{-(\ln x - \mu)^2}{2\sigma^2}\right), \quad x > \mu, \tag{7.25}$$

where $\mu \in \mathbb{R}$ is a *shift* and $\sigma > 0$ is *scale* parameter, simultaneously being its standard deviation.

The function for the standard lognormal distribution is given by

$$\mathcal{F}^{\mathrm{stLGN}}_{\sigma}(x) = \frac{1}{x\sigma\sqrt{2\pi}} \exp\left(\frac{-(\ln x)^2}{2\sigma^2}\right). \tag{7.26}$$

Its parameters can be fitted to data using the maximum likelihood estimation. Sample LGN distribution shapes are shown in Fig. 7.7.

Main statistics of the Lognormal PDF are given explicitly as below:

$$\text{mean} = e^{\mu+0.5\sigma^2}, \tag{7.27}$$

$$\text{median} = e^{\mu}, \tag{7.28}$$

$$\text{mode} = e^{\mu-\sigma^2} \tag{7.29}$$

$$\text{range} = \langle 0, +\infty), \tag{7.30}$$

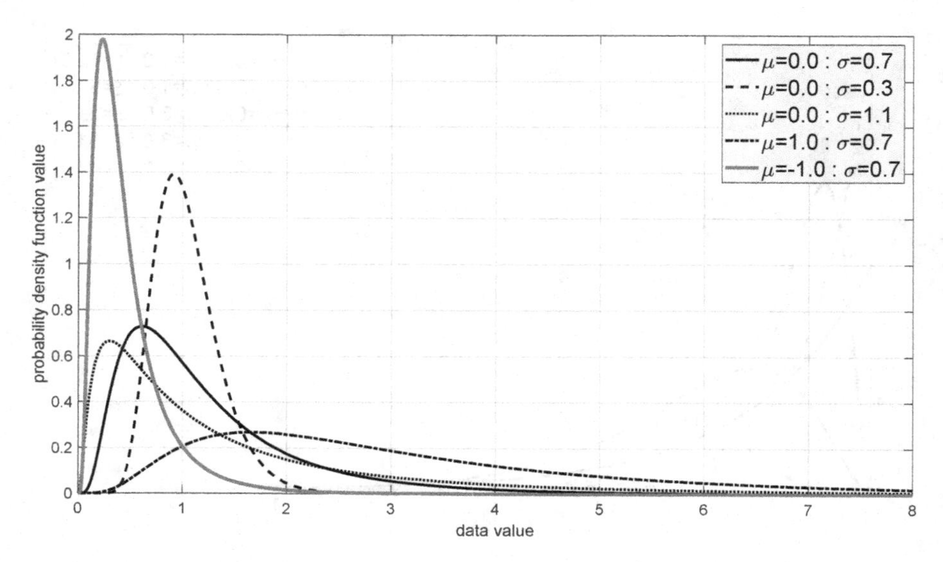

Figure 7.7: Examples of Lognormal probabilistic density functions

$$\text{standard deviation} = e^{\mu+0.5\sigma^2} \cdot \sqrt{e^{\sigma^2}-1}, \tag{7.31}$$

$$\text{coefficient of variation} = \sqrt{e^{\sigma^2}-1}, \tag{7.32}$$

$$\text{skewness} = \left(e^{\sigma^2}+2\right)\sqrt{e^{\sigma^2}-1}, \tag{7.33}$$

$$\text{kurtosis} = 3 + \left(e^{\sigma^2}-1\right)\left(e^{3\sigma^2+} + 3e^{2\sigma^2+} + 6e^{\sigma^2} + 6\right). \tag{7.34}$$

The function is frequently used in reliability analyzes to model failure times [46], in business [310], or in oceanography [359].

7.4 GENERALIZED EXTREME VALUE FUNCTION (GEV)

The generalized extreme value (GEV) distribution is a family of continuous density functions investigated in extreme value theory to combine properties of different PDFS, like Gumbel (extreme value type I), Fréchet and Weibull [196]. It was introduced by Jenkinson [172]. Its density function is given by the following equation

$$\mathcal{F}^{\mathrm{GEV}}_{\mu,\sigma,\xi} = \frac{1}{\sigma} t\left(x\right)^{\xi+1} e^{-t(x)}, \tag{7.35}$$

$$t\left(x\right) = \begin{cases} \left(1+\left(\frac{x-\mu}{\sigma}\right)\xi\right)^{-\frac{1}{\xi}} & \text{if } \xi \neq 0 \\ e^{-\frac{x-\mu}{\sigma}} & \text{if } \xi = 0 \end{cases}, \tag{7.36}$$

where $\mu \in \mathbb{R}$ denotes the *shift*, $\sigma > 0$ the *scale* and $\xi \in \mathbb{R}$ the *shape*.

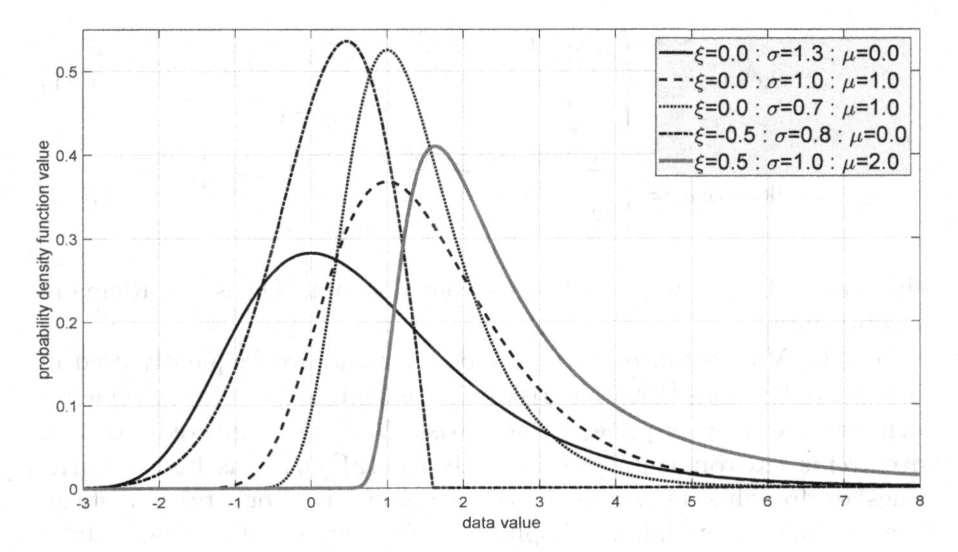

Figure 7.8: Examples of GEV probabilistic density functions

The generalized extreme value distribution parameters can be estimated using the method of moments, L-moments or with the maximum likelihood (MLE) estimation. Sample distribution shapes are shown in Fig. 7.8. Though its function shape is not strictly one-sided as it depends on the parameters selection the GEV function is frequently applied as a potential model of extreme events. Main statistics are given explicitly:

$$\text{mean} = \begin{cases} \mu + \sigma\frac{g_1 - 1}{\xi} & \text{if } \xi \neq 0, \xi < 1, \\ \mu + \sigma\gamma & \text{if } \xi = 0, \\ \infty & \text{if } \xi \geq 1, \end{cases} \tag{7.37}$$

$$\text{median} = \begin{cases} \mu + \sigma\frac{(\ln 2)^{-\xi} - 1}{\xi} & \text{if } \xi \neq 0, \\ \mu + \sigma \ln\ln 2 & \text{if } \xi = 0, \end{cases} \tag{7.38}$$

$$\text{mode} = \begin{cases} \mu + \sigma\frac{(1+\xi)^{-\xi} - 1}{\xi} & \text{if } \xi \neq 0, \\ \mu & \text{if } \xi = 0, \end{cases} \tag{7.39}$$

$$\text{standard deviation} = \begin{cases} \sigma\frac{\sqrt{g_2 - g_1^2}}{\xi} & \text{if } \xi \neq 0, \xi < \frac{1}{2}, \\ \sigma\frac{\pi}{\sqrt{6}} & \text{if } \xi = 0, \\ \infty & \text{if } \xi \geq \frac{1}{2}, \end{cases} \tag{7.40}$$

$$\text{skewness} = \begin{cases} \text{sgn}(\xi)\frac{g_3-3g_2g_1+2g_1^3}{\left(g_2-g_1^2\right)^{\frac{3}{2}}} & \text{if } \xi \neq 0, \xi < \frac{1}{3}, \\ \frac{12\sqrt{6}\zeta(3)}{\pi^3} & \text{if } \xi = 0, \end{cases} \tag{7.41}$$

$$\text{(excess)kurtosis} = \begin{cases} \frac{g_4-4g_3g_1+6g_1^2g_2-3g_1^4}{\left(g_2-g_1^2\right)^2} & \text{if } \xi \neq 0, \xi < \frac{1}{4}, \\ \frac{12}{5} & \text{if } \xi = 0. \end{cases}, \tag{7.42}$$

where $g_k = \Gamma(1-k\xi)$, γ is Euler's constant and $\zeta(x)$ is the Riemann zeta function.

The GEV distributions are very popular and are frequently used in various applications [196], like risk management, models for extreme returns or losses in asset prices, value-at-risk (VaR) and expected shortfall, in extreme environmental events analyzes [121], such as floods, hurricanes, or droughts, in estimating the maximum load or stresses a structure can hold before failure. Applications in control engineering address control performance assessment [82].

7.5 WEIBULL FUNCTION (WEI)

The formulation of general Weibull probabilistic density function [24] is given by the following equation

$$\mathcal{F}^{\text{WEI}}_{\mu,\beta,\xi}(x) = \frac{\xi}{\beta}\left(\frac{x-\mu}{\beta}\right)^{\xi-1} \exp -\left(\frac{x-\mu}{\beta}\right)^{\xi}, \; x \geq 0, \tag{7.43}$$

where $\mu \in \mathbb{R}$ is a *shift*, $\beta > 0$ a *scale* and $\xi > 0$ is a *shape*.

The Weibull function with shape $\xi = 1$ and scale parameter $\beta >$ 0is the exponential function with its scale equal to β. The case with $\mu = 0$ is named the 2-parameter Weibull function, while in case of $\mu = 0$ and $\beta = 1$ the PDF reduces to the so-called standard Weibull PDF

$$\mathcal{F}^{\text{stWEI}}_{\mu,\beta,\xi}(x) = \xi x^{\xi-1} \exp -\left(x^{\xi}\right). \tag{7.44}$$

Weibull distribution parameters can be accurately estimated using the maximum likelihood estimation. Sample Weibull function shapes are shown in Fig. 7.9.

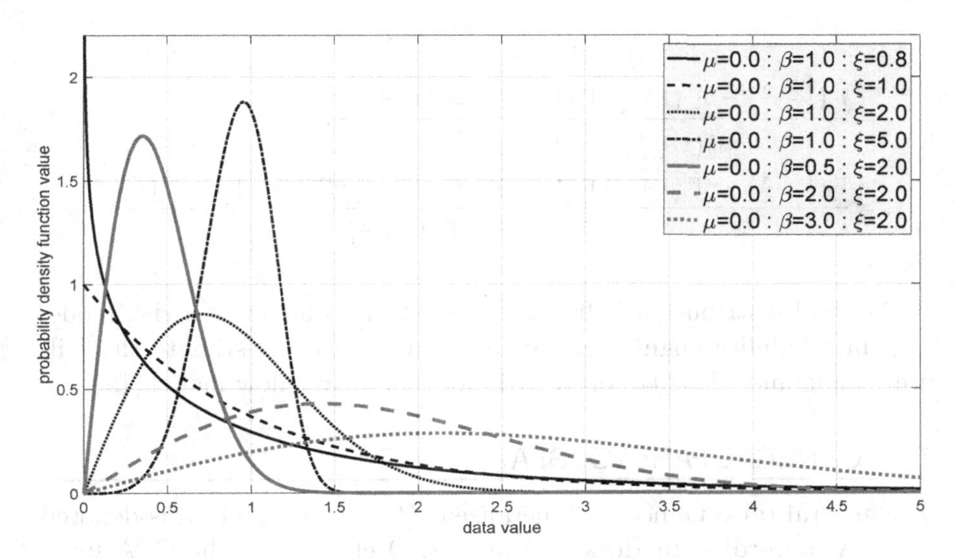

Figure 7.9: Examples of Weibull probabilistic density functions

Main statistics of Weibull PDF are given explicitly as follows:

$$\text{mean} = \beta\Gamma\left(\frac{\xi+1}{\xi}\right), \tag{7.45}$$

$$\text{median} = \beta\left(\ln 2\right)^{\frac{1}{\xi}}, \tag{7.46}$$

$$Q_1 = \beta\left(\ln 4 - \ln 3\right)^{\frac{1}{\xi}}, \tag{7.47}$$

$$Q_3 = \beta\left(\ln 4\right)^{\frac{1}{\xi}}, \tag{7.48}$$

$$\text{mode} = \begin{cases} \beta\left(1-\frac{1}{\xi}\right)^{\frac{1}{\xi}}, & \text{for } \xi > 1, \\ 0, & \text{for } \xi \leq 1, \end{cases} \tag{7.49}$$

$$\text{range} = \langle 0, +\infty), \tag{7.50}$$

$$\text{standard deviation} = \beta\sqrt{\Gamma\left(\frac{\xi+2}{\xi}\right) - \left(\Gamma\left(\frac{\xi+1}{\xi}\right)\right)^2}, \tag{7.51}$$

$$\text{coefficient of variation} = \sqrt{\frac{\Gamma\left(\frac{\xi+2}{\xi}\right)}{\Gamma\left(\frac{\xi+1}{\xi}\right)} - 1}. \tag{7.52}$$

The skewness and kurtosis can be derived from the general moment results, though the formulas are complex and not particularly helpful.

$$\gamma_3 = \frac{\Gamma\left(1+\frac{3}{\xi}\right) - 3\Gamma\left(1+\frac{1}{\xi}\right)\Gamma\left(1+\frac{2}{\xi}\right) + 2\Gamma^3\left(1+\frac{1}{\xi}\right)}{\left[\Gamma\left(1+\frac{2}{\xi}\right) - \Gamma^2\left(1+\frac{1}{\xi}\right)\right]^{\frac{3}{2}}}, \tag{7.53}$$

$$\gamma_4 = \frac{\Gamma\left(1+\frac{4}{\xi}\right) - 4\Gamma\left(1+\frac{1}{\xi}\right)\Gamma\left(1+\frac{3}{\xi}\right) + 6\Gamma^2\left(1+\frac{1}{\xi}\right)\Gamma\left(1+\frac{2}{\xi}\right) - 3\Gamma^4\left(1+\frac{1}{\xi}\right)}{\left[\Gamma\left(1+\frac{2}{\xi}\right) - \Gamma^2\left(1+\frac{1}{\xi}\right)\right]^2}. \tag{7.54}$$

Weibull distribution is frequently used in economy as a risk model [24], in reliability engineering or material science to estimate the failure/fatigue models [204], or in hydrology as a river flow model [128].

7.6 GENERAL PARETO (GPA)

The general (also named as generalized) Pareto distribution is denoted as GPA (according to Hoskings) and GPD elsewhere. The GPA function is similar to the Burr distribution and is a flexible three-parameter probabilistic density function described as follows:

$$\mathcal{F}^{\mathrm{GPA}}_{\mu,\sigma,\xi} = \frac{1}{\sigma}\left(1 + \frac{\xi\left(x-\mu\right)}{\sigma}\right)^{\left(-\frac{1}{\xi}-1\right)}, \tag{7.55}$$

where $\mu \in \mathbb{R}$ is a *shift*, $\sigma > 0$ a *scale* and $\xi \in \mathbb{R}$ a *shape* factor. The functions is defined for $x \geq \mu$ if $\xi \geq 0$ and $\mu \leq x \leq \mu - \frac{\sigma}{\xi}$. The PDF has fixed lower bound and may also be upper bounded when the shape parameter $\xi > 0$. There exists several special cases of the GPA functions:

- if $\xi = \mu = 0$ we get the exponential distribution,
- if $\xi = -1$, the GPA is equivalent to the continuous uniform distribution $U\left(0, \sigma\right)$,
- if $\xi > 0$ and $\mu = \frac{\sigma}{\xi}$ it becomes the Pareto distribution with scale equal to $\frac{\sigma}{\xi}$ and shape $\frac{1}{\xi}$.

The GPA distribution's parameters may be estimated in many ways [307], like Bayesian Reference Intrinsic (BRI) approach, maximum likelihood, probability weighted moments, the Hill estimator and a Bayesian approach using a Jeffreys prior. Sample GPA distribution shapes are shown in Fig. 7.10, while its main statistics are given as below:

$$\text{mean} = \mu + \frac{\sigma}{1-\xi}, \text{ for } \xi < 1, \tag{7.56}$$

$$\text{median} = \mu + \frac{\sigma\left(2^{\xi}-1\right)}{\xi}, \tag{7.57}$$

$$Q_1 = \beta\left(\ln 4 - \ln 3\right)^{\frac{1}{\xi}}, \tag{7.58}$$

$$Q_3 = \beta\left(\ln 4\right)^{\frac{1}{\xi}}, \tag{7.59}$$

$$\text{mode} = \mu \tag{7.60}$$

$$\text{standard deviation} = \frac{\sigma}{(1-\xi)\sqrt{(1-2\xi)}}, \text{ for } \xi < \frac{1}{2}, \tag{7.61}$$

$$\text{skewness} = \frac{2(1+\xi)\sqrt{(1-2\xi)}}{(1-3\xi)}, \text{ for } \xi < \frac{1}{3}, \tag{7.62}$$

$$\text{(excess) kurtosis} = \frac{3(1-2\xi)\left(2\xi^2+\xi+3\right)}{(1-3\xi)(1-4\xi)}, \text{ for } \xi < \frac{1}{4}. \tag{7.63}$$

The GPA distribution is very popular and is used in various applications like metallurgy [309], meteorology and wind engineering [148], financial risk assessment [17], insurance [120], medicine [44] and others.

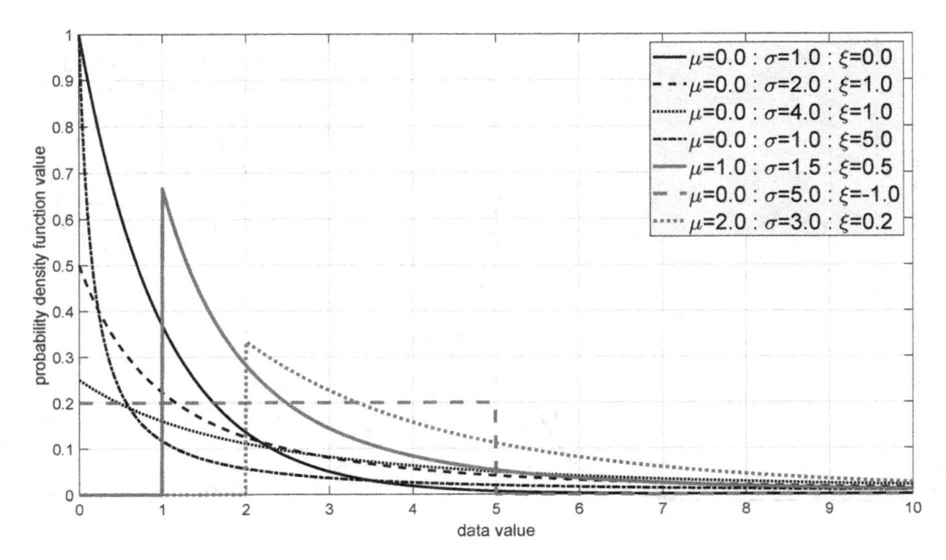

Figure 7.10: Examples of the GPA probabilistic density functions

7.7 GENERALIZED LOGISTIC (GLO)

The generalized logistic heavy-tailed distribution (GLO) is defined [153] as a three-parameter function given by the following equation

$$\mathcal{F}^{\mathrm{GLO}}_{\xi,\alpha,\kappa}(x) = \frac{\alpha^{-1}\exp\left(-\left(1-\kappa\right)y\right)}{\left(1+\exp\left(-y\right)\right)^{2}}, \tag{7.64}$$

$$y = \begin{cases} -\frac{1}{\xi}\ln\left(1-\frac{1-\kappa(x-\xi)}{\alpha}\right) & \text{if } \kappa \neq 0 \\ \frac{x-\xi}{\alpha} & \text{if } \kappa = 0 \end{cases}, \tag{7.65}$$

where $\xi \in \mathbb{R}$ is a *shift* factor, $\alpha > 0$ a *scale* and $\kappa > 0$ is a *shape* parameters. Once $\kappa = 0$ the function reduces to the logistic distribution.

The above formulation differs from the other definitions popular in the literature. It is defined as a reparametrization of the log-logistic function [5]. Its name reflects similarities with the GPA and GEV.

The most common approach to the estimation of its factors (α, ξ, κ) is to use the L-moments [153, 279], however the maximum likelihood estimation can be used as well. Exemplary shapes of the GLO distribution are sketched in Fig. 7.11. As one can see depending on the PDF parametrization we can obtain various, even sometimes non-intuitive shapes.

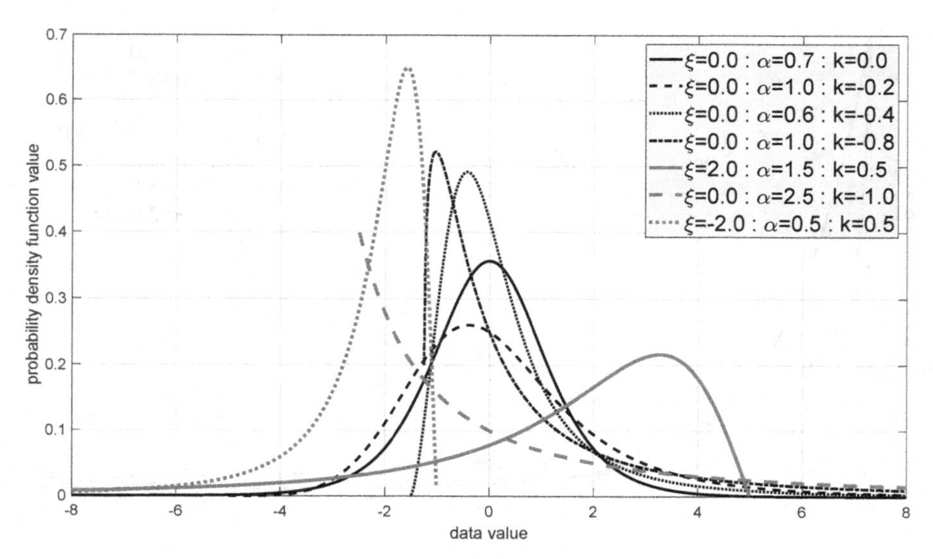

Figure 7.11: Examples of the GLO probabilistic density functions

Main statistics of the GLO distribution are given as follows:

$$\text{mean} = \begin{cases} \xi + \alpha\kappa^{-1}(1-h_1), & \text{for } \kappa \neq 0,\ \kappa > -1, \\ \xi, & \text{for } \kappa = 0, \\ \infty, & \text{for } \kappa \leq -1, \end{cases} \tag{7.66}$$

$$\text{median} = \xi \tag{7.67}$$

$$\text{range} = \begin{cases} \xi + \frac{\alpha}{\kappa} \leq x < \infty, & \text{for } \kappa < 0, \\ -\infty < x < \infty, & \text{for } \kappa = 0, \\ -\infty \leq x < xi + \frac{\alpha}{\kappa}, & \text{for } \kappa > 0, \end{cases} \tag{7.68}$$

$$\text{standard deviation} = \begin{cases} \frac{\alpha}{\kappa}\sqrt{1-2h_1+h_2}, & \text{for } \kappa \neq 0,\ \kappa > -\frac{1}{2}, \\ \frac{\alpha\pi}{\sqrt{3}}, & \text{for } \kappa = 0, \\ \infty, & \text{for } \kappa \leq -\frac{1}{2}, \end{cases} \tag{7.69}$$

$$\text{skewness} = \begin{cases} \operatorname{sgn}(\kappa)\frac{1-3h_1+3h_2-h_3}{(1-2h_1+h_2)^{\frac{3}{2}}}, & \text{for } \kappa \neq 0,\ \kappa > -\frac{1}{3}, \\ \frac{\alpha\pi}{\sqrt{3}}, & \text{for } \kappa = 0, \\ \infty, & \text{for } \kappa \leq -\frac{1}{3}, \end{cases} \tag{7.70}$$

where $h_r = \frac{r\pi\kappa}{\sin r\pi\kappa}$.

The generalized logistic distribution is mostly utilized in the extreme analysis of hydrology events to capture [153, 192].

7.8 FOUR-PARAMETER KAPPA FAMILY (K4P)

The four-parameter Kappa distribution is a generalization of other three-parameter distributions and has been introduced by Hosking [151]. It is described by the following function.

$$\mathcal{F}^{\text{K4P}}_{\xi,\beta,k,h}(x) = \frac{1}{\beta}\left[1-\frac{k}{\beta}(x-\xi)\right]^{1/k-1}\left\{1-h\left[1-\frac{k}{\beta}(x-\xi)\right]^{1/k}\right\}^{1/h-1}, \tag{7.71}$$

where $\xi \in \mathbb{R}$ is a *shift*, $\beta > 0$ a *scale* and $h > 0$ and $k > 0$ denote *shape* parameters.

The K4P distribution is defined in the following ranges:

$$\text{Range} = \begin{cases} \infty < x \leq \xi + \frac{\beta}{k}, & \text{for } k > 0, \\ \xi + \beta\frac{1-h^{-k}}{k} < x < \infty, & \text{for } h > 0, \\ \xi + \frac{\beta}{k} \leq x < \infty, & \text{for } k < 0,\ h \leq 0. \end{cases} \tag{7.72}$$

Its quantiles can be obtained using the following quartile function:

$$x^{\mathrm{K4P}}_{p|\xi,\beta,k,h} = \xi + \frac{\beta}{k}\left(-1 + \left(\frac{1-p^h}{h}\right)^{-k}\right), \tag{7.73}$$

For $k \neq 0$, the K4D reduces to other distribution, i.e. for $h = 1$ to GPA, for $h = 0$ to GEV and for $h = -1.0$ to GLO. For $k = 0$ the function reduces to exponential as $h = 1$, to Gumbel (extreme value type I) as $h = 0$ and logistic if $h = -1$. When $k = 1$, the K4D reduces to reverse exponential in case of $h = 0$ and uniform for $h = 1$. Exemplary shapes of the GLO distribution are sketched in Fig. 7.12.

Its high flexibility and many degrees of freedom allow its utilization in many situations. From that perspective we may notice its similarity to the α-stable family of distributions. The α-stable function is also characterized by four factors: shift, scale and two shape coefficients. Thus, we may consider these two distributions as the general formulations for one-sided data (K4D) and for two-sided cases (α-stable distribution).

Once we estimate K4D parameters, for instance using the L-moments method [153] or the regression [64], we may apply it to various applications, like in hydrology [191, 121] or semiconductor industry [268].

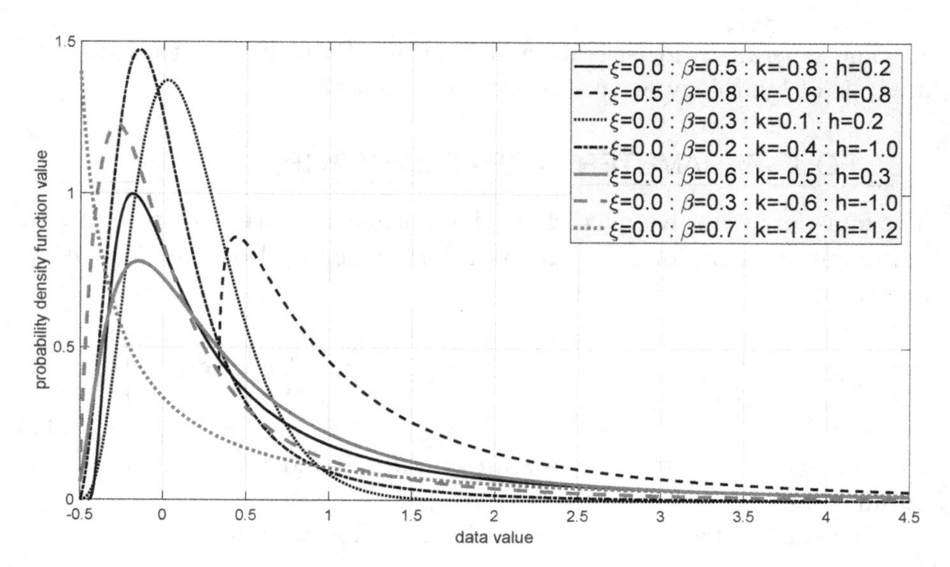

Figure 7.12: Examples of Four-parameter Kappa probabilistic density functions

7.9 EXTREME DISTRIBUTIONS FITTING EXAMPLES

Above explanations allow to use the extreme events analysis in practice. Industrial one-sided data are already introduced in the beginning of this chapter (see Fig. 7.1). Let us know try to fit extreme statistics distributions to these data. All described distribution functions are fitted to dataset histograms. Table 7.2 combines all fitting results and the estimated statistical factors of the considered distribution.

Fig. 7.13 shows how one-sided distribution capture statistical properties of the Y_1 dataset. Upper plot shows general fittin to whole data, while the lower one magnifies the peak area. Number of histogram bins

Table 7.2: Parameters of the fitted distributions

		Y1	Y2	Y3	Y4
GAM	ξ	1.51	2.65	3.21	0.57
	β	14.47	31.78	6.80	10144.21
LGN	mu	2.72	4.23	2.92	7.57
	σ	0.87	0.57	0.51	1.75
EXP	β	21.90	84.17	21.85	5792.57
WEI	μ	0.00	0.00	0.00	0.00
	β	22.93	93.92	24.61	4265.00
	ξ	1.11	1.42	1.54	0.69
GEV	ξ	0.44	0.48	0.42	0.81
	σ	7.36	23.20	5.52	1906.46
	μ	11.51	52.79	14.72	1493.77
GPA	ξ	0.09	-0.05	-0.11	0.42
	σ	19.69	88.46	24.02	3145.62
	μ	0.00	0.00	0.00	0.00
GLO	ξ	14.46	63.56	16.88	2737.30
	α	6.26	20.14	4.51	2027.97
	κ	-0.51	-0.47	-0.49	-0.58
K4P	ξ	8.54	42.03	12.43	1442.36
	β	8.82	31.05	6.58	2410.86
	k	-0.41	-0.34	-0.38	-0.54
	h	0.56	0.64	0.58	0.24

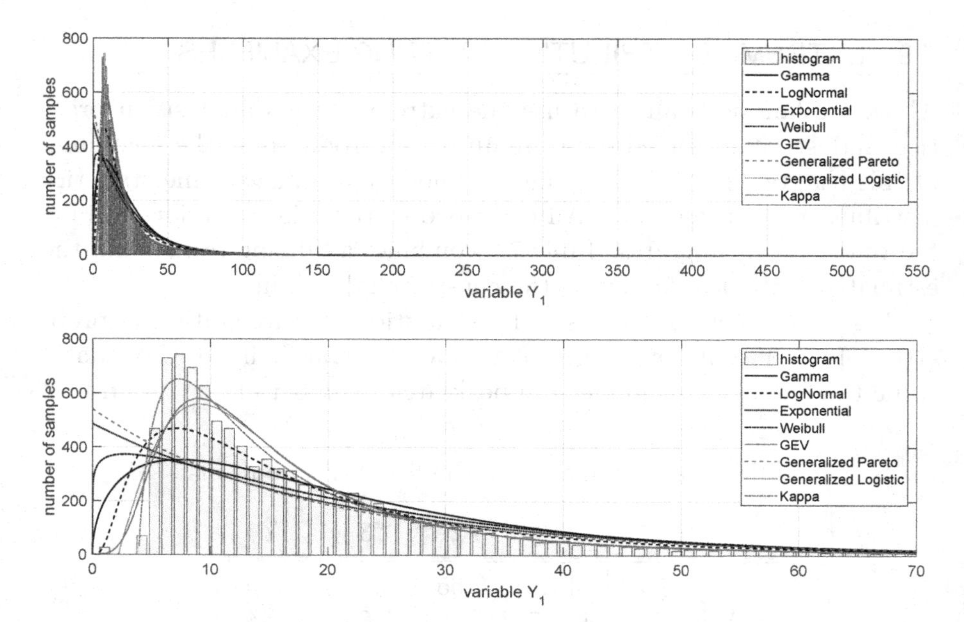

Figure 7.13: Extreme distribution fitting to the sample Y_1 dataset

remains unchanged and equals to 500. Visual comparison shows that the four parameter Kappa, GEV and Generalized logistic seem to deliver the best fit. We observe that the most challenging to be reflected is the short left tail close to zero. In further analyzes, the above visual interpretation will be justified by fitting metrics.

Fig. 7.14 shows the same analysis for the Y_2 time series. In that case the interpretation is more difficult and we deserve some numbers. We may only say that the exponential distribution seems to be the least appropriate. Further interpretations will be allowed by the introduced metrics and other statistical plots.

As we may observe in Fig. 7.15 only some of the distribution may capture the shape of the histogram, and these are the GEV, GLO and K4P. As previously, the exponential distribution is the least effective in modeling the stochastic properties for Y_3 time series.

Finally, Fig. 7.16 presents the PDF fitting efficiency for Y_4 dataset. In that case visual inspection rather fails, and we need to use other means for a determination of the appropriate PDFs.

One possible way is to evaluate some fitting residuum measure, as for instance mean square error or mean absolute error. As the data may be affected by outliers (in that case as default) we use the MAE index

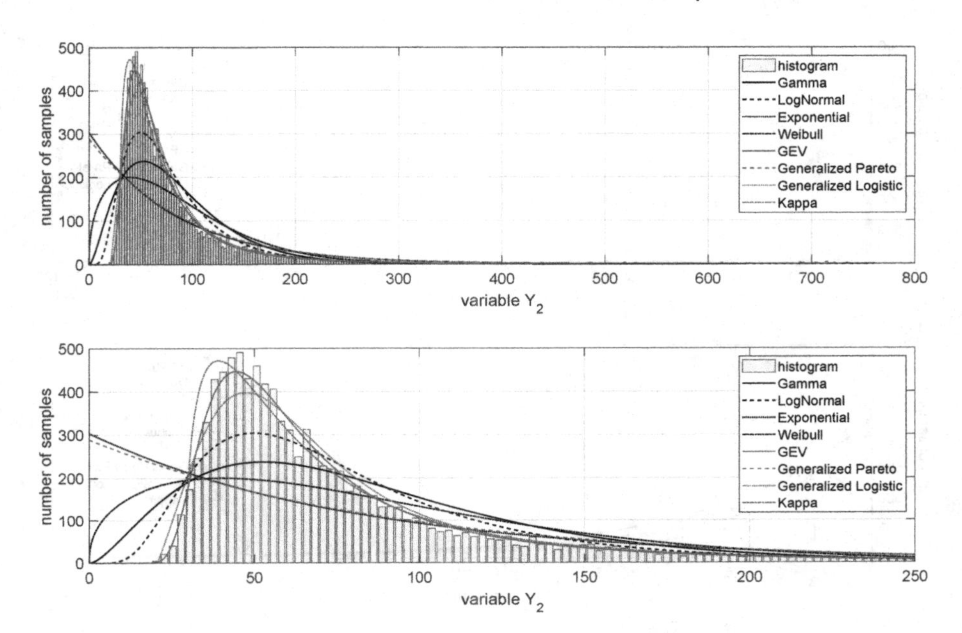

Figure 7.14: Extreme distribution fitting to the sample Y_2 dataset

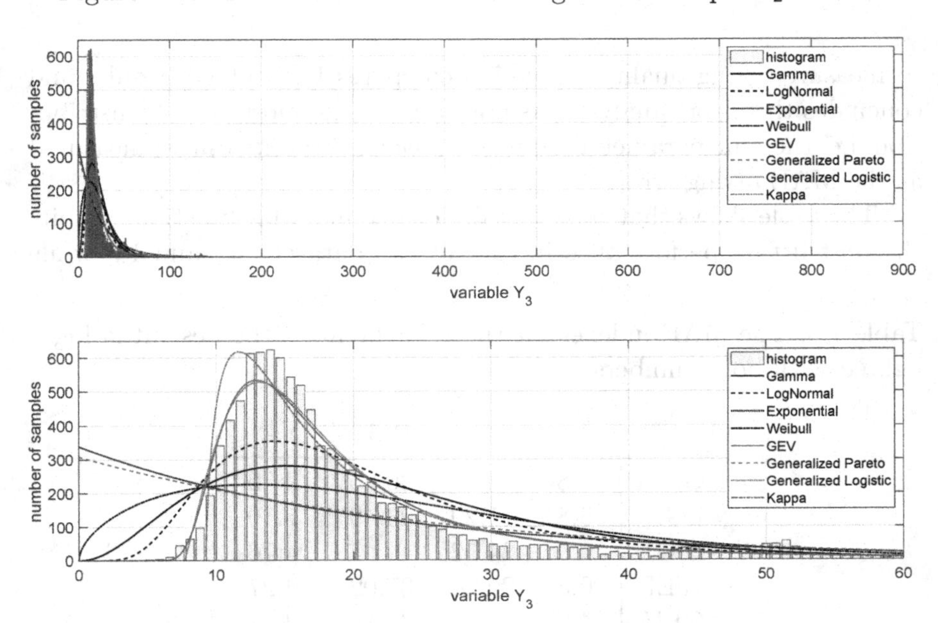

Figure 7.15: Extreme distribution fitting to the sample Y_3 dataset

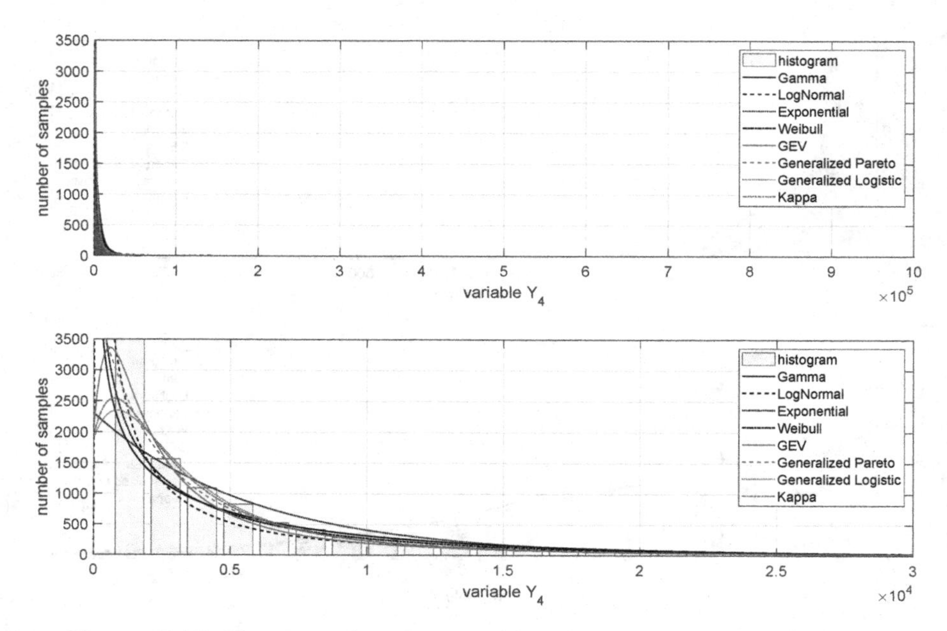

Figure 7.16: Extreme distribution fitting to the sample Y_4 dataset

to measure fitting quality. Table 7.3 compares PDF fitting results and concludes this short analysis. As one can see, the most flexible distribution, i.e. the four-parameter kappa obtains the best fit results, measured as the MAE fitting error.

The table shows that the generalized extreme value function exhibits the best fitting performance in case of two datasets: Y_2 and Y_3, while

Table 7.3: The MAE indexes for the PDF fitting – the best fit is highlighted with bold numbers

	Y1	Y2	Y3	Y4
GAM	8.84	16.52	30.86	4.89
LGN	5.81	11.29	23.49	5.19
EXP	10.96	24.52	44.44	4.76
WEI	10.32	20.24	37.02	4.27
GEV	4.22	**3.55**	**10.34**	3.11
GPA	10.87	24.38	43.88	**2.807**
GLO	4.33	5.72	10.57	2.97
K4P	**2.80**	5.19	12.38	2.811

four parameter kappa function behaves the best for Y_1 and generalized Pareto for Y_4. The K4P function is the second best in case of two other datasets. We also may notice big differences in data properties. Time series Y_4 is relatively well captured by all the considered distributions, while in case of the rest datasets there are good and bad choices. This simple and small example shows the selection of the stochastic model, i.e. the distribution matters, also in the case of the extreme value analyzes and one-sided data. Wrong choice may lead to erroneous conclusions.

7.10 CONCLUDING REMARKS

It must be noted that extreme value analysis is not only limited to a certain number of distributions. The EVA uses specific analyzes that address extreme statistics issues, but could be also taken into account in other research areas. We may enlist three following examples:

1. moment ratio diagrams (MRD) and their new variant called L-moment ratio diagram (LMRD) ,
2. flow duration plots,
3. return period analysis.

The MRDs and LMRDs are described in detail in Section 9.3. At this point, it is only necessary to mention that the diagrams visualize the relationships between selected statistical moments in two-dimensional graphs.

The flow duration relationship is a diagram that shows the percentage of time that a given variable is likely to equal or exceed some specified threshold. In most cases, these charts are used in hydrological analyses and are concerned with flows in rivers, hence their name. In its original application, it may be used to present the percentage of time river discharge can be expected to exceed a given flow value, or to show the river flow exceeds some value in some time percentage. Once a certain distribution is selected and fitted to the data the analysis can be run.

Despite the fact that flow duration analyses are deeply embedded in hydrological studies, there is no contraindication to using this concept in other engineering analyzes, such as for estimating exceedances of emission limitations. A stochastic variable is just a variable, and we should not lock ourselves into just one narrow research context. Analogous considerations can be applied to the return period analysis.

A return period, also denoted as a recurrence interval or repeat interval, is an estimated time between some events described by some stochastic variable. If a given process is stationary the return period of an event (exceeding some threshold value) is defined as the average time that occurs between two successive realizations of the considered event. Alternatively, we may evaluate the return level, i.e. what value is expected to be exceeded, once every return period, or with a given probability. It is typically based on historical data over an extended period, we usually use it in the risk analysis.

In this discussion, the ideas of flow duration and return period are only mentioned and will not be described further. Nevertheless, it is worth mentioning that such concepts exist and can be used in control engineering, such as, almost naturally, in control design for water systems (flood control) and renewable hydropower.

> As in case of two-sided cases, the selection of proper distribution function still matters. We mustn't use two-sided function in case of extreme-like statistical cases, because it simply shows ignorance. Extreme statistics deliver interesting tools to the control engineer's toolbox and it should be remembered.

GLOSSARY

EVA: Extreme Value Analysis

VaR: Value-at-Risk

BRI: Bayesian Reference Intrinsic

EXP: Exponential distribution

GAM: Gamma distribution

LGN: Lognormal distribution

GEV: Generalized Extreme Value distribution

WEI: Weibull distribution

GPA: General Pareto distribution

GLO: Generalized Logistic distribution

K4P: Four-parameter Kappa distribution

MRD: Moment Ratio Diagram

LMR: L-moment Ratio Diagram

CHAPTER 8

What happens in tails? – outliers

STORY of tails is inseparable from the outliers, as those outlying observations, anomalies in data, discordant events, or whatever we call them are responsible for tails (see Chapter 3). Generally speaking, outliers make tails, while the tail itself can be also just the result of the phenomenon, the fundamental process, which is hidden behind the data.

This book aims at the tail-aware assessment and statistics. Thus, the introduction of the statistical tail detection subject is essential. Previous sections have aimed at different picture, i.e. how the tails are connected with distributions and their factors. This section focuses on algorithms supporting the detection of outlying observations.

Outliers just exist [93] in real systems and we have to be aware of that fact. Industrial systems associated with control engineering should (or even must) take that fact into account. Outliers can appear because of various, frequently unknown reasons. We can encounter them in many situations, like in statistical investigations, data regression, classification or control. Their occurrence sometimes helps, but in general engineers do not like them. They are considered as a contamination of ideal picture. Outliers detection, labeling, taking into account or removing is a focus of many researches.

Historically, the statistics encountered them, identified and started to analyze. As the research has continued, many statistical methods and algorithms have been developed. Recently, the development of artificial intelligence (AI) and machine learning (ML), especially addressing the task of classification has brought many novel approaches. The situation is quite obvious, once we know what is normal and what is abnormal.

DOI: 10.1201/9781003604709-8

Such good and bad patterns allow to use many ML algorithms [3, 4, 236]. They can be successfully used in univariate time series analysis or multidimensional cases like picture pattern recognition. Especially, the latter area delivers dozens of efficient and promising methods.

However, there are situations that we have no knowledge that something is abnormal until it happens. This task is even more complex as we understand the fact that new and unknown observation, which does not fit to the previously observed behavior does not have to be abnormal, as our observation history is not long enough to cover all possible good situations. Such scenarios are very challenging, as we do not have enough history to learn from. Extrapolation of our ML *black box* knowledge is just difficult.

Statistical approaches address this problem from the opposite perspective. The available historical observations allow to develop the fundamental process behind the data. Mostly, distributions play this role. Their properties and analysis allow to perform knowledge extrapolation, in a similar way as the first-principle models do in modeling and identification of dynamic processes.

The issue of unusual behavior detection, especially in control systems analysis context has not been sufficiently covered by researchers, though is a daily practice in industry. Their effects are frequently neglected, if they do not cause breakdown or undesired behavior. This chapter describes the subject of statistical outlier detection from the perspective of control system assessment. Presented methods and approaches are visualized with exemplary data.

8.1 STATISTICAL OUTLIER DETECTION IN TIME SERIES

Outlying observations in time series can indicate an incorrect observation or be a representative of an endogenous (possible but very infrequent) variability in data [252, 26]. Unusual observations might occur because of human unintended errors or intentional behavior, due to failures caused by equipment, like sensors, computer systems, telecommunication links, because of sampling errors or as standardization mismatch. Identification and repair of such data is generally questionable, biased, relative and poorly justified. Data repair has to be done cautiously with external cross-checking and confirmation.

Very often we lack sufficient knowledge to correctly assess the nature of phenomena, due to its non-linearity, complexity, natural human simplifications and incomplete knowledge. Insufficient information leads to wrong interpretation of properties of the fundamental stochastic process [163]. Such unusual behavior manifests itself in form of non-intuitive and strange-looking shapes of data distributions. They can be asymmetric, heavy and long-tailed, multi-modal, flat or just of a jagged shape. Sampling of data with some correct or incorrect interval may additionally bias the analysis. The assumptions about the properties of the data that we made at the beginning of the study may be wrong, for the simple reason of lack of complete information [89]. Our errors may be due not only to insufficient information about the data generation mechanism itself, but also because there may be other mechanisms that we simply know nothing about. Data non-stationarity, trends and properties fluctuations, varying and delayed correlations, self-similarities, fractional or multi-fractal properties further complicate an already complicated picture [229, 38]. On the other hand, not every unexpected observation has to be artificial and strange from the process perspective; perhaps it is simply an extremely rare but possible behavior. Above facts make the picture complex and challenging, which may hesitate researchers and engineers, whose habit is reduced to linear, quadratic and Gaussian world.

We observe different reactions to unusual phenomena. Some of us continue their business as usual pretending that everything is alright and according to the pre-assumed *normality*. This approach seems to work if the outliers are irrelevant and entail no consequences. OK - we lose the opportunity to learn, but who cares. If we don't notice something, it doesn't mean that it doesn't exist. If we are aware of some abnormal situation, we may intentionally acknowledge and neglect them, and simultaneously learn.

Finally, the outliers might be considered as an ultimate research goal. We look after them. We consider detected data abnormality as a source of important information. Such an approach we take during leakage detection, medical diagnoses, fraud detection, leakage detection, detection ofcyber attacks, etc. [199, 236]. No matter what the situation is, we need to detect the outliers. Once we find them, we may proceed. We may remove unintended contaminating observations. We may label them for further examination if they are suspected to be legitimate. Labeled outliers might be isolated, analyzed allowing preventive actions.

The simplest situation and common understanding is that the outlier is a single strange observations, which occurs occasionally. However, the outlier does not have to be an isolated value. It can be a strange segment of the time series, just a part of it. Fig. 8.1 shows sample illustration of isolated outliers or data outlying segment.

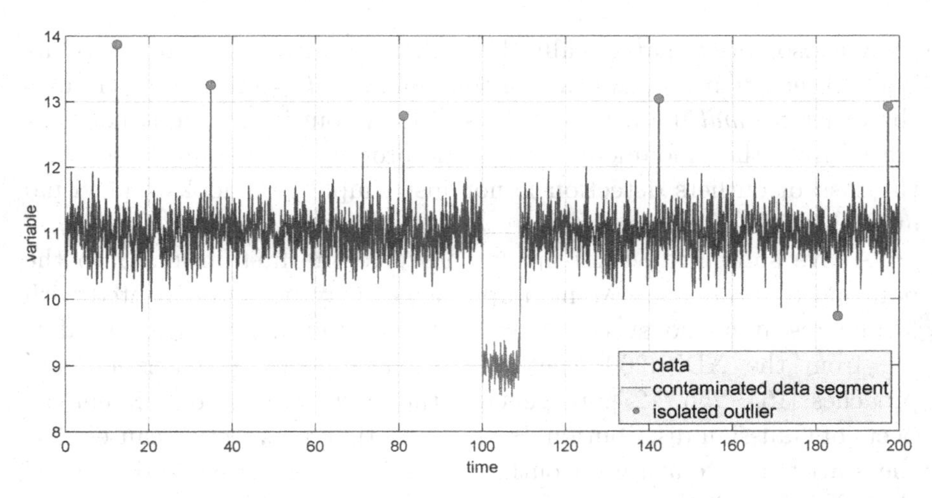

Figure 8.1: Sample data time series with labeled two types of outliers

The simplest approach is just to plot a given variable time series and do its visual inspection to look for strange-looking data occurrences (like in Fig. 8.1). We use our experience and label what we think is abnormal. Next step could be to plot data histogram. Fig. 8.2 shows the respective histogram, plot for the above sample time series. We clearly see the main *normal* data grouped around the value of 11. The contaminated segment is shown with a small peak around value of 9, while isolated outliers are just in form of small, single bins.

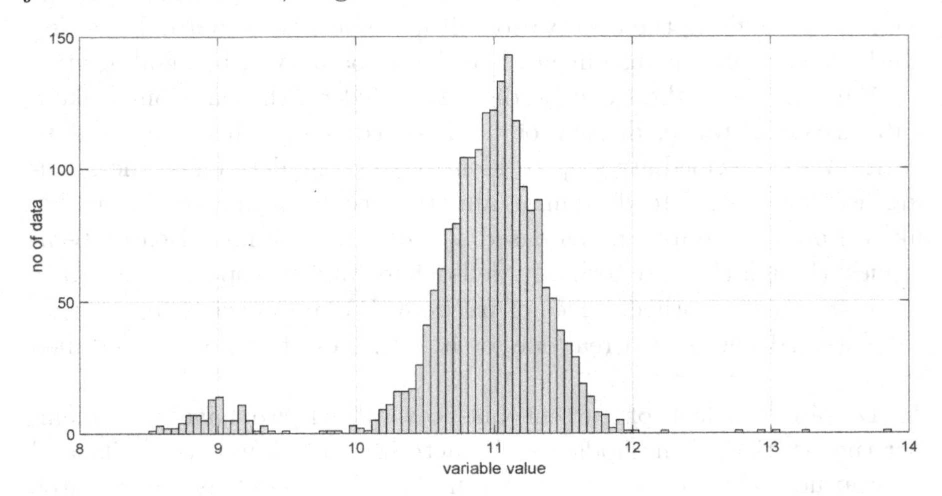

Figure 8.2: Histogram of contaminated sample data

We also notice that visually the outlier can be hardly detected. Outlying observation that appears in time moment $t = 185$ is very close to the other *normal* data, though it is still the outliers, as it is not generated from the fundamental stochastic process. Therefore, we see that the task of outliers detection is not just simple as it looks and visual analysis might not be enough.

Above simplified visual approaches give the first insight into the data ind its properties. Manual approaches may be supplemented with formal research and systematized methods. Initial investigations dating from the XIX [260] century were using various statistical approaches [164, 256, 127]. In general, the works exploited different aspects of Gaussian distribution. Such a long-time perspective causes that the statistical literature on outlier detection is extensive and detailed [138, 287, 140, 281]. However, its popularity peak is over.

Recently, artificial intelligence approaches are *en vogue* and gain the largest publicity [375, 13, 323, 334, 314, 169, 250]. One may distinguish three different types of algorithms. Supervised methods use for training historical data about normal and abnormal objects. Semi-supervised approaches are used for learning only normal nor abnormal examples. No training data is utilized in the unsupervised methods. We may also distinguish the methods by their method of reference resolution, what means the difference of the range: local versus global. Actually, might be positioned somewhere between. Next difference is associated with what kind of output the method returns. Labeling approaches give binary value, naming the observation either abnormal or normal. Scoring methods give continuous output, like the probability of being abnormal.

Finally, the method can be directly based on the data only, like a data-driven distance, density or angle-based approach or we need to perform an intermediate step to identify some model. Thus, the labeling uses the model to determine abnormality according to the model, like rational or sample model-based approaches. The model-based techniques, though they are formally well-defined and transparent introduce to the analysis another degree of freedom. We are never sure, whether the detected outlier is a real one, or it is the result of poorly identified model.

Despite the fact of current bias toward AI, we have to remember that statistical methods are characterized by formal simplicity and rigorousness. Moreover, recent findings of the non-Gaussian research and robust statistics [286, 330, 89, 280] have opened new potential perspectives. Comprehensive summaries on statistical outlier detection can be found in [341, 342, 345, 90, 215, 243, 134].

Good practice shows that the statistical outlier detection should comprise of three phases [163]:

1. labeling, i.e. flagging for further investigation,
2. treatment with robust statistical methods analyzes,
3. final test if the considered observation is abnormal.

As shown, we may find numerous methods for statistical anomaly detection. The basic fact is that statistical analysis depends on the initial assumption about the underlying distribution of the data. Majority of methods is limited to the univariate cases, which moreover assume data normal distribution model. Control performance assessment perspective generally meets the univariate assumption.

The story of statistical outlier detection started with methods that assume data normal properties, which assumption forms the basis of Z-Scores and its modified versions [163], Grubb's test [127], Tietjen-Moore test [335], minimum covariance determinant [158], extreme studentized deviate (ESD) test[282], and Thompson Tau test consisting of two steps [332].

Research on robust regression [287, 157] has shown that mean squares (Gaussian assumption) can be biased by even single outlying observations. This fact led to new research focusing on the use of robust location and scale estimators, like Z-scores with median and MADAM (median of the absolute deviations about the median [163]), Interquartile Range (IQR) [351], Hampel filter [258] or power law tail index estimates proposed by [325]. Finally, α-stable distribution can be considered as an underlying generation mechanism. This function exhibits a lot of attractive properties [88]. Research shows that this distribution model may be considered as a fundamental process behind the data, also in control engineering [77]. It includes not only scale and location but also stability factor responsible for tails and skewness coefficient.

Concluding the contribution of statistics to the outlier detection, we have to remember an idea [193] that "*the notion of outliers has to be considered not by itself but in connection with underlying scheme*". Selected statistical approaches are introduced below and their description is followed by their exemplifications.

8.2 Z-SCORES

The so-called Z-Scores approaches [163], by some authors are named MDist [199]. They seem to be the earliest and the most popular methods. They are based on the well-known property of normal distribution that once some variable X is distributed with $N(\mu, \sigma^2)$, then $Z = (X - \mu)/\sigma$ is distributed with $N(0, 1)$. Thus, we may use the Z-scores of a population of observations $x_1, x_2, \ldots, x_n$,

$$z_i = \frac{x_i - \hat{\mu}}{\hat{\sigma}} \tag{8.1}$$

as a method that detects outliers.

8.2.1 MDist-G – 3σ method

The common rule indicates the distance MDist that exceed $D = 3$ in absolute value as outliers . This method is commonly known under the name of 3σ. Such a selection determines that 99.7% of data is treated as inliers and only rare event, which occurs at approximately 1 of 370 samples is considered as an outlier.

The 3σ method is very simple and consequently attractive. However, classical estimators of mean and standard deviation are sensitive to outliers. The method is executed for the complete time series, which includes outliers. Although the method is intended to detect outliers in data sets, its operation depends on the values contaminating the data and is self-influenced. Other approaches, like robust statistics or MCD [287] should minimize that impact.

8.2.2 MDist-MAD – robust

It is known that Gaussian regression mean and standard deviation estimators are not robust and are biased even by single outlier. Thus, the researchers proposed other shift and scale estimators, which exhibit *breakdown point* exceeding zero. Data median seems to be the simplest shift alternative, while MADAM might be utilized in place of standard deviation

$$\text{MADAM} = \text{median}\left\{|x_i - \text{median}\left\{x_i\right\}|\right\}. \tag{8.2}$$

Notions of MAD and MADAM are frequently interchangeably used. Consequently, we modify MDist-G method exchanging $\hat{\mu}$ with median $\overline{x}_{\text{med}}$ and $\hat{\sigma}$ with MADAM. Let's denote as MDist-MAD a method with the score equal to 3.

8.2.3 MDist-mMAD – modified robust

Further research have shown that the Z-Scores relation used in MDist-G should be modified by a constant 0.6745 as $E\{\text{MADAM}\} = 0.6745 \cdot \sigma$ for large number of observations

$$z_i^{\text{MADAM}} = 0.6745 \cdot \frac{x_i - \text{median}\{x_i\}}{\text{MADAM}}. \tag{8.3}$$

Additionally, it is suggested to use multiplier $D = 3.5$. This modification is denoted as MDist-mMAD.

8.2.4 MDist-Hub1 and MDist-Hub2 – Robust Huber

Robust statistics researchers propose many alternative estimators for time series contaminated with outliers [157]. Let's denote M-logist estimator of shift as $\hat{x}^{\text{R}}$ and for scale as $\hat{\sigma}^{\text{R}}$. Moreover, we can utilize different Z-scores multipliers as in previous definitions.

Method denoted as MDist-Hub1 takes standard $D = 3$ and MDist-Hub2 utilizes robustified $D = 3.5/0.6745$.

8.2.5 MDist-G(tail) – Gaussian tail

One can propose further modifications of the MDist method. A possible alternative is to use the notion of the tails and the measure of the tail starting point proposed in [325]. For normal distribution the tail ξ^{G} starting points are evaluated using the following relation

$$\hat{\mu} \pm \sqrt{\frac{1}{2}\left(5 + \sqrt{17}\right)} \cdot \hat{\sigma}, \tag{8.4}$$

which allows to derive

$$\xi^{\text{G}} = \sqrt{\frac{1}{2}\left(5 + \sqrt{17}\right)} \cdot \hat{\sigma} \approx 2.13 \cdot \hat{\sigma}. \tag{8.5}$$

Following above equation, the common 3σ rule can be modified using the relaxed (more objects will be detected as outliers) threshold multipliers [89], as $D \approx 2.13$.

8.2.6 MDist-TS – power law tail

Tail starting point approach might be extended with the use of other probabilistic density functions. It is shown [325] that for a general class of symmetric distributions that follow power law (t-Student function)

$$\xi^{\mathrm{P}} = r(\alpha) \cdot \sigma = \frac{\sqrt{\frac{5\alpha + \sqrt{(\alpha+1)\cdot(17\alpha+1)}+1}{\alpha-1}}}{\sqrt{2}} \cdot \sigma, \tag{8.6}$$

where α denotes a tail index (stability exponent). $r(\alpha)$ tail factor depends on estimated power law stability exponent α. The method requires the estimation of α index. We may use maximum likelihood (MLE) estimation to fit the power-law distribution to the empirical data [51] and then obtain the stability exponent.

8.2.7 MDist-α – α-stable tail

In control engineering the assumption that normal distribution can be considered as an underlying stochastic process is not universal [77]. The research and the review of control error properties show that α-stable distribution as in Eq. (6.8) might better reflect the properties. We may identify its parameters and then to use them in the evaluation of the tail start point following the idea of crossover point proposed above.

Crossover occur for the α-stable function only for symmetric shape ($\beta = 0$), i.e. SαS case. We observe this effect when the scale is constant ($\gamma = \text{const}$) and the stability exponent α changes, as shown in Fig. 8.3. We observe four crossover points $\{a_1, a_2, a_3, a_4\}$. The inner ones, i.e. $\mathcal{F}(x) \in (a_2, a_3)$ determine the so-called "high peak" inner tunnel, while the tails (outer tunnel) occur for $\mathcal{F}(x) < a_1$ and $\mathcal{F}(x) > a_4$.

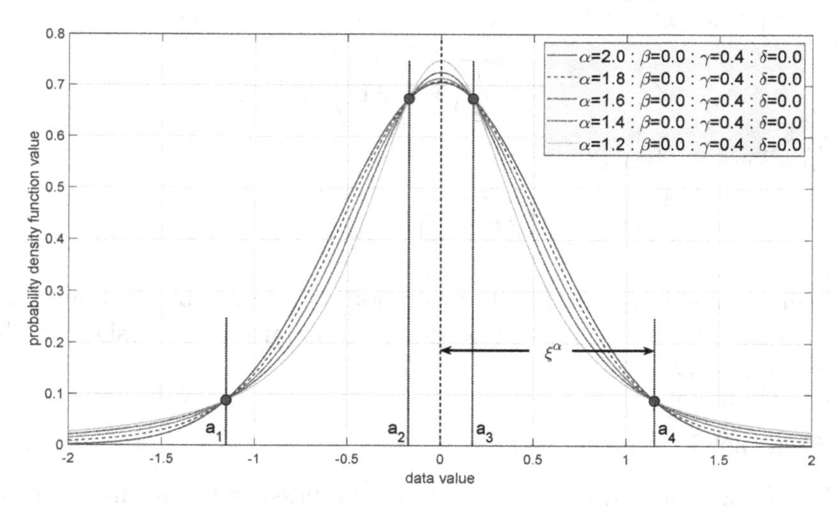

Figure 8.3: The α-stable distribution with crossovers

Thus, the values of crossover points $a_1 = -a_4$ may be used as an indication where the tail starts, so we may use them as outlier detection thresholds ξ^α. We can derive their analytical values, so they are calculated numerically. Finally, the outlier threshold must be agreed with the Z-Scores multiplier for normal case and we obtain $D_\alpha = 1.468$.

8.3 HAMPEL FILTER

Hampel filter [258] is defined as a sliding window version of classical Z-scores MDist-MAD detection. We use Hampel filter to indicate locally outlying observations, in contrary to the globally applied to whole dataset the MDist-MAD method. Ot is defined as a standard median filter with a symmetric moving window. We have to define one tuning coefficient: the width of the sliding window.

8.4 INTERQUARTILE RANGE

Majority of data does not pass the normality test and we cannot consider reliably that they follow Gaussian distribution. A possible statistic in such a case is the Interquartile Range (IQR) method [351] – see Eq. (4.27). The IQR is calculated as the difference between upper 75th and lower 25th percentiles of the data $\text{IQR} = Q_3 - Q_1$.

We may use it to detect outliers. Observations that fall below $LL = Q_1 - 1.5\,\text{IQR}$ or above $HH = Q_3 + 1.5\,\text{IQR}$ can be considered as outlying. They are often presented in a boxplot: the highest and lowest occurring values are indicated by whiskers of the box and possible outliers as individual points. The breakdown point for IQR is equal to 25%, which is better than the one for standard estimators, but still it's not fully immune to outliers.

In case of data exhibit Gaussian distribution the method detects more outliers than Gaussian MDist-G as detection points out data outside of a range $\mu \pm 2.698 \cdot \sigma$.

8.4.1 IQR-α

The IQR-α method is sometimes named the Pareto tail modeling and it is suggested for skewed data [6]. We may consider it as a generalization of the IQR. The method assumes α-stable function as an underlying outlier generating process. Observations that are larger than a certain quantile of the fitted α-stable distribution are indicated as outliers. The

IQR-α method is suggested for heavy-tailed data, but it may fail for time series that include point-wise observations that are far away from the data "high peak" inner tunnel [65].

Once we fit α-stable PDF to data, we need to determine which quantiles are taken as thresholds. There are many suggestions in the literature. Danielsson *at al.* [65] propose to take 5% quantiles IQR-$\alpha_{5\%}$, while [6] suggests to select 0.5% quantile IQR-$\alpha_{0.5\%}$.

8.5 GENERALIZED EXTREME STUDENTIZED DEVIATE TEST

The generalized extreme studentized deviate (ESD) test was proposed by Rosner [282]. We may use it for data taken from an approximately normal distribution to detect one or more outlying points. The test assumes the upper number for detected outliers. This assumption is different from the Grubb's [127] or Tietjen-Moore [335] tests, which detect an exact number of outliers. Assuming the upper constraint for the number of outliers n, the test conducts n separate tests: for one $n = 1$, two $n = 2$, and so on up to $n = N$ outliers.

The ESD tests the null hypothesis that there is no outlier versus the alternative, that there is at most N outlying observations. For a given dataset X, which consists of N observations, we define n test statistics $T_1, T_2, \dots T_n$. Each T_i is formulated as a two-sided Grubbs' statistics, defined as $X_1 = X$, $\hat{\mu}_i$ being the mean estimate of X_i and $\hat{\sigma}_i$ an estimate of X_i standard deviation

$$T_i = \frac{\max\{|x - \hat{\mu}_i| : x \in X_i\}}{\hat{\sigma}_i}, \tag{8.7}$$

and $X_{i+1} = X_i - x_i$, where $x_i \in X_i$ such that $|x - \hat{\mu} i|$ is maximized. The generalized extreme studentized deviate test is nothing but the sequentially applied Grubb's test, adjusted for the critical values based on the number of tested outliers. It holds a robustness to the effect of significant masking advantage.

8.6 MINIMUM COVARIANCE DETERMINANT

The minimum covariance determinants (MCD) [158] are considered as robust location and scale estimators, which are frequently applied to multivariate cases. Above feature enables to propose MCD as a potential outlier detection method. The resulting distances are not sensitive to the masking effect, so we can use them to identify outliers [42]. It is proposed to use the Fast MCD algorithm [284]. This algorithm searches for n observations with the smallest scatter. Determinant of the covariance matrix is a good measure of the scatter, in cases when the cluster of n the most central points is symmetrically distributed around data center.

8.7 THOMPSON TAU TEST

In general, the Thompson Tau test, denoted Th-τ is run in two steps [332]. Potential outliers are indicated first, through the calculation of the highest absolute difference between the data and their mean (or any other shift estimator) value. In the second step, the sample standard deviation is calculated is evaluated together with τ taken from T critical values of Student's T-distribution, which identify a rejection region

$$RR = \frac{t_{\alpha/2}\,(N-1)}{\sqrt{N}\sqrt{N-2+t_{\alpha/2}^2}}, \tag{8.8}$$

where $t_{\alpha/2}$ is a Student's T critical value based on the level $\alpha = 5\%$ with two degrees of freedom and N is a sample size.

We may consider various variants of the general Th-τ approach:

- the application of the classical mean estimator – Th-τ(mean),
- variant using median and MADAM – Th-τ(med),
- version with bi-weight shift and scale estimators – Th-τ(bw).

Generally speaking, as we have various mean and scale estimators, one can imagine many different variants of the algorithm.

8.8 EXAMPLES OF OUTLIER DETECTION

The outlier detection analysis is performed for the selected time series playing the exemplification role throughout this book. Five datasets are considered with their time trends shown in Fig. 4.1. Selected group of detection algorithms is tested, i.e. MDist-G, MDist-Hub1, MDist-α, ESD, MCD, IQR , IQR-$\alpha_{5\%}$, IQR-$\alpha_{0.5\%}$, Th-τ(mean), Th-τ(med) and Th-τ(bw). The results are shown in three ways: tabular, histogram with detection threshold and time trends with highlighted data pointed out as outlier. Columns [minThr] and [maxThr] denote detected min and max detection thresholds, columns [minCount] and [maxCount] detected number of outliers below minThr and above maxThr and column [Percent] shows calculated share of all detected outliers.

Table 8.1: Outlier detection results for dataset X_1

	minThr	maxThr	minCount	maxCount	Percent
MDist-G	-4.269	4.284	12	15	0.27%
MDist-Hub1	-4.290	4.299	12	14	0.26%
MDist-α	-4.277	4.297	12	14	0.26%
ESD	0.000	0.000	0	0	0.00%
MCD	-2.246	2.237	543	601	11.44%
IQR	-3.835	3.847	33	33	0.66%
IQR-$\alpha_{0.5\%}$	-3.669	3.689	49	48	0.97%
IQR-$\alpha_{5\%}$	-2.340	2.360	481	498	9.79%
Th-τ(mean)	-4.683	4.667	8	5	0.13%
Th-τ(med)	-4.683	4.667	8	5	0.13%
Th-τ(bw)	-4.928	4.959	4	3	0.07%

8.8.1 Dataset X_1

Table 8.1 summarizes detection results for X_1 time series. It is supplemented with the histogram shown in Fig. 8.4, which visualizes the placement of detection thresholds associated with each method.

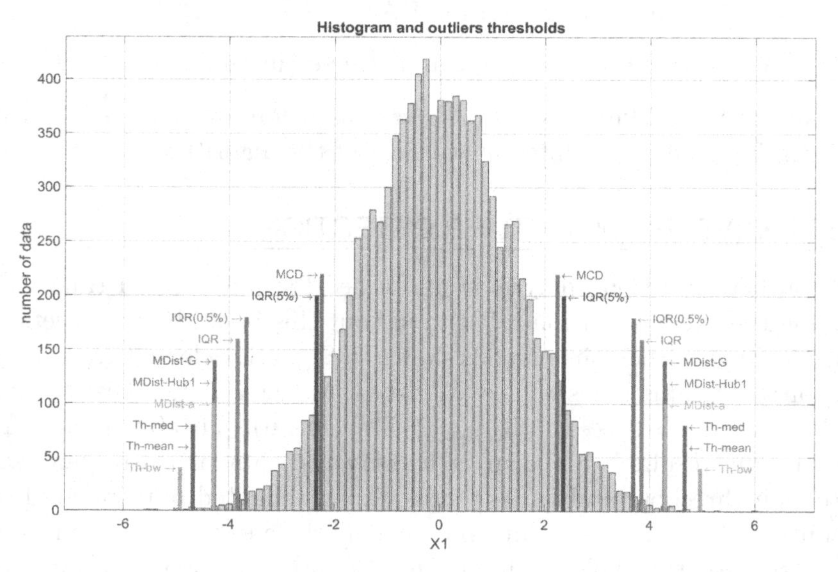

Figure 8.4: Histogram with outliers thresholds for dataset X_1

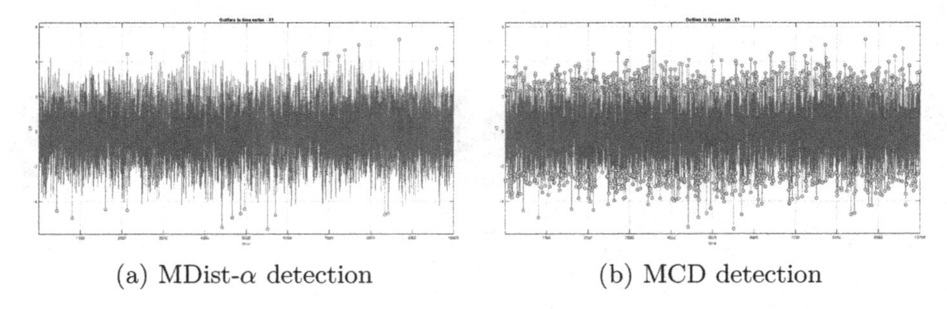

(a) MDist-α detection (b) MCD detection

Figure 8.5: Time trends with labeled outliers for X_1

We see that each approach has different sensitivity and the range of detection. The ESD algorithm seems to be the most conservative, as it does not indicate any outlying observation. The Th-τ(bw) algorithm labels the last number of observations as outlying out of those, which detect any. In contrary, the MCD tends to label many observations as outliers. Fig. 8.5 shows the effect of outliers detection in time domain. The MDist-α is quite conservative, while the MCD approach considers too many observations as outlying.

8.8.2 Dataset X_2

Table 8.2 summarizes detection results for X_2 time series, while Fig. 8.6 visualizes detection thresholds associated with each method.

Table 8.2: Outlier detection results for dataset X_2

	minThr	maxThr	minCount	maxCount	Percent
MDist-G	-12.713	12.699	6	5	0.11%
MDist-Hub1	-4.403	4.329	62	67	1.29%
MDist-α	-4.345	4.287	68	70	1.38%
ESD	6.861	6.999	22	21	0.43%
MCD	-2.280	2.202	631	654	12.85%
IQR	-3.933	3.872	94	89	1.83%
IQR-$\alpha_{0.5\%}$	-4.645	4.764	54	44	0.98%
IQR-$\alpha_{5\%}$	-2.488	2.443	481	515	9.96%
Th-τ(mean)	-4.849	4.746	50	47	0.97%
Th-τ(med)	-4.752	4.656	53	52	1.05%
Th-τ(bw)	-4.911	4.822	47	44	0.91%

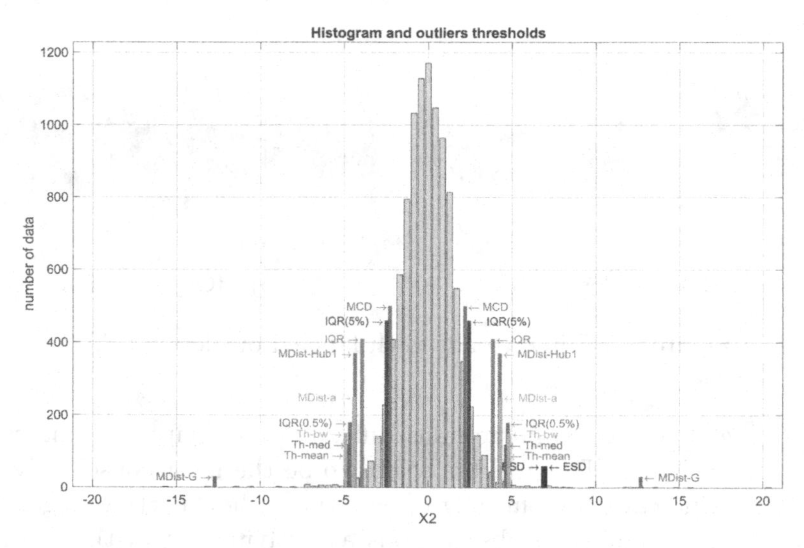

Figure 8.6: Histogram with outliers thresholds for dataset X_2

In this case we observe the effect of large unique outliers, which affect the ESD detector. In this case the MCD also is too optimistic in detection, while standard 3σ solution (MDist-G) works quite properly. The comparison of these two methods is sketched in Fig. 8.7.

8.8.3 Dataset X_3

Table 8.3 summarizes X_3 detection, while Fig. 8.8 visualizes the placement of detection thresholds associated with each method. In this case the pessimistic is the ESD approach, while the IQR-$\alpha_{5\%}$ method lies in the opposite extreme. Fig. 8.9 presents time trends for these two methods. In this case the MCD doesn't return any indication.

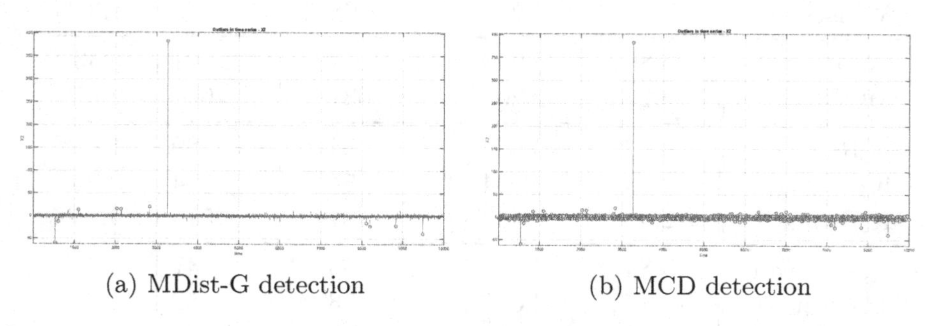

(a) MDist-G detection (b) MCD detection

Figure 8.7: Time trends with labeled outliers for X_2

Table 8.3: Outlier detection results for dataset X_3

	minThr	maxThr	minCount	maxCount	Percent
MDist-G	-0.445	0.455	44	35	0.79%
MDist-Hub1	-0.399	0.407	59	72	1.31%
MDist-α	0.000	0.000	0	0	0.00%
ESD	-0.681	0.682	15	15	0.30%
MCD	-2.240	2.243	0	0	0.00%
IQR	-0.355	0.360	91	147	2.38%
IQR-$\alpha_{0.5\%}$	-0.405	0.488	56	29	0.85%
IQR-$\alpha_{5\%}$	-0.220	0.239	547	534	10.81%
TTh-τ(mean)	-0.453	0.468	40	33	0.73%
Th-τ(med)	-0.428	0.439	49	48	0.97%
Th-τ(bw)	-0.454	0.468	39	33	0.72%

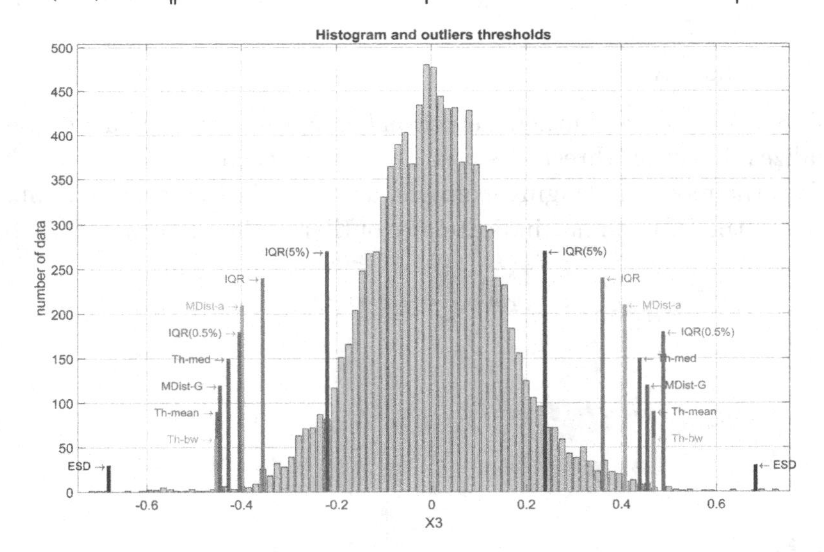

Figure 8.8: Histogram with outliers thresholds for dataset X_3

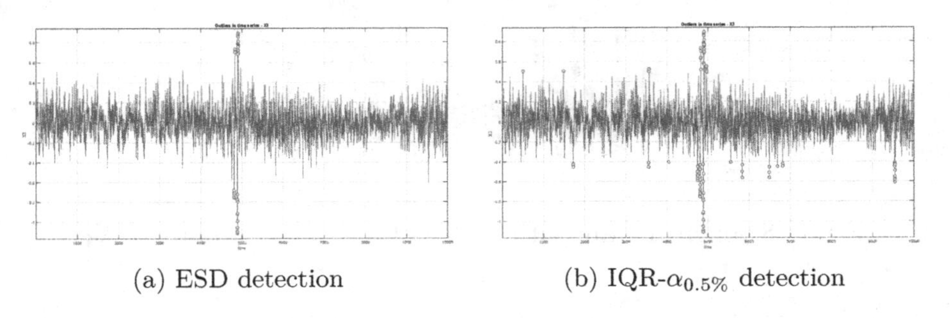

(a) ESD detection

(b) IQR-$\alpha_{0.5\%}$ detection

Figure 8.9: Time trends with labeled outliers for X_3

Table 8.4: Outlier detection results for dataset X_4

	minThr	maxThr	minCount	maxCount	Percent
MDist-G	48.867	54.711	0	4	0.04%
MDist-Hub1	48.678	54.901	0	0	0.00%
MDist-α	0.000	54.813	0	3	0.03%
ESD	0.000	0.000	0	0	0.00%
MCD	49.552	54.035	67	72	1.39%
IQR	49.000	54.600	8	15	0.23%
IQR-$\alpha_{0.5\%}$	49.194	54.383	10	28	0.38%
IQR-$\alpha_{5\%}$	50.132	53.445	437	404	8.41%
Th-τ(mean)	0.000	0.000	0	0	0.00%
Th-τ(med)	0.000	0.000	0	0	0.00%
Th-τ(bw)	0.000	0.000	0	0	0.00%

8.8.4 Dataset X_4

Table 8.4 summarizes outliers detection for X_4 time series, while Fig. 8.10 visualizes detection thresholds associated with each method.

It is the most challenging example, as shown in Fig. 8.11. Thompson tests and the ESD do not indicate any outliers. The MCD and IQR-$\alpha_{5\%}$

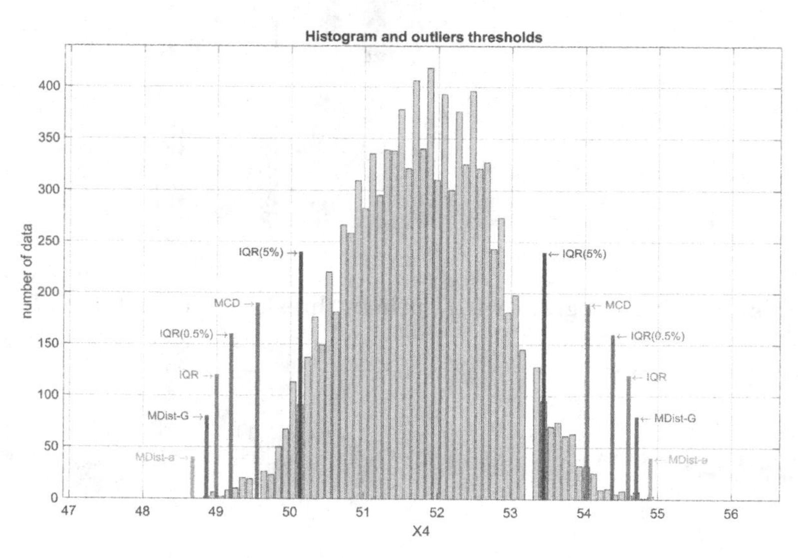

Figure 8.10: Histogram with outliers thresholds for dataset X_4

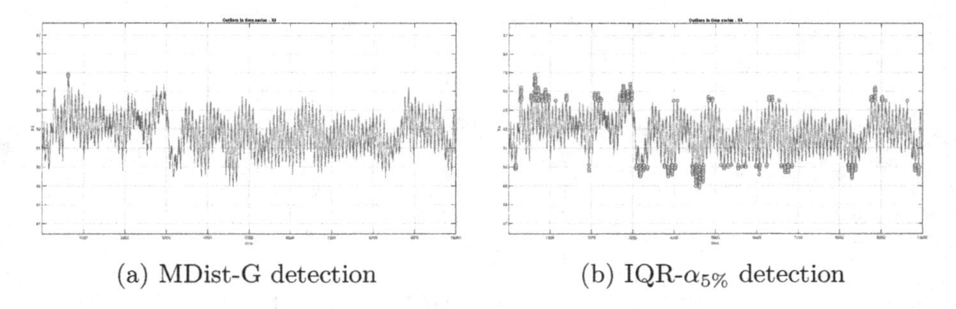

(a) MDist-G detection (b) IQR-$\alpha_{5\%}$ detection

Figure 8.11: Time trends with labeled outliers for X_4

are the most opportunistic, as they point out numerous outliers. In this case, the MDist-α algorithm detects the least number of outliers. Difficult detection is probably caused by signal non-stationarity (observed oscillations) and possible existence of a non-zero long-term trend.

8.8.5 Dataset X_5

Finally, Table 8.5 summarizes outliers detection results for X_5 time series, while Fig. 8.12 visualizes the placement of detection thresholds associated with each method.

These data are also not so well fitted for the detection as we may guess that there happened a change in the time series fundamental pro-

Table 8.5: Outlier detection results for dataset X_5

	minThr	maxThr	minCount	maxCount	Percent
MDist-G	-53.936	13.590	68	81	1.49%
MDist-Hub1	-47.506	7.167	129	142	2.71%
MDist-α	-25.996	-14.252	2580	2550	51.30%
ESD	-68.239	27.688	22	23	0.45%
MCD	-22.542	-18.060	3969	4045	80.14%
IQR	-44.527	4.085	176	206	3.82%
IQR-$\alpha_{0.5\%}$	-58.767	20.877	42	41	0.83%
IQR-$\alpha_{5\%}$	-36.241	-3.814	541	537	10.78%
Th-τ(mean)	-51.814	11.168	82	97	1.79%
Th-τ(med)	-49.245	8.973	109	126	2.35%
Th-τ(bw)	-51.054	10.602	88	102	1.90%

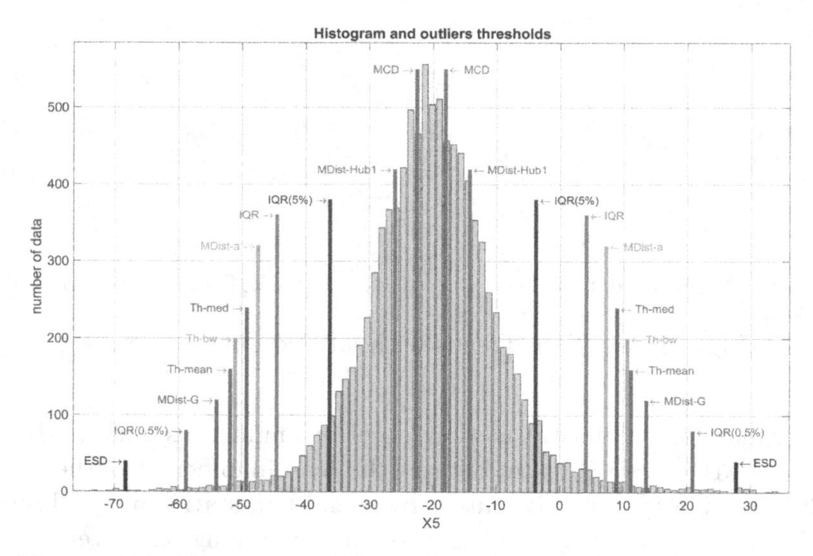

Figure 8.12: Histogram with outliers thresholds for dataset X_5

cess, as at the beginning the data are fluctuating in much more range than after sample ≈ 500.

We clearly see that in this case the MCD algorithm is so strongly biased that it indicates almost 80% of data as outliers, which is clearly wrong. Also the MDist-α method is strongly affected with the detection rate exceeding 50%. In this case the ESD algorithm seems to be the most reasonable, due to its strong reluctance in detection. In this case also the IQR-$\alpha_{0.5\%}$ approach is not so aggressive in detection. Both of them are shown in Fig. 8.13.

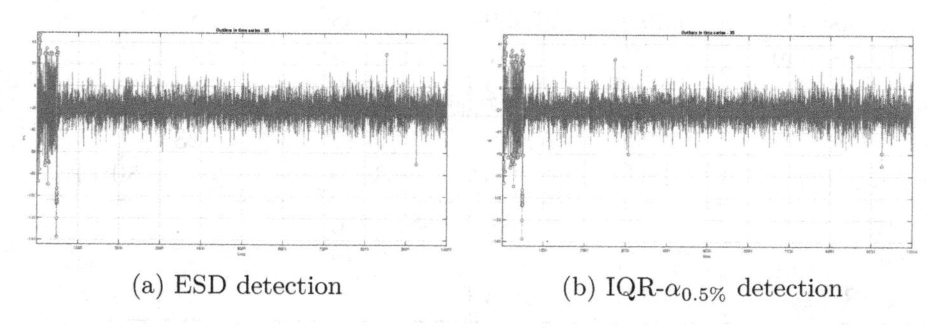

(a) ESD detection

(b) IQR-$\alpha_{0.5\%}$ detection

Figure 8.13: Time trends with labeled outliers for X_5

8.9 CONCLUDING REMARKS

This chapter is devoted to outliers and their detection statistical algorithms. Short introduction to the outlier identification task is followed by the presentation of several available statistical approaches. We show that control engineer has numerous options and his choice is not easy. The fact that we have so many options does not make the choice easier, in fact it makes it more difficult. Differences between algorithms are sometimes minor, such as the choice of the D factor in Z-scores methods or the choice of an appropriate shift/scale factor estimator.

Due to the above, the methods are either very resistant to detecting unusual values, or on the contrary, they indicate too many observations, which immediately raises doubts. These effects are clearly visible during a short comparison of data examples $X_1, \ldots, X_5$.

Generally, one should always take a lot of caution during selection analyzing data properties before. The generalized extreme studentized deviate (ESD) test seems to be the most promising, in contrary to the minimum covariance determinants (MCD) test, which has a dominating tendency to overestimate the number of outliers. The selection of the Z-scores methods depends on data character. In Gaussian case the 3σ approach is quite useful, however in other cases, especially heavy-tailed it may be wrong, because of the overestimated scale estimation. In such a situation, its α-stable counterpart, the MDist-α approach might help. Also the Interquartile Range and its IQR-$\alpha_{0.5\%}$ variant might be very useful, especially for skewed data.

There is one important issue, which should be addressed here. It is the effect of non-stationarity, which exists in X_4 and X_5 datasets. The X_4 time series is clearly oscillatory with additional slow-varying long-time trend. The X_5 data change their statistical properties, probably the scale. These effects have not been addressed yet, but Section 10.1 discusses this subject. Without prejudice to the description, it is worth mentioning here that the simplest solution may be to remove non-stationarity by building an incremental series ΔX_k and conducting the analysis for such time series

$$\Delta X\left(i\right) = X\left(i\right) - X\left(i-1\right). \tag{8.9}$$

Concluding the subject remains still open. Machine learning will not solve it automatically. We need attention and prudence. We need to learn about data and the process behind them. We need to consider several approaches before we make the final selection.

Statistical outliers identification requires a deep analysis of data properties, an overview of different approaches and a judicious choice. Automatic and superficial operation may lead either to excessive sensitivity of the algorithm, which will lead to long-term disregarding the results, or to its high resistance, which may result in unnoticed outlying observations.

GLOSSARY

AI: Artificial Intelligence

ML: Machine Learning

MLE: Maximum Likelihood Estimation

MADAM: Median of the Absolute Deviations About the Median

ESD: Extreme Studentized Deviate

MCD: Minimum Covariance Determinant

IQR: InterQuartile Range

CHAPTER 9

Visualization of statistical properties

VISUALIZATION allows to see, what is hidden behind numbers and equations. It is especially important in case of statistics, as the picture is no longer black and white. The true information remains hidden behind various shades of gray and the researcher requires different tools to interpret imprecision, uncertainty, variability and complexity. Statistical properties describe the original datasets, which may be in form of data tables (not characterized directly by time; not time-sampled) or in form of time series, i.e. sampled functions of time. Another type is a geospatial data, which is an emerging issue, however not within control engineering scope of interest. Thus, the most obvious is the presentation of data in their original form – in a tabular way. Though this form is not readable and its illegibility increases with the number of data, it is used, especially in case of a small sample data.

The next step is to present data in a graphical way using various graphs. As a picture is worth a thousand words, the presentation matters and data analytics results might be shown in different ways. Generally, the data may be visualized using the following proposed classification:

- charts and information presentation,
  - bar charts,
  - pie charts,
  - bubble charts,
  - radar plots,

DOI: 10.1201/9781003604709-9

- data points visualization,
  - scatter plots,
  - line plots and time trends,
  - step plots,
  - functional plots,
  - contour plots,
  - 3D drawings,
- representation of statistical properties,
  - histogram,
  - box-whisker plot (boxplot),
  - probabilistic density function,
  - quantile (Q-Q) plot,
  - moment ratio diagram and L-moment ratio diagram.

Each diagram highlights different features of data and should serve different purposes. Proper use will not only allow to better show the data, but also enable you to observe certain features not visible on another plot. Perhaps a book on statistics is not the place for an intricate discussion of data visualization, however, from the point of view of proper data analysis, it is important to present the results in such a way as to simplify their interpretation and enable correct conclusions to be drawn as quickly as possible. Of course, the presentation layer is closely related to the purpose. The following will be a brief discussion realized in relation to the task of evaluating the quality of control systems.

Charts, like bar, pie or bubble are generally used for the visualization of some classifications. In case of the control system we might visualize the status how the loop is utilized in different control modes, like for instance automatic (AUTO), manual (MAN), remote cascaded (RCAS) or supervised (SPV). Another classification may include the share of time the loop (actuators) spends in high/low saturation versus normal (unconstrained) operation. In multi-loop complex control systems we may present how many loops control temperature, pressure, flow, level or concentration. Actually, the utilization of the charts depends on the case, author of the analysis and recipients of the report.

Since charts are very popular and frequently used it makes no practical sense to devote more space to them. A slightly different situation applies to radar charts, which are not very often used although they allow multi-criteria visualization. Since the performance evaluation of control systems in general is multi-criteria (the classic conflict between accuracy and control time), it is worth using radar charts to visualize the performance evaluation of the control system. Radar plots have different names, like star, web or spider diagram [251].

Radar plots allow the comparison between different quantitative variables and give an opportunity to perform multi-criteria assessment. This allows to discover which loop property (reflected by certain index) scores low or high, which may give suggestions about potential improvement actions. Sample radar diagram is shown in Fig. 9.1 with comparison between four control loops being assessed by five indexes.

Multicriteria data assessment, also related to the CPA task may naturally include various statistical factors, like shift, scale, skewness or kurtosis estimators. As each of them reflects certain properties, often mutually independent this type of the presentation is very useful. Despite its readability and easy comparison, radar charts have some risks, as each of the KPIs used on the axes must be scaled to a single interval, e.g. $[0, 1]$ with a clear indication that, for example, 0 means a bad value

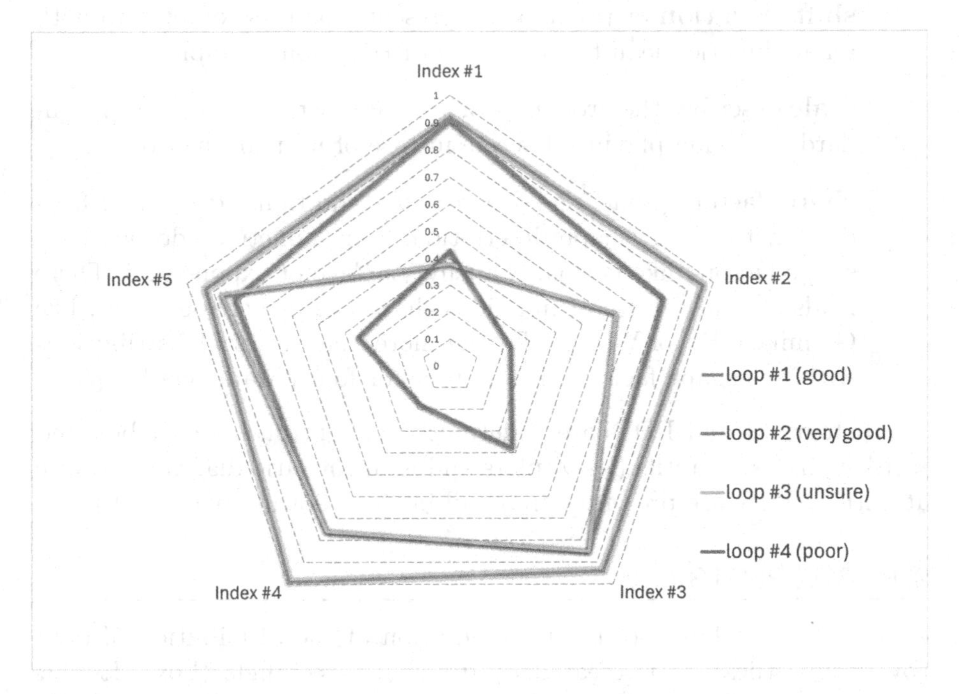

Figure 9.1: Sample radar plot for control performance assessment

and 1 a good one (or vice versa). When such scaling is necessary, the choice of scaling factors directly affects the subsequent evaluation and can often be highly discretionary in itself. Further practical aspects of the multi-criteria assessment and visualization can be found in [103, 88].

Data points visualization can be done in many ways, in for of the scattered plots, line plots, functional representation or in case of the three-dimensional visualization we get 3D graphs or contour plots. There is no specific distinction between these plots and their statistical context.

Quite a different story occurs with specific statistical plots, which directly address certain statistical properties or data representation. Histograms and probabilistic density functions are already introduced in previous sections. Histogram is a data-driven representation of the data distribution within its range. Drawing histogram requires only data, nothing else. A probabilistic density function is a function graph that is designed to model and visualize the distribution of data in the form of a parameterized function. Thus, it is a statistical model that allows the representation of data in parameterized form and the analysis and comparison in space of these parameters. The factors of the probability density function can have the following meanings:

- **shift** (**location** or **position**) represents the location of data with an arithmetic mean being normal distribution example,
- **scale** describes the broadness of the data distribution, with a standard deviation playing this role in case of normal function,
- **shape** factors (none, one or two) offer additional degrees of freedom in the PDF shape description. Some functions do not have shape factors, like normal, Cauchy, Laplace or exponential. There is also a wide group of functions that have one shape factor, like Gamma, GEV or Weibull. Furthermore, there are still distributions with two shape factors, like α-stable or four parameters kappa.

Histograms and PDFs are widely used in control, however box and whisker charts, quantile (Q-Q) plots and moment ratio diagrams require attention. They are presented in detail in the following paragraphs.

9.1 BOX PLOTS

A box plot is a kind of visual presentation of the distribution of data (by its quartiles). It is created by drawing a rectangle (box). Its left or bottom side (depends on whether the chart is presented vertically

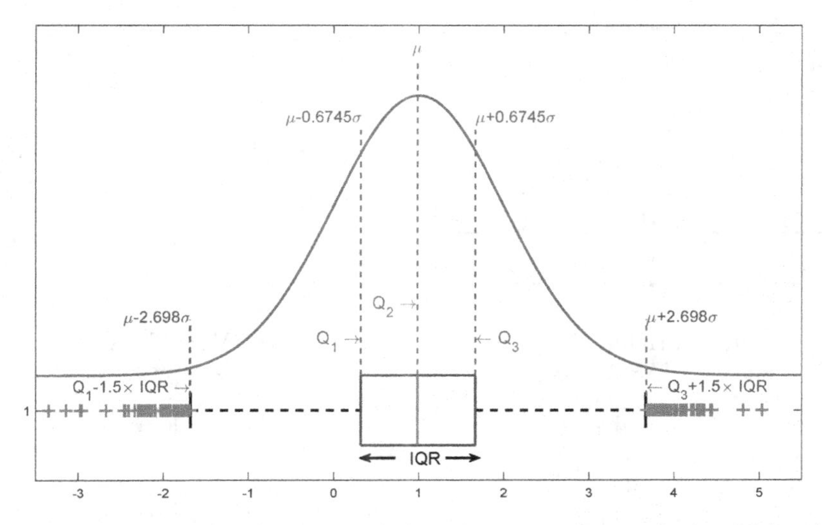

Figure 9.2: Boxplot for normal distribution – example for $N(1,1)$

or horizontally) is determined by the first quartile Q_1 and the opposite side by the third quartile Q_3. While, inside the rectangle there is a line defining the median $\overline{x}_{\text{med}}$ (the second quartile Q_2). The lines extending parallel from the boxes are called "whiskers". They are used to indicate the variability beyond the lower and upper quartiles. By default, the whiskers are one and a half times the value of the quartile spread. Fig 9.2 shows what it means if we take into account normal distribution $\text{N}(1,1)$.

In its simplest form, the left/lower end of the left/lower segment determine the smallest value in the set, while the opposite end is the largest value. Outliers are plotted as single red pluses that follow whiskers. While box plots seem primitive compared to a histogram or density plot, they have the advantage of taking up less space. This is useful when comparing distributions between multiple groups or data. It's easy to read:

- lower/upper quantile,
- median,
- whether the data are symmetric,
- how closely the data are grouped,
- how many outliers there are and what values they have,
- what is the skewness of the data and in which direction.

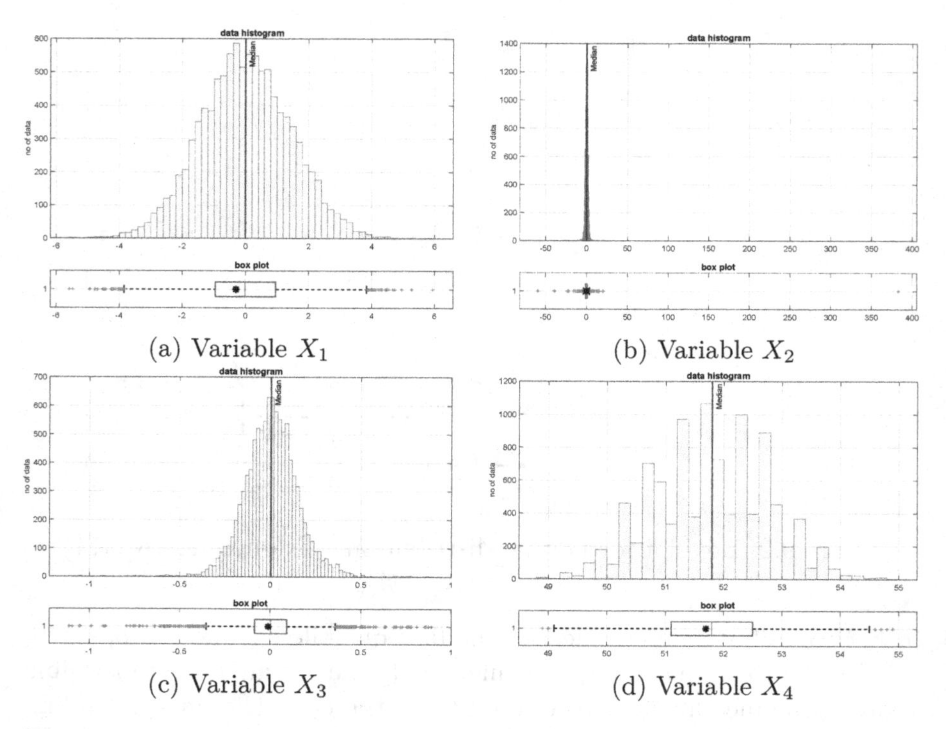

Figure 9.3: Histogram and boxplot for exemplary data – black star marks the center of the tallest bin

However, the main advantage of the box plot that it does not depend on the data statistical model (its distribution) and it might be used as one of the first analysis steps delivering better insight into the stochastic properties, which characterize the data. The boxplot can be used to any type of the data. Moreover, we may plot more than one boxplot in a single chart, what allows not only the analysis of a single dataset, but also to compare numerous ones.

Fig. 9.3 presents respective box-whisker plots for sample variables $X_1 \dots X_4$, while Fig. 9.4 presents the remaining chart for the X_5.

Boxplot visualizes in a very direct way the relationships already found for the data under consideration. A single, simple drawing offers very powerful capabilities. In particular, it is easy to see which part of the data is treated as the main one, the proportion of outliers and where they are located in relation to the rest of the data. For example, in the case of the X_2 variable, we can immediately see that there is one drastic outlier with a value of about 380, which "stretches" the histogram plot.

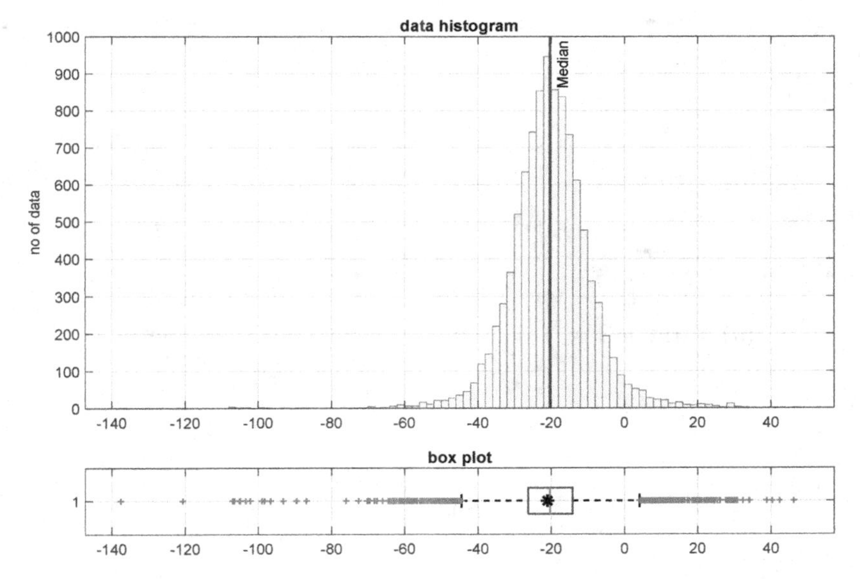

Figure 9.4: Histogram and boxplot for sample data X_5 – black star marks the center of the tallest bin

With a high degree of probability, this observation can be identified as an erroneous and most likely simply removed. We see that the boxplot may be used in the labeling of the outliers.

We can also see that the highest bar of the histogram does not necessarily coincide with the median. In addition, the degree and type of asymmetry in the data can also be estimated. The analysis of the "box" itself allows us to determine the asymmetry of the main part of the data (the basic element of the impact of the process), while the way the abnormal values are distributed can show that they are the ones that affect the tails and can cause their asymmetric distribution.

Boxplots are universal and do not depend on data properties. We may freely use them also in asymmetric, one-sided extreme events analysis. Fig. 9.5 presents respective box-whisker plots for extreme analysis data examples $Y_1 \ldots Y_4$. Actually, the observations are analogous to the previous ones, and they only confirm the versatility of the box plot and indicate that its use will allow the analyst to better see into the depths of the analyzed data.

However, there is another feature of box charts that we have not pointed out so far, which allows them to be used even better – this is the ability to conduct comparative analyzes of different datasets. Fig. 9.6

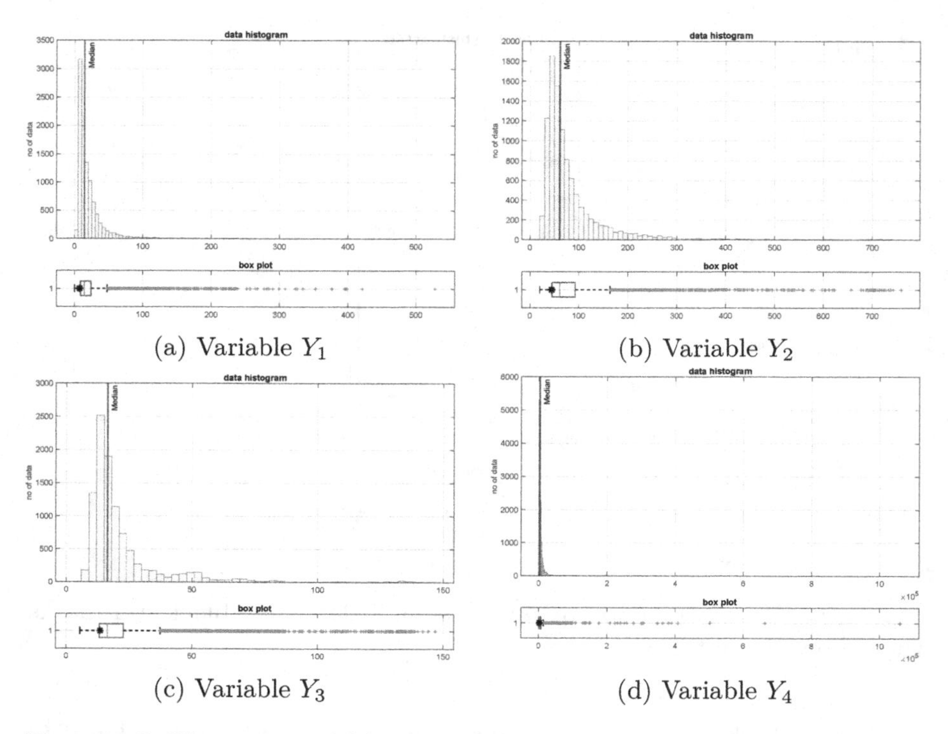

(a) Variable Y_1

(b) Variable Y_2

(c) Variable Y_3

(d) Variable Y_4

Figure 9.5: Histogram and boxplot for data $Y_1 \ldots Y_4$ – black star marks the center of the tallest bin

compares all sample data used in this book in a single plot. The upper chart puts in one plot all the datasets, however one of them exhibits extremely large values, what deteriorates the comparative analysis, and only shows different domains of values for the variable Y_4. Lower diagrams compare datasets with the Y_4 time series excluded. We may simply compare the share, location and distribution of the outlying observations, datasets shift and their scale size, which is reflected by the box size, their height in case of vertical boxplots.

It's a tad unfortunate that box plots are so sparingly used in analyses of control engineering applications. It seems that they could have helped significantly, for example, in comparing between the performance of individual controllers in multi-loop systems.

9.2 Q-Q PLOTS

Graphical methods are becoming increasingly popular for verifying whether observations come from a given distribution. This is another

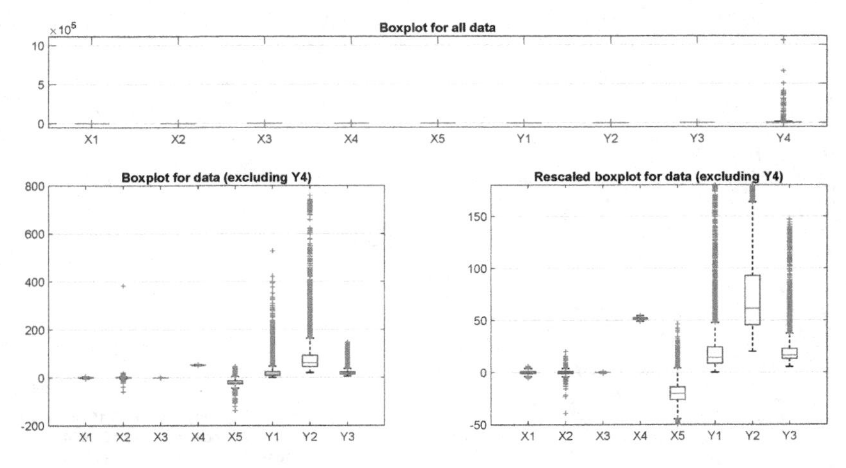

Figure 9.6: Comparison of data with the boxplot

approach to the task of finding appropriate distribution fitting empirical observations. For example, quantile plots, or Q-Q plots, are effective for this. We can also use these graphs to verify whether observations from two samples have the same distribution.

To obtain a quantile plot, we draw values of the quantile function for empirical distribution formed from the observations on one side against the theoretical quantile function values on the other side in an $\mathbb{R} \times \mathbb{R}$ Cartesian square. $\mathcal{F}^{-}(x)$ is a generalized inverse function. If $\mathcal{F}(x)$ is a distribution, then the function $Q_{\mathrm{F}}(x)$ is a quantile function of $\mathcal{F}$

$$Q_{\mathrm{F}}(x) = \mathcal{F}(x) . \tag{9.1}$$

At the same time, we define the empirical version of the quantile function by considering the generalized function from the empirical distribution defined as $Q_{\mathrm{e}}(x) = Q_{\mathrm{F}_{\mathrm{e}}}(x)$. Therefore, the Q-Q plot is a set of points $\{Q_{\mathrm{F}}(x), Q_{\mathrm{e}}(x)\}$. So the Q-Q graph forms a non-decreasing step function, often plotted just in form of the scattered plot.

The general philosophy of quantile graphs is to observe the linearity of the graph, which is easy to see. For example, let's consider the data obtained using $\mathrm{N}(0,1)$ normal distribution. We generate randomly four datasets with a different number of points $\{50, 200, 100, 10000\}$. The resulting plot is shown in Fig. 9.7. It gives three different takeaways.

All datasets fit better or worse to the normal distribution. We see that the more observations we use, better fitting is obtained. Visual assessment is one thing, but for quantitative comparison we may calculate

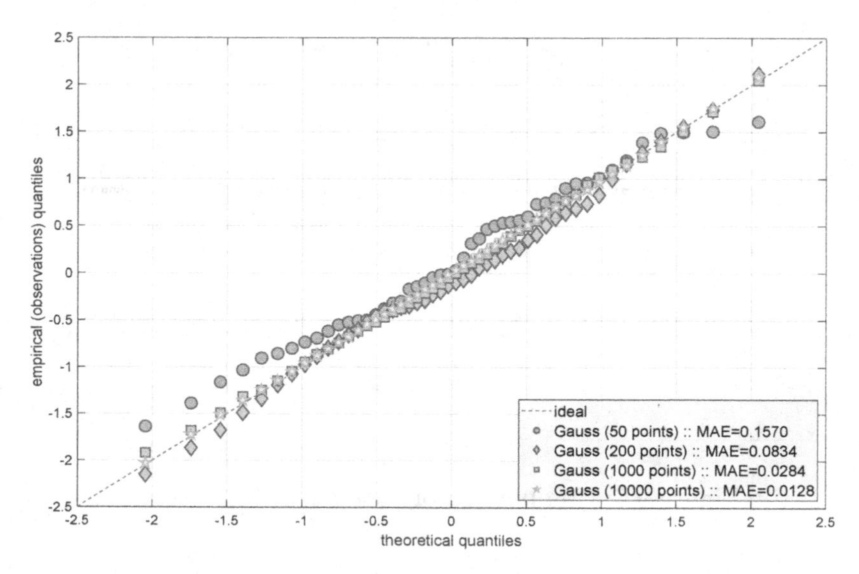

Figure 9.7: Q-Q plot for sample observations from N$(0, 1)$ distribution

the residuum versus the known N$(0, 1)$ PDF. The Q-Q plot not only visualizes distribution fitting, not only compares various datasets, but also can be used to evaluate how good the fitting is. For that different indexes can be used, but the MAE seems to be an obvious choice. These indexes can be utilized for PDF fitting, as we may compare empirical observation with the theoretical quantiles derived for various distributions.

Fig. 9.8 depicts Q-Q plots for sample variables $X_1 \ldots X_4$, while Fig. 9.9 presents the remaining chart for the X_5. These plots show, which PDF are more appropriate in reflecting empirical data properties. Not only we can say that some PDF is good or bad. In case of poor fitting we may see, where matching fails: in the peak region or in tails.

As we look at variable X_1 we see that almost all functions can fit except Laplace and Cauchy, which is obvious as all functions except those two include with their parameters a normal distribution. Variable X_2 addresses the opposite situation. Due to the extreme outliers and long tails normal-like distributions totally fail, while α-stable and generalized Student's t- distribution capture data properties. We notice the improving effect of robust estimation as it allows to improve normal fitting dramatically, in contrary to mean and standard deviation.

Other variables are quite similar to the X_1, however with different scale of the effect. Especially, the X_3 and X_5 definitely is not heavy-tailed as Cauchy distribution is not appropriate at all. We can say that

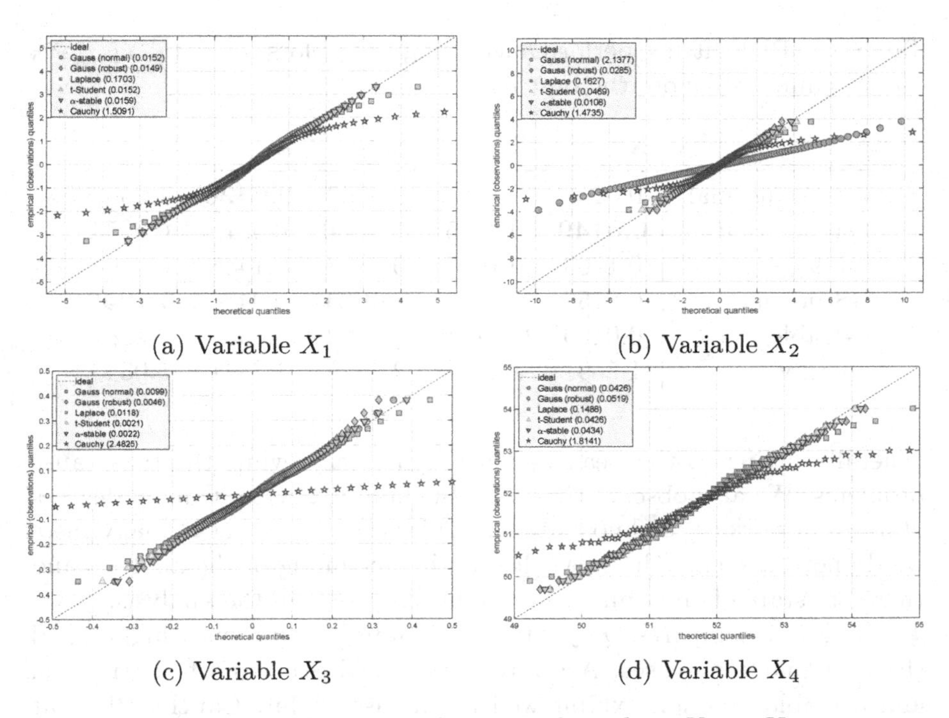

(a) Variable X_1 (b) Variable X_2

(c) Variable X_3 (d) Variable X_4

Figure 9.8: Q-Q plots for exemplary data $X_1 \ldots X_4$

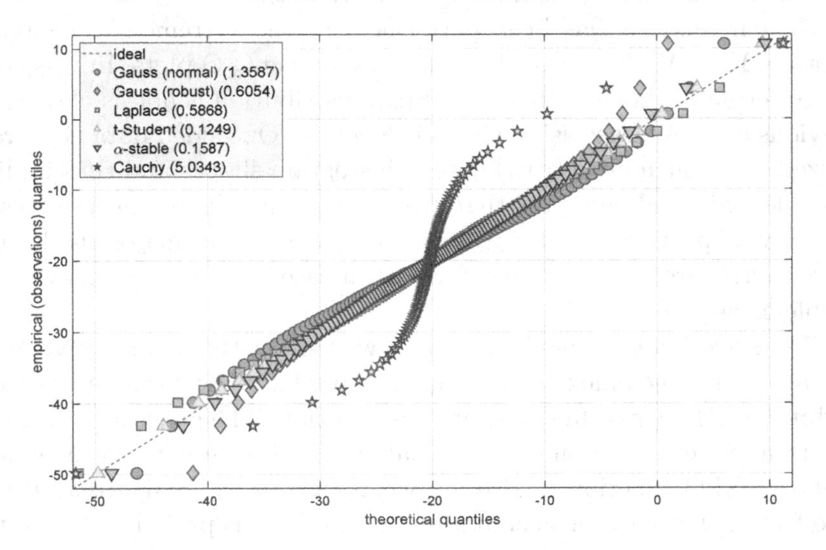

Figure 9.9: Q-Q plot for exemplary data X_5

Table 9.1: PDF fitting performance using Q-Q plots for data $X_1 \dots X_5$ – bold numbers denote the best fit

	X1	X2	X3	X4	X5
Gauss (normal)	0.0152	2.1377	0.0099	**0.0426**	1.3587
Gauss (robust)	**0.0149**	0.0285	**0.0046**	0.0519	0.6054
Laplace	0.1703	0.1627	0.0118	0.1488	0.5868
t-Student	0.0152	0.0469	**0.0021**	0.0426	**0.1249**
α-stable	0.0159	**0.0108**	0.0022	0.0434	0.1587
Cauchy	1.5091	1.4735	2.4825	1.8141	5.0343

generally fitting to the peak region is quite easy, while the tails cause problems. We also observe that α-stable family, due to it many degrees of freedom works well almost always. This is good news, as in many cases we do not have enough knowledge at the beginning of the analysis and therefore stable family might be the initial choice in the analysis.

Table 9.1 compares Q-Q fitting residuum indexes, with highlighted the best matching cases. Actually, generalized Student's t-distribution and α-stable function exhibit well in all cases, while Cauchy PDF, at least in the considered examples, generally fails.

Similar analysis has been performed for the extreme-like technical datasets $Y_1 \dots Y_4$. Fig. 9.10 shows the respective Q-Q plots. In this case, the selection of the favorite or universal distribution is not as clear as in previous case. At least using visual inspection. One-sided data is characterized by a much greater richness of histogram shapes and thus finding a single universal function that would work in all, or most, cases is not easy. Especially, the shape of the long tail creates major challenges, however the properties around zero, as in case of Y_2 and Y_3, may cause problems as well.

The selection criteria change, once we look at the quantitative number of the fitting index, as shown in Table 9.2. The "winner" is clear and it is the four parameter kappa distribution. The generalized logistic function succeeded in one case (variable Y_4), but we have to remember that the GLO function is also a special case of the K4P distribution. The fact that it has not been finding in that case is probably due to the estimating mechanism.

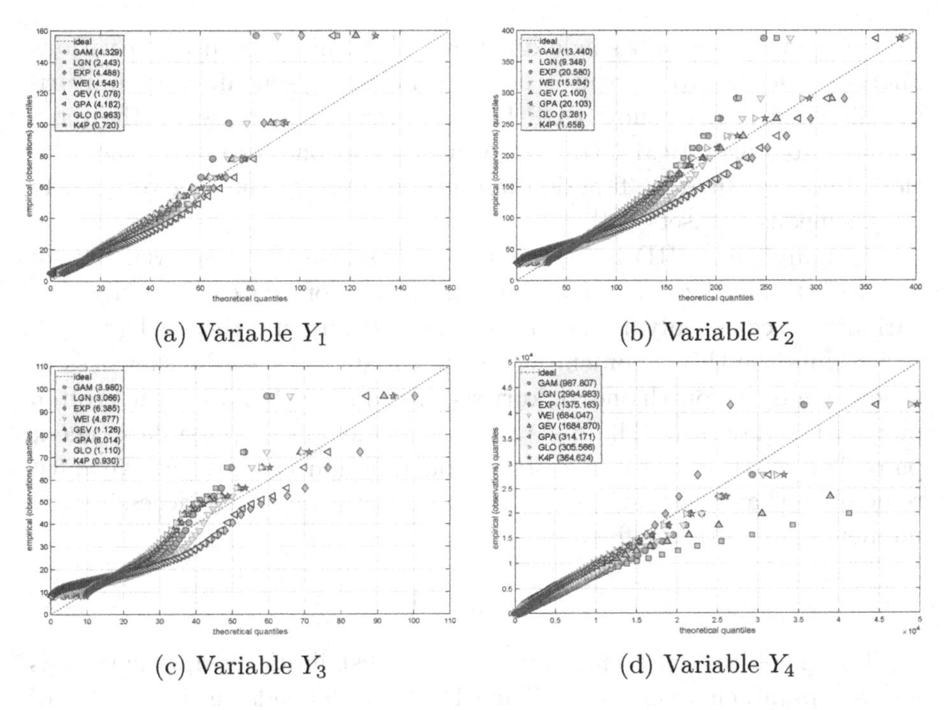

(a) Variable Y_1 (b) Variable Y_2

(c) Variable Y_3 (d) Variable Y_4

Figure 9.10: Q-Q plots for exemplary data $Y_1 \ldots Y_4$

9.3 MOMENT RATIO DIAGRAMS – MRD

Moment ratio diagram (MRD), known from works of Craig [57] has been initially mentioned by Karl Pearson works in early 19th century. They aim at graphical presentation of data statistical properties of the con-

Table 9.2: PDF fitting performance using Q-Q plots for data $Y_1 \ldots Y_4$

	Y1	Y2	Y3	Y4
GAM	4.3289	13.4395	3.9795	987.81
LGN	2.4433	9.3483	3.0665	2994.98
EXP	4.4880	20.5802	6.3854	1375.16
WEI	4.5481	15.9340	4.6767	684.05
GEV	1.0783	2.1005	1.1256	1684.87
GPA	4.1816	20.1025	6.0144	314.17
GLO	0.9632	3.2805	1.1097	**305.57**
K4P	**0.7198**	**1.6583**	**0.9296**	364.62

sidered time series in a single diagram. They might be used in various analyzes such as to measure the proximity between different univariate distributions, to show PDF function versatility, to select the most appropriate theoretical PDF to given observations, to clarify the relations between different function families, to classify them or to identify homogeneous datasets [34].

Actually, the MRD is a graphical representation of a given pair of selected standardized moments in Cartesian coordinates. Actually, two variants are commonly used [337]. The most common is the MRD(γ_3, γ_4), which puts the third moment γ_3 (or its square γ_3^2), i.e. the skewness as abscissa and the fourth one γ_4, kurtisis, as ordinate. This diagram is often plotted upside down. The diagram should include all possible moment pairs $\{\gamma_3, \gamma_4\}$ that a given distribution function can attain. We must remember that there exists theoretical constraint of the accessible by moments pairs area, defined as

$$\gamma_4 - \gamma_3^2 \geq 1. \tag{9.2}$$

The graphical representation of a given distribution can take various forms: a point, curve or region. The PDF locus depends on the number of shape factors of the respective theoretical PDF. The distributions that lack shape factor (like Gauss, Laplace or exponential) are represented by single point. Normal distribution is a point $(0, 3)$ and Laplace $(0, 6)$.

The PDF functions that are parameterized with one shape parameter, like generalized Students's t-distribution, GEV or Weibull are represented by a curve. Regions refer to distribution functions having two shape factors, like α-stable and K4P.

The second classical formulation of the moment ratio diagram was proposed by Cox and Oaks [55]. The MRD(γ_2, γ_3) relates the skewness γ_3 as the ordinate to the variance γ_2 as the abscissa. This diagram is location and scale-dependent. Typical layouts of MRD(γ_3, γ_4) and MRD(γ_2, γ_3) are shown in Fig. 9.11. Unfortunately, the MRDs are almost unknown among control engineers. Their use is sporadic [43].

Moment ratio diagrams for exemplary data are sketched below. Fig. 9.12 presents the most common MRD(γ_3, γ_4) diagram, while Fig. 9.13 the MRD(γ_2, γ_3). We observe that classical MRDs have disadvantages due to unscaled data and wide variability in the underlying values of moments. Therefore, the zoomed-in data are shown to present

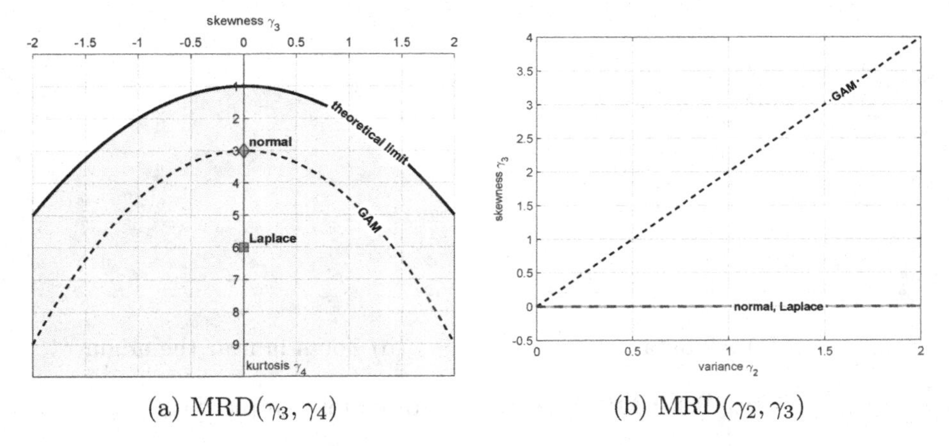

(a) MRD(γ_3, γ_4)

(b) MRD(γ_2, γ_3)

Figure 9.11: MRDs with theoretical PDFs, accessible area in gray

what happens close to the origin. The MRD(γ_3, γ_4) confirms that variables X_1 and X_4 are quite close to be considered normally distributed, while the others lie far away from the point $(0, 3)$, which reflects normal distribution – their distributions are not Gaussian.

Moreover, all the one-sided data are highly skewed to the right, which is visible with their high positive skewness values. Generally, the comparison using unscaled moments is ineffective with a very limited applicability to the very similar datasets. Though moment ratio diagrams offer high potential, their applicability requires much attention in scaling and it's expected that the use of L-moments should help in that issue.

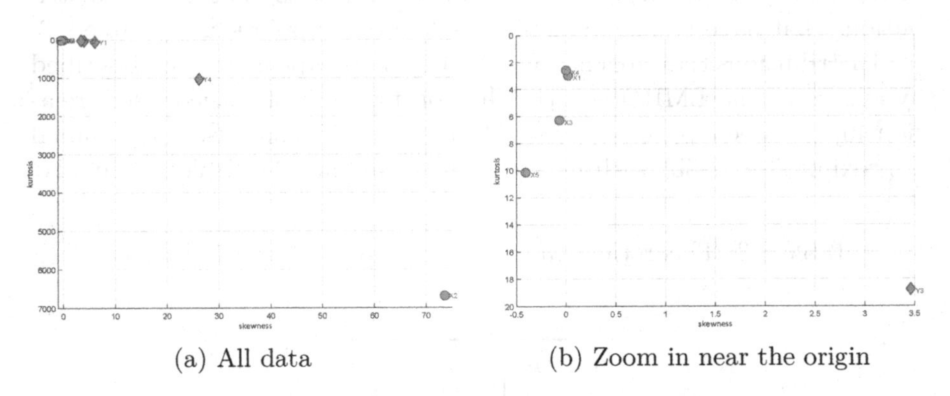

(a) All data

(b) Zoom in near the origin

Figure 9.12: MRD(γ_3, γ_4) diagrams for sample data

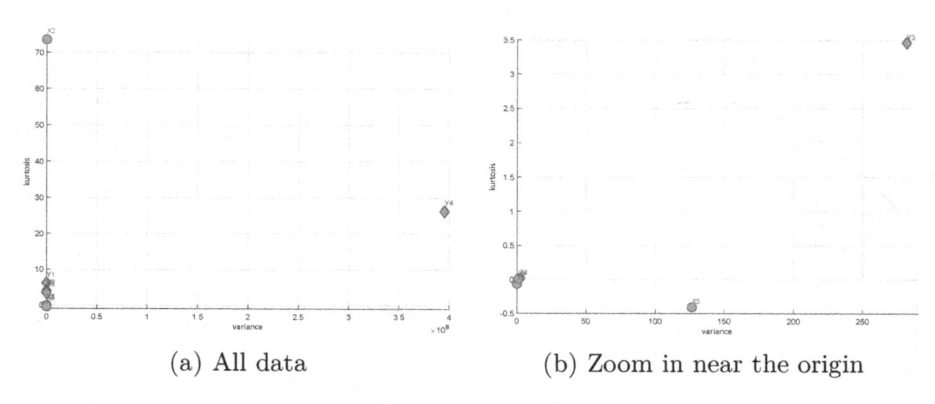

(a) All data (b) Zoom in near the origin

Figure 9.13: MRD(γ_2, γ_3) diagrams for sample data

9.3.1 L-Moment Ratio Diagrams – LMRD

As we have shown L-moments [149] propose a new interpretation to the background notion. Thus, it is only natural to expand the concept of MRD and introduce a new type of diagram exchanging moments with L-moments. As the result we obtain the L-moment ratio diagrams, denoted as the LMRDs. It appears that they quickly got high popularity among extreme value analysis (EVA) community [340, 121]. Similarly to the MRD, the LMRD supports the identification of an appropriate theoretical distribution to given empirical observations [259]. The diagram that compares L-Kurtosis τ_4 with the L-skewness τ_3, denoted as LMRD(τ_3, τ_4), is the most common. Fig. 9.14 presents the blank LMRD(τ_3, τ_4) with various theoretical distributions. Theoretical distributions that have no shape factor are defined by points as in Table 9.3.

Distributions that are characterized by one shape factor are described by curves in the LMRD(τ_3, τ_4). The construction of a theoretical relationship between τ_3 and τ_4 for selected distributions uses polynomial approximations [153] with a_i coefficients sketched in Table 9.4. It also

Table 9.3: Theoretical values for τ_3 and τ_4 of selected PDFs

	τ_3	τ_4
uniform	0.0	0.0
normal	0.0	0.1226
Laplace	0.0	17/22
EXP	1/3	1/6

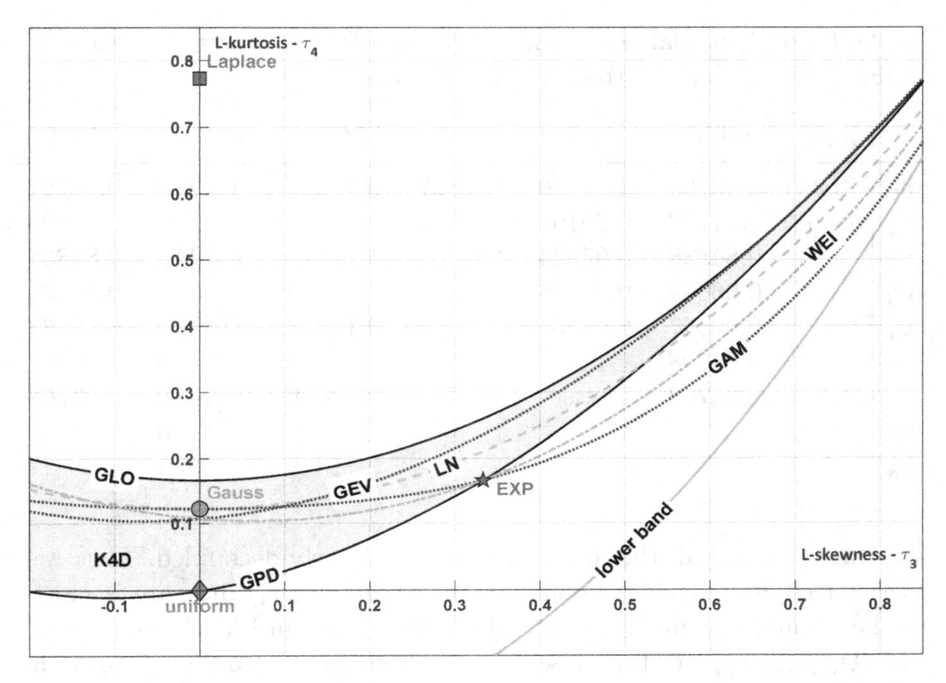

Figure 9.14: Blank LMRD(τ_3, τ_4) diagram with theoretical PDFs

includes the theoretical limit polynomial coefficients

$$\tau_4 = \sum_{i=0} a_i \tau_3^i. \tag{9.3}$$

The four parameter kappa distribution is described by two shape parameters, and therefore it is reflected by the region lying between GPD and GLO curves.

By the analogy to MRD(γ_2, γ_3), we may relate the third L-moment to the second one. We get two options, i.e. the extension comparing the L-skewness with L-Cv (τ_2) in the formulation of the LMRD(τ_2, τ_3). However, the LMRD(l_2, τ_3) might be used as well. Research shows that its functionality is useful in control applications [95].

The LMRDs are highly popular and frequently used in life sciences, especially in hydrology and climatology [191, 121, 130]. The idea to take an advantage of LMRDs in control engineering uses an assumption that properly controlled loop process variable and respective control error should behave according to the Gaussian properties – they represent a sequence of serially uncorrelated random variables. It reflects the situation that given controlled system is linear, symmetric, stationary, noises

Table 9.4: Polynomial coefficients of τ_4 as a function of τ_3 for selected PDFs – "limit" denotes theoretical limit curve

	limit	GEV	GPD	GLO	LGN	GAM	WEI
a_0	-0.25	0.10701	0	0.16667	0.12282	0.1224	0.10701
a_1	0	0.1109	0.20196	0	0	0	-0.11090
a_2	1.25	0.84838	0.95924	0.83333	0.77518	0.30115	0.84838
a_3		-0.06669	-0.20096		0	0	0.06669
a_4		0.00567	0.04961		0.12279	0.95812	0.00567
a_5		-0.04208			0	0	0.04208
a_6		0.03763			-0.13638	-0.57488	0.03763
a_7					0	0	
a_8					0.11368	0.19383	

are Gaussian and all the disturbances are properly decoupled. Thus we expect that its associated point on the LMRD(τ_3, τ_4) diagram is close to the normal distribution point $(0, 0.1226)$, while in LMRD(τ_2, τ_3) and LMRD(l_2, τ_3) is placed as close as possible to the plot origin $(0, 0)$ with zero L-skewness $\tau_3 \approx 0$.

L-moment ratio diagrams for sample data are sketched below. Fig. 9.15 depicts common LMRD(τ_3, τ_4) diagram.

We clearly see the advantage of the LMRD over the MRD plot. The L-moments are scaled and belong to certain known limitations. The plot is consistent and there is no need for any scaling. The difference between two-sided data X_i and one-sided extreme-like time series Y_i is clearly visible. Also it is shown that the distance from some reference point, like for instance the point reflecting normal distribution can be considered as a similarity measure to some theoretical distribution (normal in this case) or to the good control (in control engineering interpretation).

Diagrams that incorporate the relation of skewness versus scaling are addressed in the following paragraphs. Fig. 9.16 presents the relation of skewness versus L-Cv in LMRD(τ_2, τ_3), while Fig. 9.17 the skewness as a function of l_2 in LMRD(τ_2, τ_3). Both diagrams share the disadvantage of required scaling, which makes their utilization inconvenient in comparison with LMRD(τ_3, τ_4).

Respective research starts to address these issues in control engineering in their direct use [95] or in the context of control performance

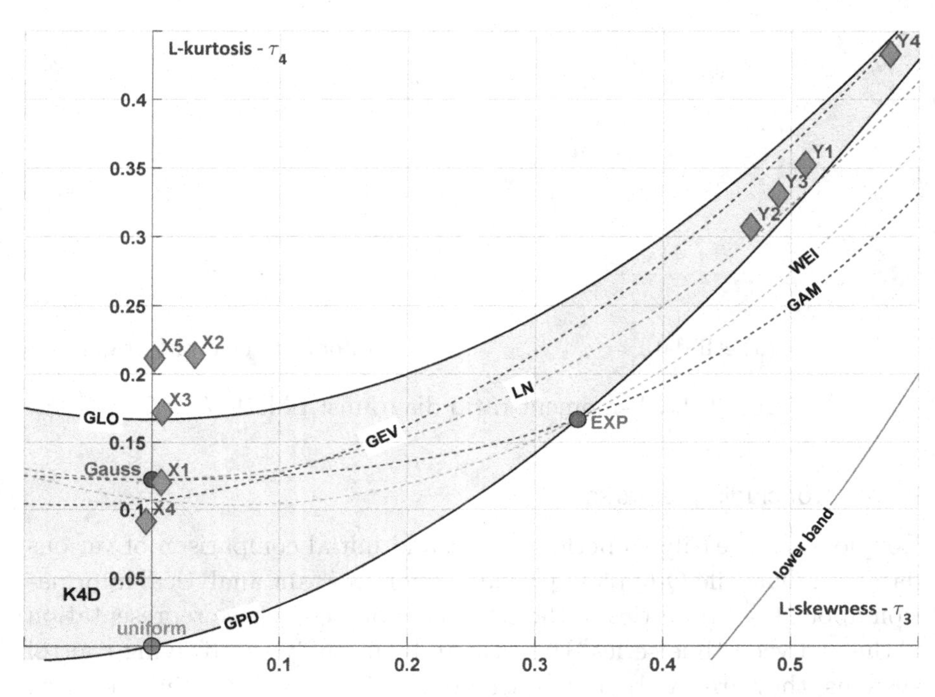

Figure 9.15: LMRD(τ_3, τ_4) for exemplary data

sustainability [98]. The general idea of moment ratio diagrams might be extended in two directions: regression model residuum analysis or ratio diagrams, which may include not only moments but other statistical factors, like the ones of the α-stable distribution or other measures.

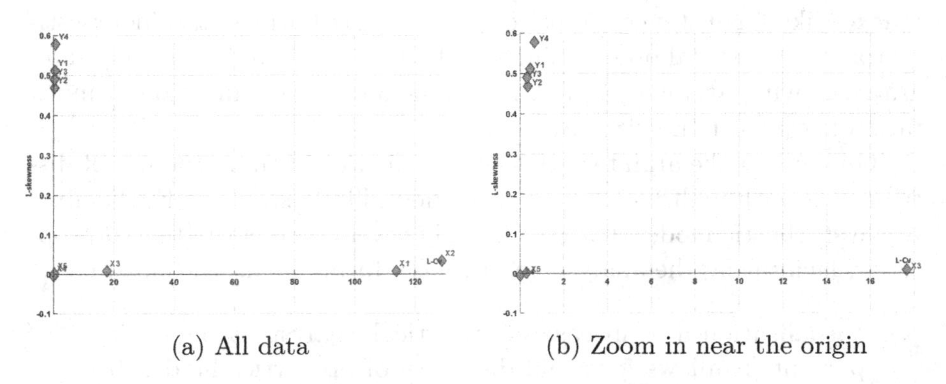

(a) All data (b) Zoom in near the origin

Figure 9.16: L-moment ratio diagrams: LMRD($L - Cv, \tau_3$)

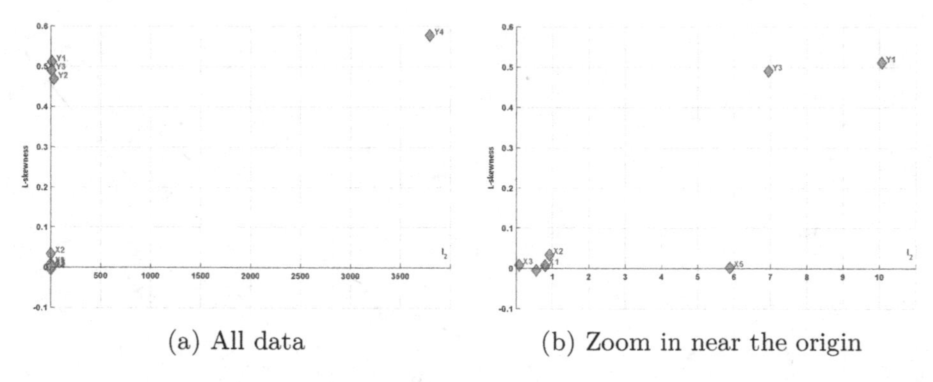

(a) All data (b) Zoom in near the origin

Figure 9.17: L-moment ratio diagrams: LMRD(l_2, τ_3)

9.3.2 Concluding remarks

Boxplots allow easily to perform draft and initial comparison of various datasets. Quantile Q-Q plots provide very important analytical information about the properties of the stochastic process whose representation is the analyzed time series. Unfortunately, in analyses involving control systems, they are vastly underestimated and are used very infrequently. And this is a pity, since they can allow rapid validation of hypotheses on interference and process noise, for example, in order to design appropriate filters.

Moment ratio diagram gives further insight into statistical data properties. They also can be applied to the multicriteria regression analysis, especially in case of the machine learning approaches, due to the high number of possible methods [182]. Their extension with the mix of other indexes like for instance factors of various distributions, tail index estimators, or fractional order estimates of auto-regressive fractionally integrated moving average (ARFIMA) filters goes in the direction of index ratio diagrams (IRD) [181, 91].

One may make an interesting observation concerning the use of distributions with multiple degrees of freedom, like α-stable or K4P deliver single stochastic model that significantly covers different potential properties, facilitating the analysis in any case during identification stages.

Visualization also in case of statistical analysis is highly important. It allows for rapid discovery of properties hidden behind the numbers. Control engineering mostly uses time trends.

However, histograms, boxplots, Q-Q plots and moment ratio diagrams may improve the analysis and as such should be put into the engineer's toolbox.

GLOSSARY

Q-Q: Quantile plot

MRD: Moment Ratio Diagram

LMRD: L-Moment Ratio Diagram

EVA: Extreme Value Analysis

ARFIMA: Autoregressive Fractionally Integrated Moving Average

IRD: Index Ratio Diagram

CHAPTER 10

Research derivatives

STATISTICS apart from the above, let's call standard issues, brings much more. It is due to the fact that the data properties are often more complex than assumed or expected. Generally, the notions of moments and probabilistic density functions assume that data are stationary. However, this assumption is hardly met in industrial practice.

Since researchers rarely note this fact and, consequently, practitioners equally rarely take it into account, it is worth referring to stationarity in the data and methods for its detection. At the same time, a related aspect related to oscillatory behavior and periodic trends is considered more extensively.

Another topic that could practically be presented in a separate book is the issue of regression. In contrast, in this paper it will be described as briefly as possible, referring only to statistical properties in the data and their impact on regression tasks itself.

As related, the concepts of tail index, long-term memory, homogeneity or causality are introduced. These topics practically extremely rarely appear in industrial control engineering, and we show that they are worth remembering and knowing what they can contribute to data analysis and our knowledge of the problem under consideration. The above selection may be considered debatable, but nevertheless industry experience indicates their hitherto overlooked or unnoticed potential.

10.1 STATIONARITY

The popular understanding of stationarity means that we treat a time series as stationary when its statistical properties do not change over time.

DOI: 10.1201/9781003604709-10

Formally, a given time series (or rather a stochastic process that generates the time series, with both concepts closely related) is called stationary (more formally: weakly stationary) if it has time-independent mean value, variance and auto-correlation. In the case of weak stationarity for a given time series X_i, the expected value $\mathbb{E}(X_i) = \mu$, the variance $\sigma^2 < \infty$ and the covariance $\mathrm{cov}\,(r_t, r_{t-l}) = \gamma_l$ (which depend only on l)3 are assumed to be constant due to shifts in time.

The series X_i is called strictly stationary if the joint distribution of the multivariate variable $(x_{i_1}, x_{i_2}, \ldots, x_{i_k})$ is identical to the distribution of the variable $(x_{i_1+h}, x_{i_2+h}, \ldots, x_{i_k+h})$ for each h and any finite sequence of indices $(i_1, i_2, \ldots, i_k)$. Strict stationarity means that the joint distribution of a multivariate variable is constant due to shifts in time. This definition implies that its probability density function is constant in time. This is a very strong assumption, difficult to verify in practice.

The definition of strict stationarity includes the definition of stationarity in a weak sense. The opposite statement is not necessarily true. Practically, industrial reality considers the weak stationarity, which implies that statistical properties of a time series such as the mean or standard deviation do not change over time. Real-time series are usually non-stationary .

At that point it's worth to notice that the white noise time series, i.e. $\epsilon_i \sim \mathbb{N}\left(0, \sigma^2\right)$ and $\mathrm{cov}\left(\epsilon_i, \epsilon_j\right) = 0$, for $i \neq j$ is a stationary process $x_i = \epsilon_i$. In contrary, a random walk process $x_i = x_{i-1} + \epsilon_i$ is non-stationary . Its expected value is constant, while the variance is not constrained in time.

Before further description, it's worth to introduce the concept of integration. We denote the first-order integration operator as $\Delta x_i = x_i - x_{i-1}$, the second order as $\Delta^2 x_i = \Delta\left(\Delta x_i\right) = x_i - 2x_{i-1} + x_{i-2}$ and finally the n-th order

$$\Delta^n x_i = \underbrace{\Delta \ldots \Delta}_{n} x_i. \tag{10.1}$$

If the X_i time series is stationary, than it's zero-order integrated

$$x_i \sim \mathbb{I}\left(0\right). \tag{10.2}$$

If the ΔX_i time series is stationary, than it's the first order integrated

$$x_i \sim \mathbb{I}\left(1\right) \tag{10.3}$$

and in general if the $\Delta^n X_i$ time series is stationary, than it's the n-th order integrated

$$x_i \sim \mathbb{I}\left(n\right). \tag{10.4}$$

Following the above, we may introduce two types of stationarity. Incremental stationarity means that the original time series is not stationarity, while the increments are stationary $x_i \sim \mathbb{I}(n)$. Trend stationarity means that the time series is the sum of the deterministic trend and the stationary stochastic process (e.g. white noise) $x_i = \alpha + \beta i + \epsilon_i$. Therefore, a given process is said to be trend stationary (or stationary around a trend) if it can be considered as a linear combination of some stationary process and one or more processes having a trend. Thus, once we remove the trend, it would become stationary.

There are methods that allow to assess stationarity better than observing the data. Several statistical tests have been created for this purpose. The presence of a unit root indicates that the time series is non-stationary. Let's consider the first-order auto-regressive process

$$x_i = \alpha x_{i-1} + \epsilon_i. \tag{10.5}$$

The t-student's test is the simplest to check if $\alpha = 1$, as in case of $\alpha = 1$ the process X_i is a random walk. If so, the α estimator is biased and doesn't fit the t-student's distribution. Thus, we check if $\delta < 0$ for

$$\Delta x_i = \delta x_{i-1} - \epsilon_i. \tag{10.6}$$

Following the above we build the hypotheses:

$$\begin{aligned} \mathcal{H}_0 &: \alpha = 1 \Longleftrightarrow \mathcal{H}_0 : \delta = 0, \\ \mathcal{H}_1 &: \alpha < 1 \Longleftrightarrow \mathcal{H}_1 : \delta < 0, \end{aligned} \tag{10.7}$$

while the null hypothesis denotes the non-stationarity. This test is called the Dickey-Fuller test (ADF) [73]. The test exhibits low power in case of the random part auto-correlation. Moreover, it assumes that the generating process AR(1) has no free element. The augmented Dickey-Fuller test (ADF) changes the test regression into

$$\Delta x_i = \gamma x_{i-1} + \sum_{s=1}^{P} \alpha_s \Delta x_{i-s} + \epsilon_i. \tag{10.8}$$

As one can see it allows to take into account deterministic components, i.e. the free element and the linear trend. Critical values for the ADF test differ from the t-student's statistics. They are evaluated numerically and may differ depending on the utilized software [69].

In practice, the ADF-GLS test variant is used. It was developed by Elliott, Rothenberg and Stock [105] as a modification of the augmented Dickey–Fuller test. Actually, the ADF-GLS test is an augmented Dickey–Fuller test, except that the data is transformed using a generalized least squares (GLS) regression before running the test. It exhibits higher power comparing to the ADF test once the autoregressive root is large, but still less than one. The DF-GLS test gives higher probability to reject the false null of a stochastic trend during situations when the sample data stems from a time series that is close to the integrated one. It tests for an auto-regressive unit root in the detrended time series, while the basic GLS deterministic components estimates are utilized to get detrended formulation of the original data.

An alternative test, like Kwiatkowski-Phillips-Schmidt-Shin test [202] (KPSS), which unlike unit root tests, provides straightforward test of the null hypothesis of trend stationarity against the alternative of a unit root. To achieve that the time series x_i is assumed to consist of three components: a deterministic time trend, a random walk and a stationary residual:

$$x_i = \beta i + (r_i + \alpha) + \epsilon_i, \tag{10.9}$$

where $r_i = r_{i-1} + u_i$ is a random walk with the initial value $r_i = \alpha$ serves as an intercept and u_i is an independent and identically distributed variable from $\mathcal{N}(0, \sigma_u^2)$.

Following the above assumptions we build the hypotheses:

$$\begin{aligned} &\mathcal{H}_0 : x_i \text{ is trend stationary} \vee \sigma_u^2 = 0, \\ &\mathcal{H}_1 : x_i \text{ is a unit root process.} \end{aligned} \tag{10.10}$$

In some cases the simplified test version is used that does not utilize the trend component is also used. It is applied to test level stationarity. In some situations, we may use stationarity as an alternative tool to detrend time series, in case to detect and remove long-term trends to emphasize short-term fluctuations.

We frequently apply both tests together to check for the stationarit y and its type. We obtain four options:

1. DF-GLS and KPSS conclude non-stationarity.

2. DF-GLS and KPSS conclude that series is stationary.

3. DF-GLS concludes stationarity and KPSS the opposite.
4. DF-GLS concludes non-stationarity and KPSS the stationarity.

In the first and second case, the result and interpretation is straightforward. Case 3 implies that data is difference stationary and the differencing is needed to make data strictly stationary. The last case implies the need for detrending.

10.1.1 Final comments on stationarity

Generally, stationarity testing is quite easy and it is strange that stationarity tests are practically unknown and unused in control engineering. And yet it seems that they should be almost one of the first steps when analyzing data. After all, virtually all approaches to modeling or control assume stationarity.

In real industrial solutions, stationarity should not be assumed blindly, without prior verification. At the very least, such errors can lead to misinterpretations and, in worse cases, to faulty models or control strategies. And then it is difficult to find the cause when it is found at the very beginning of the reasoning by making wrong assumptions.

10.2 OSCILLATIONS AND THEIR DETECTION

Oscillations, regardless of their negative significance for the operation of a given process, are still a common occurrence. And neither the awareness of process instability nor the accelerated wear and tear of actuators (valves, dampers, pumps, fans, etc.) changes this. Oscillations that exceed the sense of safety of operators most often cause that the control loops that are most affected by this phenomenon are switched to the manual control mode. Of course, artificially forcing the immutability of the actuator is no solution, as it cures the effect and not the cause. It happens though they are most often well visible in the loop time trends.

Interestingly, oscillations are most often not the origin of undesirable phenomena, but are their result and highly visible symptom. And the cause can be a poorly (too aggressively) tuned control loop, integration wind-up in a PI/PID type controller, a malfunctioning actuator, e.g. due to friction, or an uncoupled disturbance.

Clear oscillating time series is a sine wave signal. It exhibits characteristic U-shaped histogram as shown in Fig 10.1. In practical assessment the oscillating (sinusoidal) features of process data are well detectable

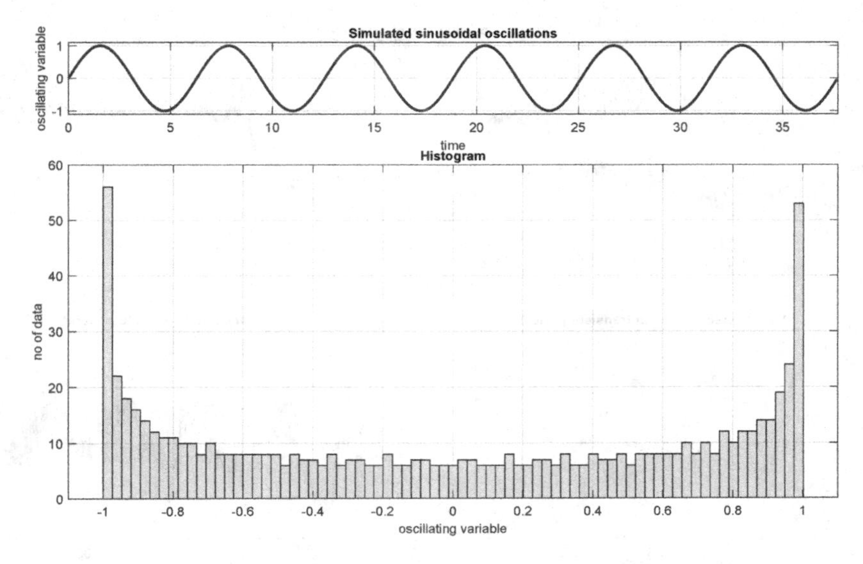

Figure 10.1: Sine wave time series and its histogram

by simple visual inspection of data time trends. Fig. 10.2 shows an exemplary time series of the oscillating variable. We may clearly notice the oscillations. They exist all the time. During transient period, when the operating steady state changes we observe clear damped oscillations, which continue to exist during the steady operating regimes.

We must be aware that the histogram should not be applied to the whole data as they are biased by a trend, which most probably is due to the changing operating regimes. Example of such improper histogram (in sense of its further analysis and interpretation) is sketched in Fig. 10.3. We observe numerous peaks in the plot (at least three dominating), which reflect different operating points. We have to be really cautious as they might be easily confused with histogram of oscillatory signal.

Therefore, we must select a period of the relatively steady operating regime for the oscillatory analysis, as shown in the example in Fig. 10.4, and do the histogram analysis for such data.

The question is how should we detect and assess the oscillatory behavior. Visual inspection is a very first step and in many cases it's quite enough. It allows to notice clear oscillations and even to measure its period and/or the amplitude. For the most part, the mere fact of detecting oscillations is enough, while additional information on their frequency will allow their tracking and possible analysis.

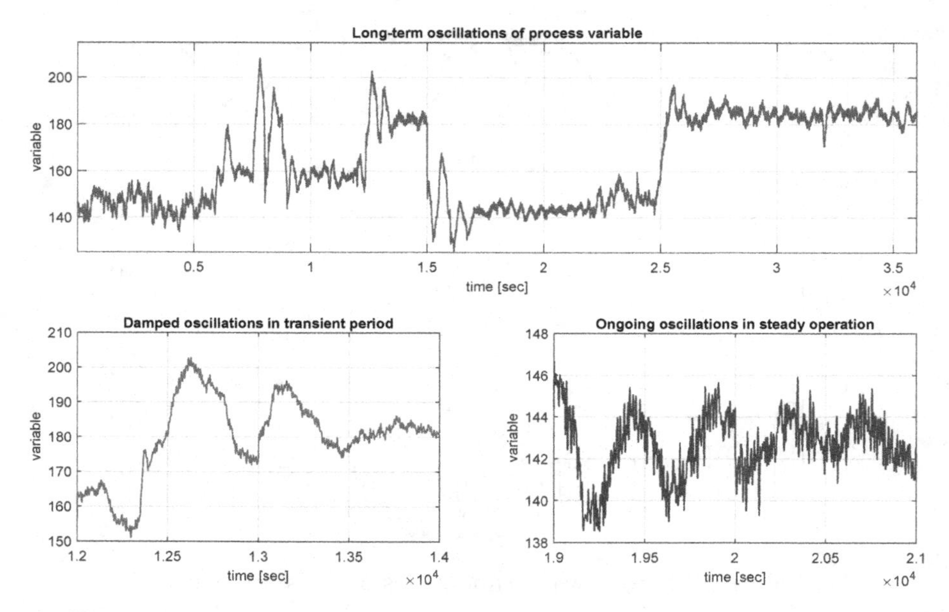

Figure 10.2: Industrial example of the oscillating process variable

Nevertheless, research has not left this issue in oblivion and we can find some interesting formal approaches. Apart from the visual investigation we may distinguish the following methods' categories:

- time domain approaches using various criteria like the mean absolute error (MAE) [141, 333],
- detection of spectral peaks in the frequency domain using Fourier transform, however this method is limited to linear systems with low noise-to-signal ratio and is seldom in process control [185],
- approaches using the auto-covariance function [238],
- methods that use wavelet plots [232].

More detailed discussion on the methods may be found out in the relevant literature, however as one can observe the methods are quite old. A very interesting approach maybe found in the novel approaches that consider more general aspect, i.e. the signal decomposition. As we have already observed in the stationarity analysis, data often include the trend, while stationarity tests assume the existence of some stochastic process and noise as well. Thus, a given time series may consist of a

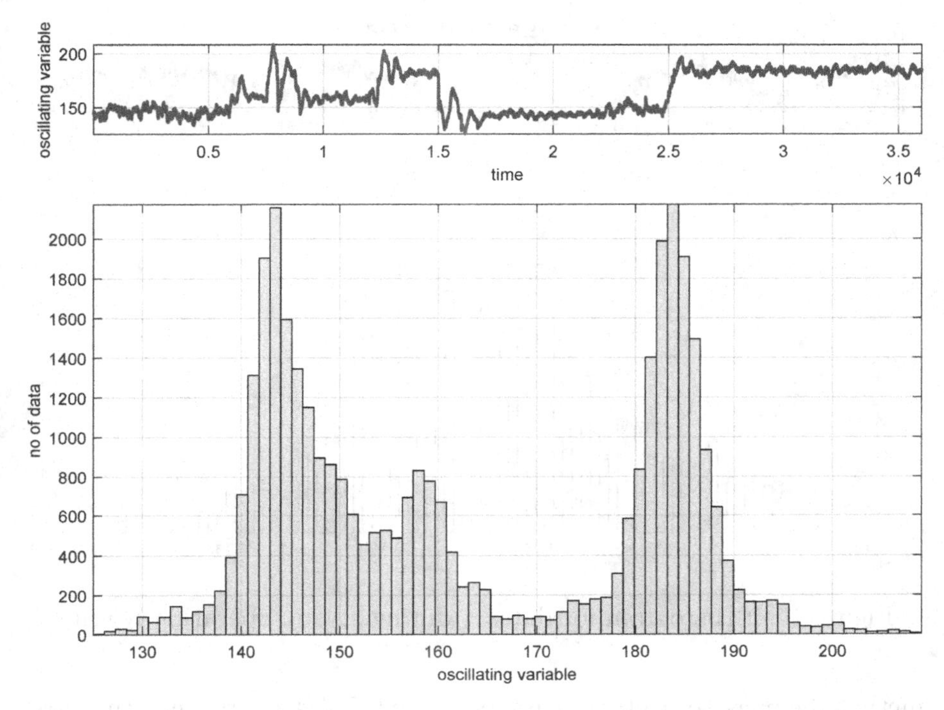

Figure 10.3: Histogram shape for whole sample dataset

trend, stochastic process, a noise and the oscillations. If we decompose signal into these elements, we will automatically detect oscillations.

Empirical mode decomposition (EMD) aims at the decomposition of the whole time series into a given number of components, named as intrinsic mode functions (IMF). Authors in [156] describe that the IMF should follow the following features:

1. the number of extremes (minima or maxima) should be equal to the number of zero crossings or they may differ at most by one,

2. the mean of local minima and maxima envelopes equals to zero.

The EMD uses time scale between successive local extremes to define intrinsic oscillatory modes. It formulates the baseline low-frequency trend to be eliminated from the time series leaving only the components of higher frequencies. However, due to the frequent noise and signal intermittency in real industrial data, the mode mixing and mode splitting (MS) effects may appear. To avoid this effect the ensemble empirical mode decomposition (EEMD) [357] has been developed. The EEMD

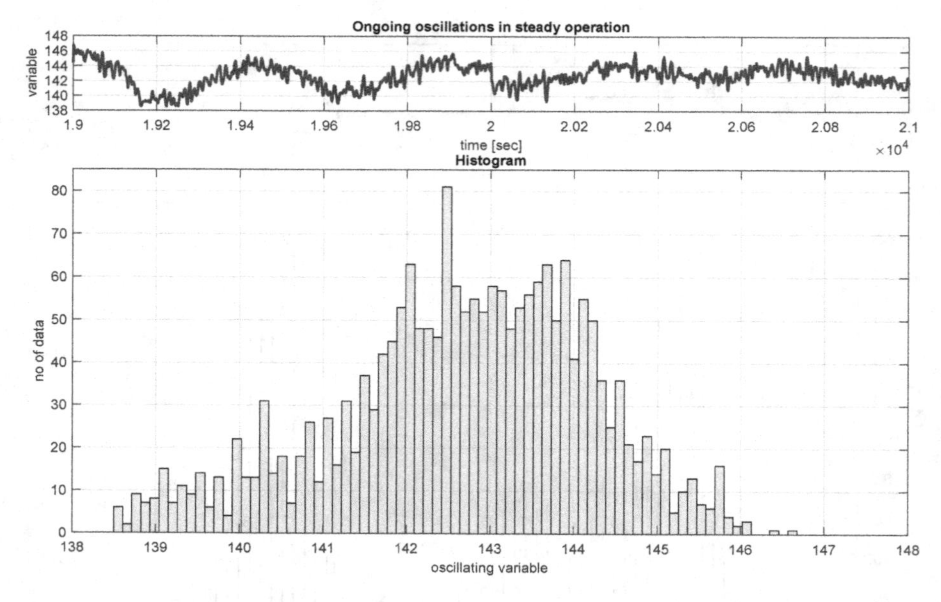

Figure 10.4: Histogram for dataset during steady operating regime

method belongs to a class of noise-assisted EMD approaches that are very powerful in time-frequency analysis [275, 231]. They aim at alleviating mode mixing caused by noise and signal intermittency. The EEMD can be successfully applied to the task of noise and oscillation identification, which is within our interest.

Median ensemble empirical mode decomposition (MEEMD) is a variation of the basic EEMD with the use of median operator in place of the mean estimator to ensemble noisy IMF trials [206]. This algorithm is just a practical extension to the EMD and justified choice for industrial analyzes. The EMD method was developed so that data can be examined in an adaptive time-frequency–amplitude space for nonlinear and non-stationary signals [231]. It decomposes the input signal into a few IMFs and a residue. The given equation is as follows

$$I(n) = \sum_{m=1}^{M} IMF_m(n) + Res_M(n), \tag{10.11}$$

where $I(n)$ is the multi-component signal. $IMF_m(n)$ is the Mth intrinsic mode function, and $Res_M(n)$ represents the residue corresponding to M intrinsic modes. The proposed median EEMD (MEEMD) version uses the median in form of

$$\text{median} = \begin{cases} IMF(t)[(N+1)/2] & \text{– when k is odd,} \\ \frac{IMF(t)[(N/2)]+IMF(t)[(N/2)+1]}{2} & \text{– when k is even,} \end{cases} \tag{10.12}$$

where $IMF(t)$ denotes the ordered IMF list at time instant t, which is obtained from N independent noise realizations. Considering a real-valued data time series $x(t)$ and a predefined noise signal component ϵ, the general formulation of MEEMD procedure is shown in Algorithm 1. The results of the MEEMD application to the considered data are shown in Fig. 10.5.

We see the decomposition of the original data into four IMFs $IMF_1 \ldots IMF_4$ and the resulting residuum ϵ. Each IMF has an increasing frequency. Please note that at first the data were detrended with the zero-order polynomial, i.e. the mean value has been removed. The decomposition shows how contaminated with different frequencies the signal is. And in general, we have to be aware of that fact. Thus, compared to

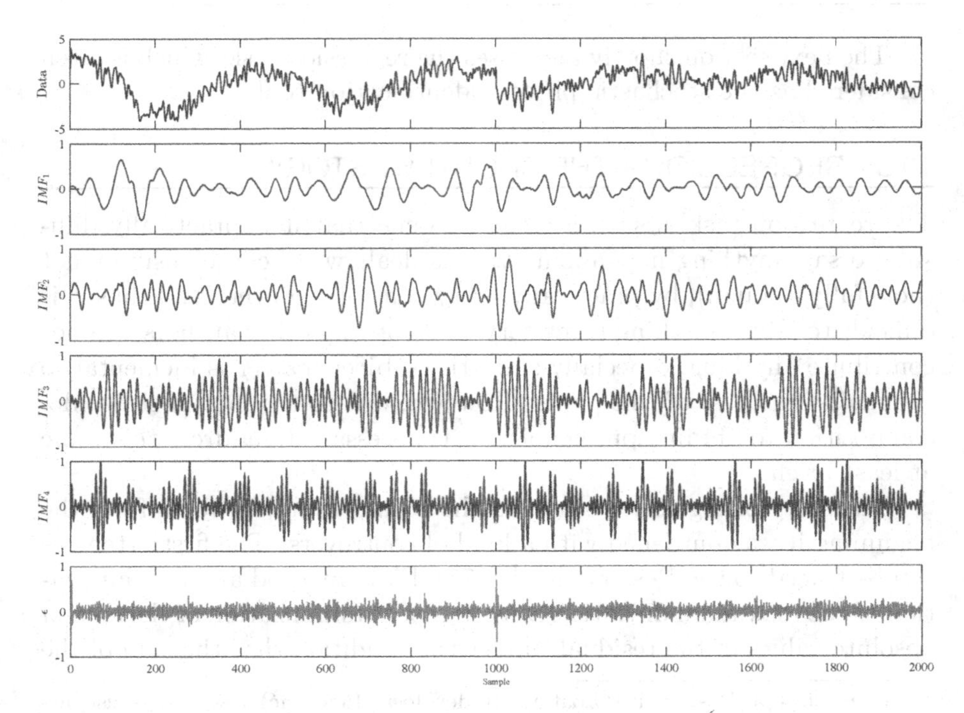

Figure 10.5: The MEEMD signal decomposition for sample data during steady operating regime

other methods for pure oscillation detection, component decomposition techniques are much more valuable.

In summary, we can see that the original signal is most often a composite of several different phenomena: a long-term trend, oscillations of one or more frequencies, the stochastic process mentioned earlier and not addressed by EMD class methods, and residual noise, which is assumed to be a stationary and independent variable, like a white noise.

Algorithm 1 MEEMD algorithm

1. Generate the ensemble $y_n(t) = x(t) + \epsilon\omega_n(t)$ for $n = 1, ..., N$, where $\omega_n(t) \sim \mathcal{N}(0, 1)$;
2. Decompose every member of $y_n(t)$ into M_n IMFs using the standard EMD, to yield the set $(d_m^n(t))_{m=1}^{M_n}$;
3. Assemble same-index IMFs across the ensemble using the median operator to obtain the final IMFs within MEEMD;
for instance, the mth IMF is computed as $d_m(t) = \text{median}(d_m^1(t), d_m^2(t), ..., d_m^N(t))$.

The next section shortly addresses the regression task, which is often considered as the stochastic process identification tool.

10.3 REGRESSION – THE SHORTEST STORY

The regression task is such a common issue that it is practically difficult to say anything new about it. It is dealt with, consciously or not (hopefully, though), by practically every control engineer. And thus it is difficult to add something revelatory, and copying other authors will not contribute anything, especially since the subject matter is incidental to the main narrative of this paper. However, due to the fact of signal decomposition to obtain "pure" noise, it is necessary to address this topic at least briefly.

The story of regression starts in the 18th century, however from the beginning it was burdened with a kind of controversy.The first attempts can be traced to the Boscowich's [1] (1711-1787) method and a minimization of the maximum deviation consisted of a minimization of the sum of absolute values of the residuals under the condition that the sum of the

[1]Roger Joseph Boscovich (Croatian: Ruđer Josip Bošković) was a physicist, astronomer, mathematician, philosopher, diplomat, poet, theologian, Jesuit priest, and a polymath from the Republic of Ragusa.

residuals should be equal to zero. Thus, translating into current naming he tried to solve the fitting (regression) task with the ℓ_1 norm.

An algorithm for finding the minimax residual was given in 1783 by Laplace, and in 1789 he simplified his earlier procedure. Another method was proposed by Euler and Lambert, according to which the estimates should be the quantities which minimize the absolute value of the largest deviation. As one can see the initial research was addressing the minimization of the residua absolute values, not the squares.

The version with the ℓ_2 norm appeared in works of Legendre (1752-1833), who published in 1805 a memoir, *Nouvelles méthodes pour la determination des cometes*, in which he introduced and named the method of least squares (LS). And the controversy started, as Gauss (1777-1855) published in 1809 a book, *Theoria motus corporum coelestium in sectionibus conicis solem ambientium*, where he discussed the method of least squares and, mentioning Legendre's work, stated that he himself had used the method since 1795. Unsurprisingly, Legendre was offended by Gauss's statement. In the following years, Gauss tried to produce evidence for his claim but had only little success. The astronomer Olbers included in a paper in 1816 a footnote asserting that Gauss had shown him the method of least squares in 1802 and Bessel published a similar note in a report from 1832.

Nowadays, the Gauss is considered as the father of the least squares method, which used to be and often still is the core of all the regression approaches. And the contributions of Legendre in case of the ℓ_2 norm and his ℓ_1 predecessor (Boscovich, Laplace) are hidden in the darkness of memory and history.

The method of least squares can be used both for static and dynamic data. Let's assume the simplest situation of two-dimensional data $y = f(x)$. A given exemplary data are generated using the relationship

$$y(i) = -0.05 \cdot x^2(i) + 0.8 \cdot x(i) + 1 + \epsilon(i), \tag{10.13}$$

where $\epsilon(i)$ comes from normal distribution $\mathcal{N}(0, 0.2)$ and $i = 1, \ldots, N$. We assume that we are fitting the polynomial of order r

$$\hat{y}(i) = \sum_{j=0}^{r} x^j(i) \cdot \Theta_j = \sum_{j=0}^{r} \phi_j(i) \cdot \Theta_j = \boldsymbol{\phi}^T(i)\boldsymbol{\Theta}. \tag{10.14}$$

As we see the model is linear according to the regression variable $\phi_j(i) = x^j(i)$ with parameters denoted as Θ_j. The solution to that task is obtained through the minimization of mean square performance index

$$\min_{\boldsymbol{\Theta}} V(\boldsymbol{\Theta}, i) = \min_{\boldsymbol{\Theta}} \frac{1}{N} \sum_{i=1}^{N} \left(y(i) - \boldsymbol{\phi}^T(i)\boldsymbol{\Theta} \right)^2$$
$$= \min_{\boldsymbol{\Theta}} \frac{1}{N} \sum_{i=1}^{N} \left(y(i) - \hat{y}(i) \right)^2 . \tag{10.15}$$

For the vector description we introduce the following notation:

$$\boldsymbol{X} = [x(1), x(2), \ldots, x(N)]^T \tag{10.16}$$
$$\boldsymbol{Y} = [y(1), y(2), \ldots, y(N)]^T \tag{10.17}$$
$$\boldsymbol{\Phi}^T = \left[\boldsymbol{\phi}^T(1), \boldsymbol{\phi}^T(2), \ldots, \boldsymbol{\phi}^T(N) \right]^T \tag{10.18}$$
$$\boldsymbol{P} = \left(\boldsymbol{\Phi}^T \boldsymbol{\Phi} \right)^{-1} . \tag{10.19}$$

Finally, we obtain the solution in form of the

$$\hat{\boldsymbol{\Theta}} = \left(\boldsymbol{\Phi}^T \boldsymbol{\Phi} \right)^{-1} \boldsymbol{\Phi}^T \boldsymbol{Y} = \boldsymbol{P} \boldsymbol{\Phi}^T \boldsymbol{Y} . \tag{10.20}$$

It exists and is optimal according to the following assumptions:

1. The considered regression model is linear in the coefficients and in the error element.
2. Error term has a zero value of the population mean.
3. All independent variables (inputs, $x(i)$ in the considered case) are uncorrelated with the error element.
4. Error observations are independent, i.e. uncorrelated with each other having constant variance (stationarity, no heteroscedasticity). Optionally, we could say that the error term is normally distributed white noise.
5. None of the independent variable is a perfect linear function (with Pearson's correlation coefficient equal to +1 or -1) of other independent variables.

At first glance, these assumptions are not threatening, but nevertheless carry quite serious consequences, which will be particularly evident in the case of dynamic version. Let's come back to the considered example Eq. (10.13). Fig. 10.6 presents to the right sample data out of that model and to the left exemplary polynomial fitting models.

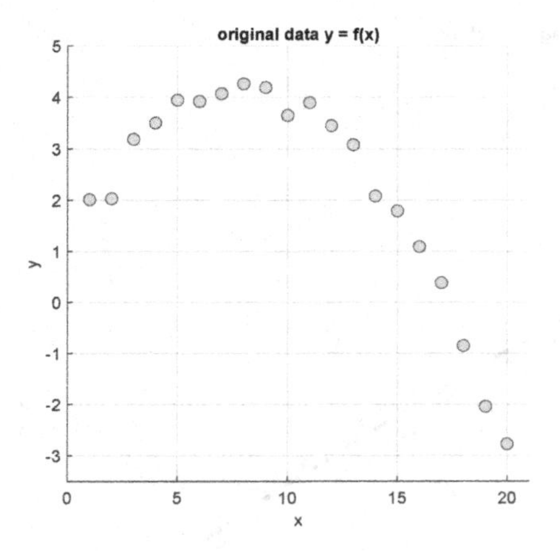

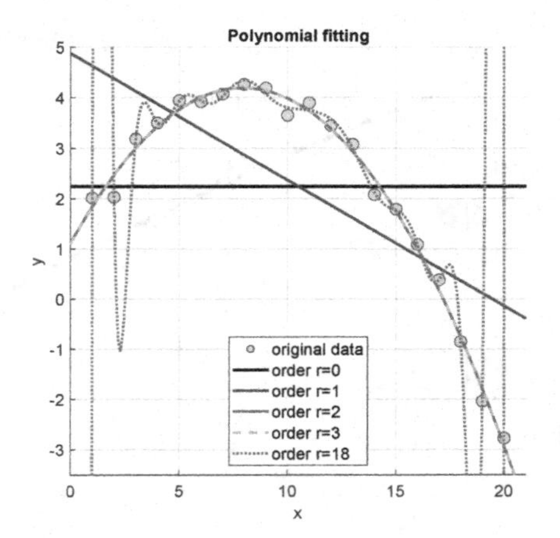

Figure 10.6: Least squares polynomial fitting in static case with different orders

As we can see the estimates $\hat{y}(i)$ fitness depends on the order r. Generally the case of $r = 2$ should be optimal, however it seems that increasing of the order for $r > 2$ improves the estimates. however if we go to far we end up with model overfitting. Therefore, despite the temptation to use high order models we should not do it. To exclude such a situation we use the second dataset, called the validation one that should be used to ultimately validate the model fitness. Fig. 10.7 visualizes that effect.

The same approach might be applied to the dynamic data allowing the identification of the linear regression models. The most common are the so-called auto-regressive moving average with auxiliary input (ARMAX) structures. More information on that subject can be found in numerous books, like for instance [219, 165, 326].

Above linear identification has its nonlinear versions using nonlinear ARMAX (NARMAX) models [18, 354] or any type of the machine learning approaches, like for instance Gaussian process regression (GPR) [300], ridge regression (RR) [209], regression trees [62], random forest regression (RFR) [146] and many, many others.

However, at that moment three issues are selected for further description, as they are less frequently addressed:

– model robustness and the effect of outliers,

– process non-stationarity and resulting need for the adaptation,

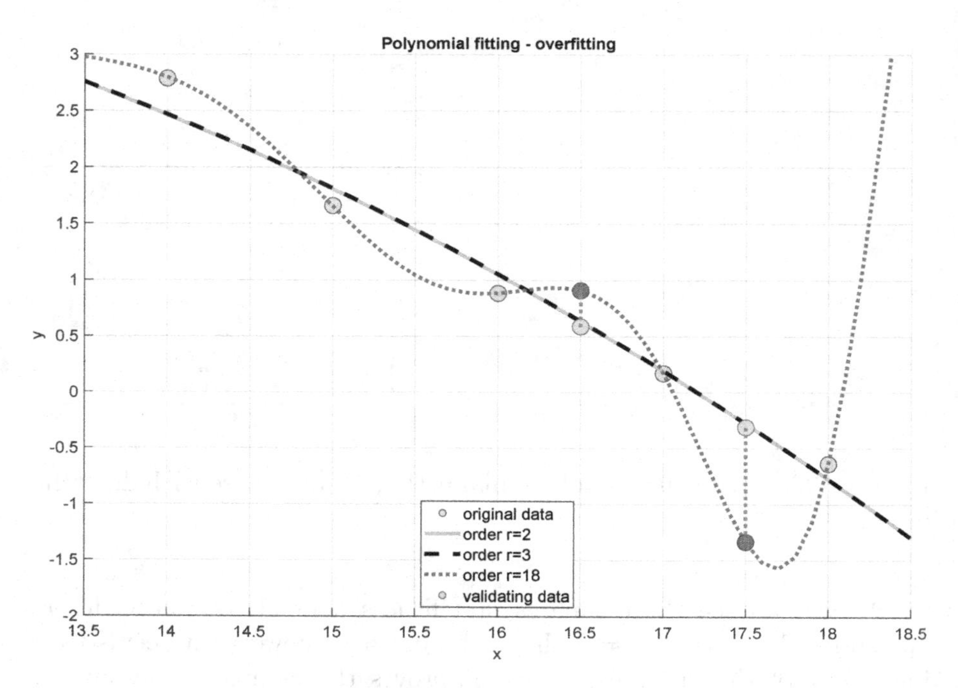

Figure 10.7: Overfitting effect and validation data

– effect of the model on method (assessment procedure, controller performance, etc.).

These aspects are not directly associated with statistics, but may indirectly affect conducted statistical analyzes. The effect of outlier s is a clear example of such an issue. As it has been already mentioned, industrial data are often contaminated with the outlying observations. We have already shown that the outliers may affect the estimation of the statistical moments, like for instance variance. If the variance is affected, therefore the means square error and the least squares estimation as well. Either way they are equivalent.

The ℓ_2 metrics and the mean square error exhibit zero breakdown point, and even single outlier may seriously affect the least squares estimation. Fig. 10.8 visualizes this effect. In the felt side diagram we observe ideal situation without any outlying observations. Everything is clean and unbiased. The LS estimated line almost exactly fits the real linear relationship.

In contrary, the right plot shows the relation, when one observation (for $x = 6$) is significantly affected (the red dot). We observe that the

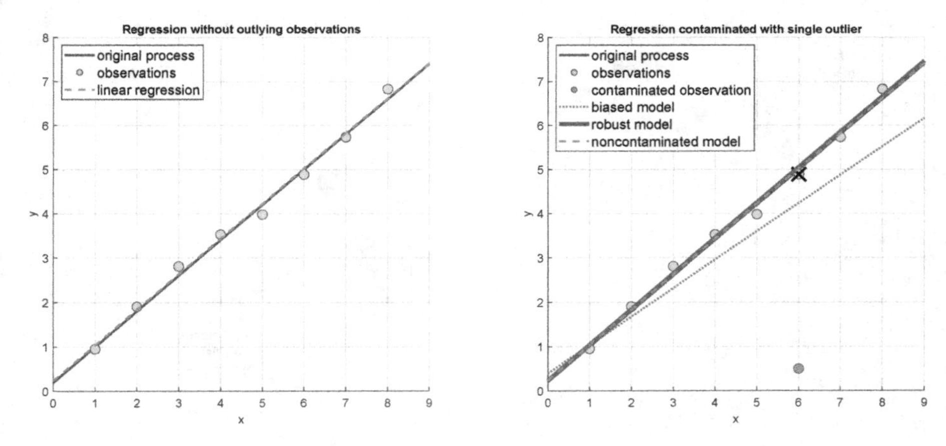

Figure 10.8: Effect of outliers and robust regression

resulting estimated line is noticeably different and the true relationship is not reflected correctly.

Though this example is simple, it clearly shows what dangers are associated with the presence of outliers and how they can skew the task of estimation and identification [287]. How can we solve the problem in case of the contaminated data or in the case of data about which we are not sure?

First of all – check data!!! Do not make blind moves and don't take any shortcuts. Secondly, If we have a significant belief that the data are contaminated and include outlying observations, let's use **robust regression**. It is not difficult. The idea is simple. We need to change the ℓ_2 mean square performance index with the other one, which is robust. Even simple change to the ℓ_1 mean absolute error may improve the situation, though its breakdown point is still zero.

Another method is to use the trimming. Least trimmed squares (LTS) is a statistical technique that provides robust regression method based on minimizing the sum of squared residuals. The idea is extremely simple. We need to remove selected number $s = 1, 2, \ldots$ of the minimum and maximum observations. Even simple removal of the smallest and the largest observations ($s = 1$) helps. Another possible simple method is to use the least median of squares (LMS) index V_{LMS}

$$\min V_{\mathrm{LMS}} = \min\left\{\text{median}\left(y(i) - \hat{y(i)}\right)^2\right\}. \tag{10.21}$$

Apart from the above simple methods, nothing prevents us from taking advantage of the full wealth of robust statistics and M-estimators as described in Section 4.3.

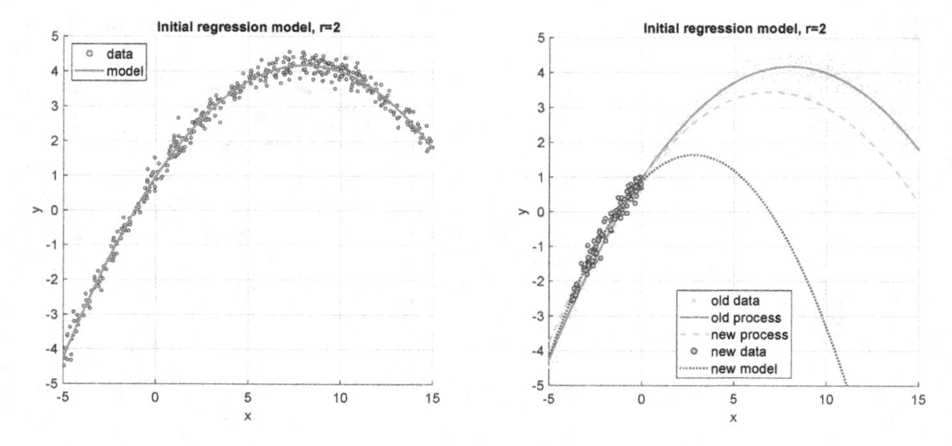

Figure 10.9: Model adaptation with least squares

The second issue that requires some attention is the model non-stationarity. Although the subject seems to be well known, it still raises doubts in practice. It is not easy to apply least squares to the non-stationary data – to self-adapt the model. There exists simple solution to that – the recursive least squares estimation (RLS). Therefore we can solve the problem in two ways: apply RLS or just repeat the LS using new data that appear.

Let's consider sample example. We take first 400 points generated from that process and we obtain almost ideal 2nd polynomial curve (see Fig. 10.9 right plot).

Next, let's assume that the process somehow has changed and now we get new model relationship (light blue curve in the plot)

$$y(i) = -0.05 \cdot x^2(i) + \mathbf{0.7} \cdot x(i) + 1 + \epsilon(i). \tag{10.22}$$

Thus, we measure new data and they appear from the limited operating region of $(-4.0, 0.0)$, for instance due to the change in the process operation. We get new set of 100 data points sketched in blue on the right of Fig. 10.9. Using these points we perform LS estimation and we get new, adapted model plotted in dark blue color. We explicitly see that this curve seriously differs from the real model. It is due to the fact that new data cover only small section of the whole process domain. This effect is named that the local change affects global performance. We have to remember that with high probability similar effect will happen with the RLS version.

To address this issue we have to have a chance to always cover with new data the whole process domain or to remember in some buffer data from uncovered parts of the domain. The other solution is to use methods, for which local change affects only local behavior, as we get in case of the kernel regression [355, 200] approaches or radial basis function networks [40, 30], or other local methods. Though the subject exists in the research with numerous approaches it's frequently forgotten in the industrial practice and thus has been treated as worthy of mentioning.

10.3.1 Concluding remarks on model-based analysis

Finally, we need to comment the whole idea of the model-based analysis. We may find numerous model-based approaches to various analyzes: outlier detection, clustering, classification, control performance assessment, control, and many others. Such analytical techniques are very powerful, as they allow to relate the results to some baseline – to the **model**.

Thus, these methods consist of at least two phases: model identification and the right domain algorithm. We obtain the result. And once the results are wrong we get into the dilemma:

> *What is wrong: the result (model, classification, assessed control performance) or the model?*

This is not an easy question. The model-based methods incorporate an additional degree of freedom of making errors, which as such are hidden behind the right method. Authors frequently propose dozens of new methods that use the model without thinking about its risks, the need to identify, maintain, adapt, resilience, etc.

We should remember about that. If we select the method, let's choose the one, which minimizes the chance of making mistakes: less steps the better. If we may use the model-free approach, let's at least try.

This is one of the important elements, in which statistics, though considered out of fashion, offers mathematical rigor and proven baseline, something more than overused machine learning methods

10.4 TAIL INDEX

Shortly saying, the tail index ξ measures the tail significance versus the rest of data. It is a measure of the tail fatness. The notion of the tail index also originates from the extreme value theory. Let's take the definition that the tail index ξ is some distribution shape coefficient that determines how heavy the tail is, with its higher value corresponding to a thinner tail. Though one may find formal definition [115] we will stay with a narrative description.

The computation of tail indices usually relies on statistical methods, such as basic Hill estimator, which estimate tail indices based on extreme values of sample data. In control engineering, tail indices might be utilized to assess the extreme system behaviors, like the analysis of failure rates or response times. This is particularly important during the risk analysis, especially once extreme events cannot be neglected.

The tail index is correlated with a distribution and its coefficients, like standard deviation and mean, but it provides unique information about its shape, focusing on the tails. This measure delivers an insight into the overall properties of the distribution and is especially important when dealing with heavy-tailed distributions or extreme values.

The tail index estimators are explained by the following formula

$$1 - \mathcal{F}(x) \sim L(x)\, x^{-\alpha}, x \to \infty, \tag{10.23}$$

where α is the tail index . Assuming that $\alpha > 0$, $L(x)$ is a slowly varying function, i.e. for every $x > 0$ as $t \to \infty$

$$\lim_{t \to \infty} \frac{L(t)}{L(tx)} \to 1. \tag{10.24}$$

For convenience we use $\xi = \alpha^{-1}$ as the extreme value index (EVI).

The most popular tail index estimator $\hat{\xi}$ was proposed by Bruce M.Hill [144], in form of the quasi-maximum likelihood estimator. The Hill estimator is a commonly used inspired by the Pareto distribution. The authors suggest using the Pareto distribution to approximate the kth largest observation from a sample. Let's consider a sample, which contains n independently and identically distributed observations, denoted $X_1, X_2, ..., X_n$. The Hill estimator is defined as

$$\hat{\xi}^H(k) = \frac{1}{k} \sum_{i=0}^{k-1} \log\left(\frac{X_{n-i}}{X_{n-k}}\right). \tag{10.25}$$

Though it's a maximum likelihood estimator (MLE), it is considered as a semi-parametric algorithm, as it only assumes certain behavior in the tails rather than making general assumptions about the entire function. The selection of the estimation parameter k is crucial, because it directly affects the estimator performance. If k is too large, the variance of the estimator increases, leading to biased estimates. Conversely, once k is too small, the estimator's bias increases, leading to inaccuracies. The use of Hill estimator of the tail index aims at a balance between the effectiveness, while minimizing the estimation error. Please note that the Hill index may even to take on negative values, which correspond to distributions with a finite right endpoint.

Huisman et al. [159] use a more general form of the tail, introduced by Hall [131]

$$1-\mathcal{F}(x) = Cx^{-\alpha}\left[1 + D_1x^{-\alpha} + \ldots + D_mx^{-m\alpha} + +o\left(x^{-m\alpha}\right)\right], \quad (10.26)$$

as $x \to \infty$.

They have noted that if the above assumption is satisfied for$m = 1$ the bias of Hill estimator is almost linear according to k. As small values of k give lower bias and higher variance, they propose to estimate $\hat{\xi}(r)$ with the Hill estimator $r = 1, \ldots, k$, and next run a regression for

$$\hat{\xi}(r) = \beta_0 + \beta_1 r + \epsilon(r), r = 1, \ldots, k. \quad (10.27)$$

Therefore, the $\hat{\xi}^{HU} = \hat{\beta}_0$ is a bias-free estimate of the tail index. Actually, the literature shows dozens of various tail index estimators. Fedotenkov, in his comprehensive summary [111] describes more than one hundred estimators. As the estimation itself is not an ultimate goal for a control engineering during his/her analysis, quite a basic approach should suffice. Author uses the Huisman estimator $\hat{\xi}^{HU}$ during his projects, nonetheless one may use others as well, if only it's justified.

The visualization of the tail index estimation is often presented in forms of the graphs that show the Hill index estimation convergence according to the number of the highest observations used or just by an index value. Two sample plots prepared using the considered visualization data are shown below. Fig. 10.10 presents the Hill index diagram for X_3 dataset, while Fig. 10.11 for Y_1.

Finally, Table 10.1 shows the summary of Hill and Huismann tail index estimators evaluated for sample datasets, two-sided and one-sided.

Table 10.1: Summary of tail index estimates for sample datasets

tail index	X_1	X_2	X_3	X_4	X_5	Y_1	Y_2	Y_3	Y_4
$\hat{\xi}^H$	3.53	2.77	2.69	126.4	-0.41	1.83	2.45	2.82	1.71
$\hat{\xi}^{HU}$	6.95	3.50	4.70	179.7	1.00	2.17	3.24	4.26	1.53

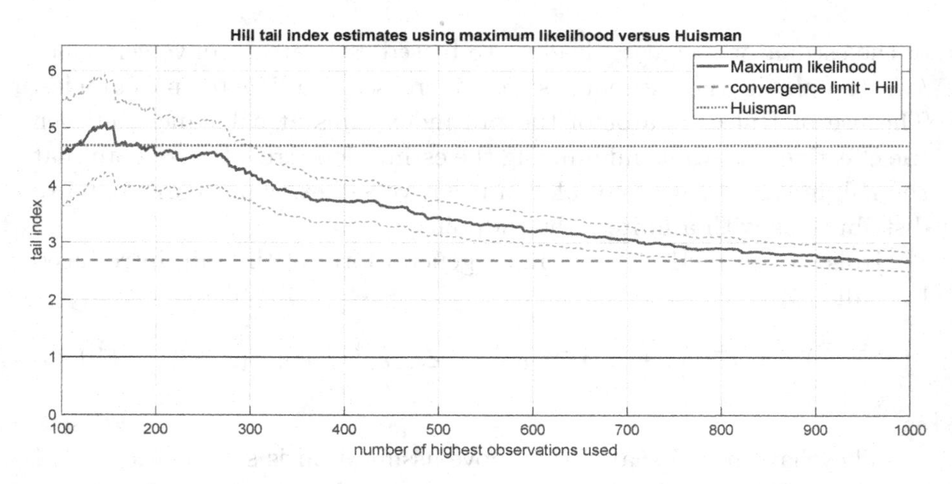

Figure 10.10: Tail index estimates for X_3 dataset

We may notice that we obtain negative Hill estimator for the X_5, which may confirm the hypothesis about distributions' finite right endpoint, which may be confirmed by strange, in some regions looking artificial, and clearly non-stationary behavior.

10.4.1 Concluding remarks on tail index

The application of tail index in the control engineering is very fresh with only a few publications [75, 76]. The application of tail indices in control

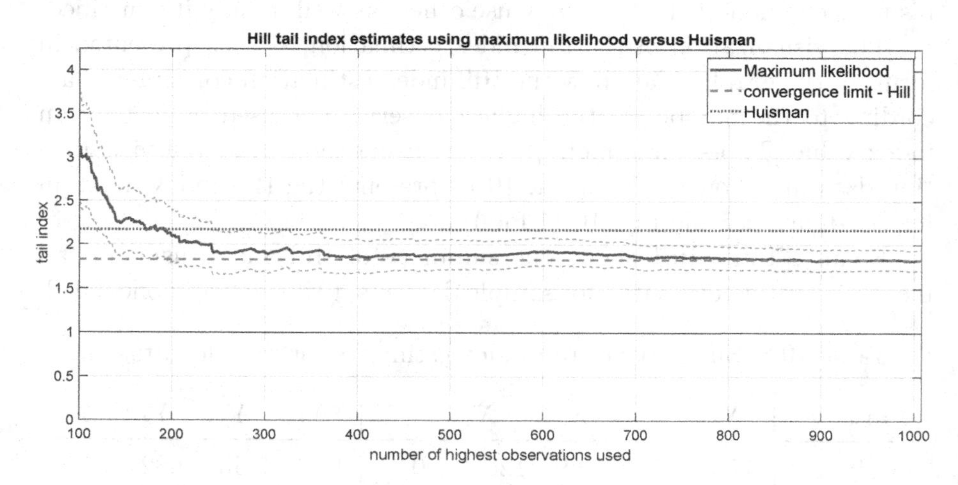

Figure 10.11: Tail index estimates for Y_1 dataset

engineering might involve modeling and the analysis of extreme events or anomalies in a system. From that perspective there is a potential area of interest in control performance assessment, which is investigated in the application part of that book.

10.5 INFORMATION MEASURE – ENTROPY

Statistical analysis of a given time series at reconstruction of the properties of the fundamental stochastic process behind the data. Entropy, which is considered as an information measure in data could be interpreted as a potential approach. There is a research showing that entropy can be successfully applied in control engineering to the task of control performance assessment. Entropy is used as the CPA index, as research shows that it corresponds to the entropy value [367].

Entropy is a concept, which is familiar in physics (thermodynamics), information theory, and mechanics. The notion of information entropy is used to measure the uncertainty of any given random variable. The information entropy is usually named Shannon's entropy after Claude E. Shannon, who was working on the fundamentals of the information theory [305]. From a physical perspective, Shannon's theory does not relates to physics, but it shares a lot of intuitions with thermodynamics and statistical mechanics [376]. It relates to the quantity of uncertainty of an event in its intuitive understanding, Therefore, it's concomitant with a given probabilistic density function. It defines entropy as a measurement of the mean information content connected to some random result. Such a definition allows an intuitive distinction mathematically precise. The Shannon's entropy should satisfy the following conditions:

1. The measure is supposed to be continuous, i.e. any small change in any of the probability value has to cause a small change in the according entropy value.

2. It gets maximal if all the observations are equally likely. However, the entropy change varies frequently according to the number of observations [376].

3. If the realization is certain, the entropy should equal to zero.

Let's assume the probability density distribution of a given variable X with its probable realizations $x_1, \ldots, x_n$ is described by function $\mathcal{F}(x)$.

Thus, the Shannon's entropy is defined as

$$H(x) = -\sum_{i=1}^{n} p(x_i) \log_2 \left(\frac{1}{p(x_i)} \right) = -\sum_{i=1}^{n} p(x_i) \log_2 p(x_i) \geq 0, \tag{10.28}$$

where $p(x_i)$ is the probability of the ith realization of X. As the outcomes probability directly relates to the distribution function values, we may simply rewrite the definition as

$$H(x) = -\sum_{i=1}^{n} \mathcal{F}(x_i) \log_2 \mathcal{F}(x_i) \geq 0, \tag{10.29}$$

Apart from basic Shannon's definitions, there are other ones, like:

- the differential entropy, H^{DE} defined as:

$$H^{\mathrm{DE}} = -\sum_{i=1}^{n} \mathcal{F}(x_i) \ln \mathcal{F}(x_i). \tag{10.30}$$

- the rational entropy, H^{RE} defined as:

$$H^{\mathrm{RE}} = -\sum_{i=1}^{n} \mathcal{F}(x_i) \log \left[\frac{\mathcal{F}(x_i)}{1 + \mathcal{F}(x_i)} \right]. \tag{10.31}$$

The first literature reference of the entropy-based control performance assessment applied it as a minimum information entropy (MIE) benchmark and the design of minimum information entropy CPA index [237, 363]. The minimum MIE bound for Gaussian and nonlinear non-Gaussian system is identically defined

$$H_{\mathrm{min}} = \ln \left(\sqrt{2\pi \exp(1)} \sigma \right) \tag{10.32}$$

where exp(1) is Euler's number. The upper bound equals to

$$H_{\mathrm{min|upper}} = \ln \left(\sqrt{2\pi \exp(1)} \sigma_{\mathrm{MV}} \right), \tag{10.33}$$

where σ_{MV} denotes the standard deviation of the minimum variance system [135]. Please note that variance and entropy are equivalent in Gaussian case. Thus the CPA index, which takes into account the MIE is formulated as

$$\eta^{H} = \frac{\exp(H_{\mathrm{min}})}{\exp(H_{\mathrm{act}})}. \tag{10.34}$$

Knowledge of the process delay enables to apply the tight minimum information entropy upper bound $H_{\min|\text{upper}}$ and allows to formulate the control measure as

$$\eta^H = \frac{\exp\left(H_{\min|\text{upper}}\right)}{\exp\left(H_{\text{act}}\right)}. \tag{10.35}$$

The notion of the Shannon's entropy exhibits some shortcuts. It may receive negative or even negative infinite values for the continuous random variable, when used as the benchmark. It may potentially limit its utilization as the benchmark measure. Thus, the rational entropy might be used instead [47, 368, 369]. This entropy's formulation exhibits similar properties, while being consistent. For a given random variable X and its probability density function denoted by $\mathcal{F}(x)$, the rational entropy is given by Eq. (10.31). Using the rational entropy definition two additional CPA indexes are proposed

$$\eta^{H_1} = 1 - \frac{\arctan\left(\left|H^{\text{RE}}_{\text{act}} - H^{\text{RE}}_{\min}\right|\right)}{\arctan\left(\max\left\{\left|H^{\text{RE}}_{\text{act}} - H^{\text{RE}}_{\min}\right|\right\}\right)}, \tag{10.36}$$

$$\eta^{H_2} = \frac{H^{\text{RE}}_{\max} - H^{\text{RE}}_{\text{act}}}{H^{\text{RE}}_{\max} - H^{\text{RE}}_{\min}}, \tag{10.37}$$

where $H^{\text{RE}}_{\max}$ is the maximum rational entropy for a given interval $[a, b]$. Authors show that such benchmarking indexes are effective and feasible covering the CPA uncertainties [368].

Apart from the above, we may try to use entropies Shannon, differential and rational or any other directly as the performance measures through their application to the control error times series [88]. Table 10.2 presents the respective entropy values evaluated for the sample control error datasets X_i. We may notice that the relationship between sets is held with all the considered entropy versions. We also notice that control errors that are close to normal distribution are described with relatively higher entropy values.

Table 10.2: Summary of entropy measures for control error data X_i

entropy	X_1	X_2	X_3	X_4	X_5
Shannon H	5.659	0.731	4.880	5.309	4.530
rational H^{RE}	1.713	0.459	1.487	1.610	1.387
differential H^{DE}	3.922	0.507	3.383	3.680	3.140

10.5.1 Concluding remarks on tail index

The notion of entropy, similarly to the tail index is relatively rare in control engineering applications, though there exist works on its use to the task of control performance assessment, both in the benchmarking approach and directly. Despite that fact, there is a noticeable potential of their use, as they do not require any modeling (we use only histogram) and are independent of any assumptions about data properties (no Gaussian assumption).

10.6 MEMORY, PERSISTENCE, FRACTALITY, FRACTIONALITY

The notions of the long-range memory, persistence, fractality or fractionality take their origins in different research areas, although they address similar aspects of data properties that are not captured and accounted for classical statistical concepts like distributions or moments.

Fractal and persistence-based approaches got the highest popularity in economics [264, 228, 324, 229, 319] and social sciences [299, 198]. We may also meet these methods in medicine [263, 220] and selected engineering areas, like telecommunication [15, 211], informatics [212], wind energy [188] or prediction [96]. In control they are seldom taken into account, however some works appeared [321, 67, 68, 97].

Data persistence (this is our interest) is the continuance of a result after its cause disappeared. This term in time series context directly relates to the idea of long-term dependencies and memory in data [234]. Simply saying, we have a persistent process if the effect of infinitesimally small change influences the future behavior of the time series for a long time. The longer the influence exists, the longer its memory [262]. Thus, the integration is an example of highly persistent process, as the information, which once appears will no longer fade away. From that perspective, the fractionally integrated time series [308] form an interesting version of persistent processes.

We can classify persistence in terms of range: short- or long-range, and in terms of its strength: weak or strong. The persistence reflects the correlations between adjacent values in the dataset. It may be can weak, strong or nonexistent, like in white noise. Single observations in data may affect other observations that are nearby in time (short-range) or far away (long-range). Persistence over time also relates to the notion of non-stationarity, like trends or cycles. We have to be aware that persistence might have unexpected negative effects, like it can produce spurious correlation We may quantify it many ways, using autocorrelation function, but also using spectral analysis and power spectrum, Hurst exponent, fractional order, fractal correlation or fractal dimension. As shown above the notions of time series memory (long or short), persistence, fractality and fractionality interpenetrate each other.

The notion of fractals, Latin *fractus*, i.e. fragmentary or broken, self-similar or being *infinitely subtle* take their beginnings from the works of B. Mandelbrot [227] in 70's. Fractal analysis incorporates notions of self-similarity, self-affinity, power-law and scale-invariance [297, 110].

The concept of self-similarity underlies behind fractals and power law. It means the invariance to changes in scale or size [299]. We interpret it as a symmetry shaping of the environment. An alternative approach to the term of randomness [324] powers fractal analysis, though the mathematics noted such a behavior long before. Previous works of Hilbert, Peano, Cantor, Koch, Julia or Sierpiñski might have motivated Mandelbrot in his development of fractal theory.

British engineer H.E. Hurst was analyzing historical Nile discharge levels [160]. Initially, everybody assumed that the flows should be independent of their history. Hurst proved it to be otherwise. He proposed persistence measure, named after him the Hurst exponent defined as the asymptotic property of the so-called rescaled range R/S diagram.

$$E\left[\frac{R(n)}{S(n)}\right] = cn^H, n \to \infty, \tag{10.38}$$

where n denotes number of points, S standard deviations for time period n, R data range of length n, c as positive constant and H – Hurst exponent. We calculate Hurst exponent H with logarithms

$$\ln\left\{E\left[\frac{R(n)}{S(n)}\right]\right\} = \ln\, c\, +\, H \ln\, n. \tag{10.39}$$

We plot this relationship in double logarithmic scale of $E\left[\frac{R(n)}{S(n)}\right]$ against n that allows to estimate H as the fitted line slope. Sample R/S plot is sketched in Fig. 10.12. We see two possible options. The left one presents almost independent realization for sample dataset X_1, which have been previously shown to be close to Gaussian. It is confirmed with the Hurst exponent equal to 0.51. In contrary, the second plot for X_4 persistent data with Hurst exponent equal to *0.79*.

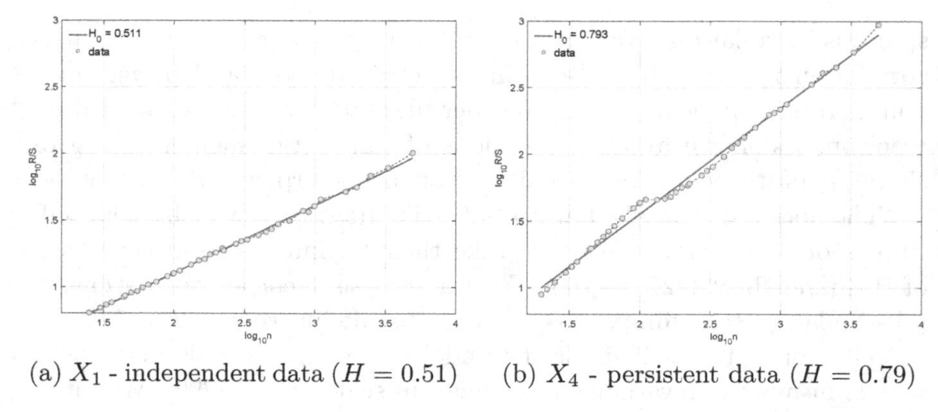

(a) X_1 - independent data ($H = 0.51$) (b) X_4 - persistent data ($H = 0.79$)

Figure 10.12: The examples of the R/S diagrams

Small value of n will make the result anti-persistent and the value of Hurst exponent will be deceiving. Results for low values of n might be also affected with data discretization, and thus one should select it with caution. On the other side a large value of n together with short data set may produce too few samples for correct self-similar process H estimation. Therefore, the number of observations and the selection of minimum n value matters. Hurst exponent is very important parameter for fractal and persistence analyzes. Its value may address the following properties of a given time series:

- Negative values $H < 0$ point out over-differenced white noise.
- Values in $0 < H < 0.5$ denote anti-persistent (ergodic) data, i.e. values returning toward an average. Data decrease in the past suggests increase in future and vice versa.
- If Hurst exponent is $H = 0.5$ then data exhibit Brownian motion properties. All observations are statistically independent. We get stochastic uncorrelated process, like a Gaussian process.
- Larger values in $0.5 < H < 1$ denote persistence in time series, i.e. data increase/decrease in the past implies also further increase/decrease in future. It is named as fractional Brownian motion, i.e fractional white noise with long range dependence (LRD).
- Even larger results $1 \leq H < 1.5$ denote fractional Brownian noise without any dependency in time: fractional integration of white noise, with $d \in [0.5, 1)$.

What is the most interesting from the CPA perspective, the Hurst exponent relates to control quality. Research shows [320, 78, 88] that persistent behavior of the control error signal $0.5 < H < 1$ reflects sluggish control, the anti-persistent one $0 < H < 0.5$ the aggressive settings, while independent control error signal realization $H = 0.5$ addresses just good control performance. We must note that this relationship is very constructive as it not only says what is good and what is bad, but also points out why.

Actually, control engineering limits Hurst exponent to $H \in (0, 1)$, however its evaluation can be problematic. It may appear that Hurst exponent is greater than 1 due to strong data complexity or internal data corruptions (periodicity or trends). In such cases, conventional approaches fail.

The research points out a hypothesis about the equivalence between the stability factor α of the α-stable distribution and the Hurst exponent . Hurst index is equivalent to the inverse of the α-stable distribution stability factor $H = 1/\alpha$ [264]. There are few assumptions on data properties that limit this hypothesis, like varying long-range correlations of small or large fluctuations. Literature shows other associated terms like fractal correlation

$$C = 2^{(2H-1)} - 1 \tag{10.40}$$

or fractal dimension

$$D = 2 - H. \tag{10.41}$$

The R/S plot is the original algorithm for H evaluation. It allows to find out not only single exponent, but also shows more features: multiple scaling and crossover. During the years researchers investigated other algorithms [226] listed in historical order:

1. periodogram method [119],
2. Higuchi (fractal dimension method) algorithm [143],
3. Haslett and Raftery estimation [136],
4. boxed or modified periodogram method [137],
5. the Haar wavelet and maximum likelihood estimation [183, 356],
6. aggregated variance method [28],
7. variance of residuals method by Peng [261],

8. absolute value method [328],
9. differential variance [331],
10. Abry and Veitch's wavelet-based analysis [1],
11. Whittle's maximum likelihood estimator originating from periodogram [327],
12. diffusion entropy method [126],
13. Detrended Fluctuation Analysis (DFA) originally suggested by Peng [261] followed by Weron [350],
14. Koutsoyiannis iterative method [198],
15. Kettani and Gubner method [189],
16. Geweke Porter-Hudak (GPH) estimator [119].

.

None of those methods is universal and each one has its operational range [173, 184, 52, 190, 114, 79]. One may call its selection as *hit and miss* game [52]. In the industrial data we should be highly cautious. We must be aware that any data preprocessing (filtering, detrending, removal of periodicity) affects estimation.

Complex process behavior may cause multiple scaling exponents in the same range. The ability to find them out and address properly is crucial. There are dedicated approaches to capture this effect. The simplest one is just through observation of slopes in the R/S diagram [79]. It's followed by dedicated solutions: wavelet leaders [168] and high-order correlation function [14]. Moreover, the analysis should also take into account crossover points as they carry on information as well [187]. The visualization of this multiple scales and crossover is shown in Fig. 10.13.

Table 10.3 shows summary of different fractal factors evaluated for sample control error data X_i. We see that signals X_1 and X_2 exhibit quite independent realizations with Hurst exponent very close to the 0.5 value. Moreover, multiple scaling is marginal. Thus, we may conclude that the associated loops are properly tuned and disturbances are decoupled.

In contrary, the other three signals are characterized by multiple scales. In such a case we take into analysis the shortest Hurst exponent, i.e. H_1. Two of the loops (X_3 and X_4) exhibit significantly persistent scaling $H_1 = 0.8238$ and $H_1 = 0.9891$, respectively. It means that the

Table 10.3: Summary of fractal factors for control error data X_i

entropy	X_1	X_2	X_3	X_4	X_5
H_0	0.5107	0.4830	0.6053	0.7928	0.5897
H_1	0.5015	0.4895	0.8238	0.9891	0.4153
H_2	0.4831	0.5292	0.7362	0.8156	0.5118
H_3	0.5112	0.4802	0.5669	0.8659	0.8346
cr_1	1650	1650	33	100	75
cr_2	2475	2475	90	165	660

associated loops are tuned in a sluggish manner, while control error X_5 is anti-persistent ($H_1 = 0.4153$) and the loop is slightly aggressive.

Those effects address dichotomy of scaling behavior that is not predicted by existing models, but empirically identified in time series [218]. They appear due to the various reasons, like additive effects of nonlinearities and influences caused by various sources with different delays

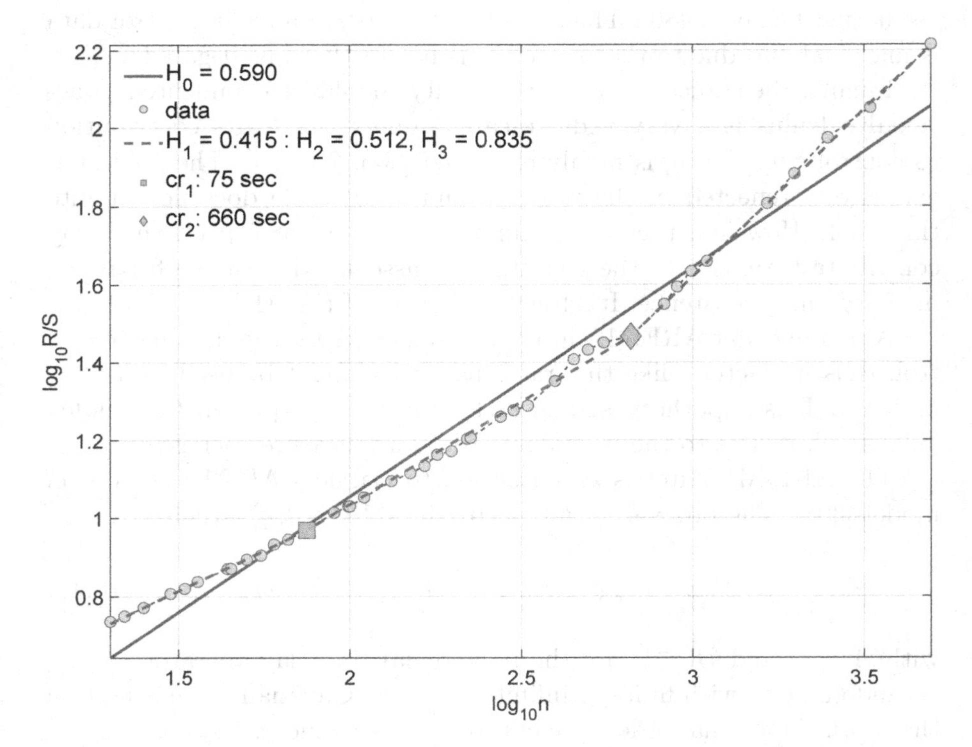

Figure 10.13: Multiple scaling in R/S plot - data X_5

[211], disturbances affecting at various delays or the nature of control system itself, i.e. steady state and transient operations.

There is also another fractal issue that might be noted, i.e. the multifractality. It's used to analyze turbulent flows [230]. Understanding of multi-fractal effects gives further insight into the dynamics of a given time series. It exhibits attractive properties, which can reproduce various aspects of data: heavy tails, long-term dependencies and multi-scaling. The analysis consists of the decomposition and characterization of information at all scales. There are different algorithms: box-counting [292], discrete wavelet multifractal model [167], wavelet transform modulus maxima [242], wavelet leaders [304] and multi-fractal detrended fluctuation analysis (MF-DFA).

Control issues that can be assessed include existence of cascade effects that may exist in complex systems and surely affects control, unknown cross-dependencies associated with un-decoupled disturbances (inappropriate use of feedforward), poor control (bad tuning), superposition of contribution of many sources or human impact and possibly oscillating behavior [80]. The analysis of multifractality in control data is marginal and due to that reason it is not further investigated.

Finally, the subject of the fractionality should be commented. Fractional calculus is a very wide research context [308]. Its contribution to control engineering is highly exploited [336, 244, 94]. The subject is extensively reflected in the research and as a whole does not fit into this book. However, there is a minor, but important aspect, that may contribute here, due to the fact that it's associated with the fractality, memory and persistence. It's fractional order of the ARFIMA filter.

Assuming that ARFIMA filters may model data from nonlinear control, thus its factors, like the fractional order might be used as a CPA measure. This hypothesis should as for the Hurst exponent CPA index [83], which relates to the fractional order and measures persistence.

The ARFIMA filter is an extension to classical ARIMA regression models[308]. The process x_k is named ARFIMA(p, d, q), when

$$A_p(z^{-1}) \cdot x_k = B_q(z^{-1}) \cdot \left(1 - z^{-1}\right)^{-d} \epsilon_k, \tag{10.42}$$

with $A(z^{-1})$ and $B(z^{-1})$ are the polynomials in delay operator z^{-1}, ϵ_k is random noise with finite or infinite variance. Gaussian noise is used in this work. Fractional order d refers to process memory.

For $d \in (0, 0.5)$ the process exhibits long memory or long-range positive dependence (persistence). The process has intermediate memory (anti-persistence) or long-range negative dependence, when $d \in (-0.5, 0)$. The process has short memory for $d = 0$; it is stationary and invertible ARMA. ARFIMA(p, d, q) time series is calculated by d-fractional integrating of a classical ARMA(p, q) process. The d-fractional integrating using $\left(1 - z^{-1}\right)^{-d}$ operator causes the dependence between observations, even as they distant in time.

The Geweke Porter-Hudak (GPH) estimator utilizes a semi-parametric method to calculate fractional memory coefficient d_{GPH} for a given ARFIMA process x_k [119]:

$$x_k = \left(1 - z^{-1}\right)^{-d} \epsilon_k, \tag{10.43}$$

Next, ordinary least squares algorithm estimates the $\hat{d}$ from the

$$\log\left(I_x(\lambda_s)\right) = \hat{c} - \hat{d}\left|1 - e^{i\lambda_s}\right| + \text{residual}, \tag{10.44}$$

calculated for fundamental frequencies $\left\{\lambda_s = \frac{2\pi s}{n}, s = 1, \ldots, m,\ m < n\right\}$, m is the largest integer in $(n-1)/2$ and $\hat{c}$ is a constant. Discrete x_k Fourier transform we define as

$$\omega_x(\lambda_s) = \frac{1}{\sqrt{2\pi n}} \sum_{k=1}^{n} x_k e^{ik\lambda_s}. \tag{10.45}$$

Application of the least squares algorithm to Eq. (10.43) yields to the final formulation

$$\hat{d} = \frac{\sum_{s=1}^{m} x_s \log I_x(\lambda_s)}{2\sum_{s=1}^{m} x_s^2}, \tag{10.46}$$

where $I_x(\lambda_s) = \omega_x(\lambda_s)\omega_x(\lambda_s)^*$ being a periodogram and $x_s = \log\left|1 - e^{ik\lambda_s}\right|$. The GPH estimation evaluates d_{GPH} without explicit specification of the ARMA model polynomial orders. The research uses implementation proposed in [28].

Fractional order CPA performs well according to the rules, similar as for the Hurst exponent : $d = 0$ indicates good control, $-0.5 > d > 0$ aggressive and $0 > d > 0.5$ sluggish, however with observed disclaimer. Theoretical relationship $H = d + \frac{1}{\alpha}$ should account for the unknown stability exponent α. Thus, proper assessment should also consider the appropriate estimation of the stability index α, which biases the d_{GPH} estimator value toward realistic representation of the control quality.

Table 10.4: ARFIMA fractional orders for control error data X_i

entropy	X_1	X_2	X_3	X_4	X_5
d_{GPH}	0.038	0.027	-0.028	0.765	0.055

Table 10.4 shows results, which for sample control error data X_i. The results are quite consistent with previous Hurst exponent analysis, with one exception, i.e. control error X_4, for which we obtain $d_{\text{GPH}} = 0.765$, which is larger than 0.5. We may suspect that it denotes no time dependency and/or fractional behavior. We may suspect that is could be connected with data discretization and/or unknown artificial effects of data pre-processing (averaging, deadbands, interpolation) done during system data archiving. This example shows how important is the analysis done by different means, as these effects were not disclosed before.

10.6.1 Concluding remarks on memory dependence in data

The notion of data memory, and its representation in form of persistence properties, fractality or fractionality are interconnected and frequently exist in time series originating from control loops. Though they exist, can be measured and properly interpreted, are seldom used in practice. This work, among others, aims at its popularization. It is shown how this measures like Hurst exponent and fractional order may improve the work of the control assessment engineer.

If we look at the signal internal memory from the other perspective, the anti-persistent time series is mean-reverting, persistent is trending and for $H = 0.5$ $(r = 0)$ the signal is purely random. Therefore, the persistence measure may be used as indicators for data trends as well.

It is also shown that each step of the analysis may disclose different features of data and thus it's always worth to perform multi-criteria analyzes evaluating multiple indexes.

10.7 HOMOGENEITY

Homogeneity analysis is fresh issue in control engineering research, while it's used in life sciences. In that context it aims at regional flood frequency analysis approaches, precipitation analyzes and other hydrological issues [27, 253]. Apart from many approaches, the one proposed by Hosking, which uses L-moments and LMRD plots seems to be the most promising [153].

Extension of the homogeneity concept toward control engineering seems to be very attractive as it may allow to investigate new issues. We may consider the homogeneity in two perspectives: space and time. Once we have a complex multi-loop control system finding homogeneous loops might help in loop benchmarking and the validation of complex control structure design. We may even support the process of complex system design and the selection of proper controlled variables and loops, by application of the homogeneity analysis to process variables. This issue is challenging [318] and as for now we do not have methodologies, apart from the best practices. Time perspective allows to introduce and to measure loop sustainability [98].

We do the homogeneity tests in two-dimensional LMRD(τ_3, τ_4) plane spanned by L-skewness and L-kurtosis as in Fig. 9.14. We start testing from visual inspection, which further is supported by formal methods [153]. Let's assume that terms μ_V and σ_V denote mean and standard deviation of the test statistics V presented in Eqs. (10.47...10.49)

$$V_1 = \sqrt{\frac{\sum_{i=1}^{N} n_i \left(\tau_2^i - \tau_2^R\right)^2}{\sum_{i=1}^{N} n_i}}, \tag{10.47}$$

$$V_2 = \frac{\sum_{i=1}^{N} n_i \sqrt{\left(\tau_2^i - \tau_2^R\right)^2 + \left(\tau_3^i - \tau_3^R\right)^2}}{\sum_{i=1}^{N} n_i}, \tag{10.48}$$

$$V_3 = \frac{\sum_{i=1}^{N} n_i \sqrt{\left(\tau_3^i - \tau_3^R\right)^2 + \left(\tau_4^i - \tau_4^R\right)^2}}{\sum_{i=1}^{N} n_i}, \tag{10.49}$$

where n_i is single data i time series length, $\tau_2^i, \tau_3^i, \tau_4^i$ data L-moments and $\tau_2^R, \tau_3^R, \tau_4^R$ grouped average L-moments.

The homogeneity statistics values H_i, which are calculated by Eq. (10.50) indicate time series, which are acceptably homogeneous, in case of $H_i < 1$, possibly homogeneous for $1 \leq H_i < 2$, and definitely heterogeneous, when $H_i \geq 2$

$$H_i = \frac{V_i - \mu_V}{\sigma_V}. \tag{10.50}$$

This definition may suit control engineering case, however the following discordance analysis, i.e. the opposite task formulation is even more feasible and exactly fulfills control problem requirements.

The above homogeneity statistics H_i does not indicate time series that causes heterogeneity. Hosking proposed [153] discordance measure

D_i, which uses L-moments $\tau_2^i, \tau_3^i, \tau_4^i$. It identifies unusual time series and their distributions, which generally are discordant with the whole set of all datasets. In the original formulation, time series relate to hydrological discharge station measurements. In control engineering case, datasets for variables coming out of the loops are considered as the data sources. We evaluate the D_i measure for each dataset separately remembering that datasets are in form of three-dimensional vectors $\boldsymbol{u}_i = [\tau_2^i, \tau_3^i, \tau_4^i]$, $i = 1, \ldots, N$

$$D_i^2(\boldsymbol{u}_i) = (\boldsymbol{u}_i - \bar{\boldsymbol{u}})^T \boldsymbol{S}^{-1} (\boldsymbol{u}_i - \bar{\boldsymbol{u}}), \tag{10.51}$$

$$\bar{\boldsymbol{u}} = 1/N \sum_{i=1}^{N} \boldsymbol{u}_i \tag{10.52}$$

$$\boldsymbol{S} = \frac{1}{(N-1)} \sum_{i=1}^{N} (\boldsymbol{u}_i - \bar{\boldsymbol{u}}) (\boldsymbol{u}_i - \bar{\boldsymbol{u}})^T \tag{10.53}$$

where $\bar{\boldsymbol{u}}$ is time series mean and $\boldsymbol{S}$ the group (cluster) covariance matrix.

Discordance measure D_i aims at indicating how far a certain vector $\boldsymbol{u}_i$ lies from the cluster center. High values indicate that a given dataset doesn't follow the leading pattern. This task is analogous to outlier detection [90], however performed in a multivariate space with the utilization of Mahalanobis distance [70]. The only issue, which must be addressed is a definition of the threshold that delimits discordant datasets. Once vectors $\boldsymbol{u}_i$ are drawn from multivariate Gaussian population, the measure D_i^2 approximately follows three degrees of freedom χ^2 distribution. Thus we usually apply the square root of 0.975 quantile for χ^2 distribution threshold: $d_0 = \sqrt{\chi^2_{3,0.975}} \approx 3.06$.

We need to remember that mean and covariance estimators, which are originally applied, might be eventually biased by outliers [287], which would deteriorate the quality of the whole analysis. Few outliers can attract mean $\bar{\boldsymbol{u}}$ and inflate the covariance protecting their detection. This result is called the masking effect. To solve that issue, we may take into account robust statistical estimators with higher breakdown point, preferably equal to 50% [246]. Many estimators might be applied, however fast minimum covariance determinant (FAST-MCD) method [284] has been selected. The algorithm identifies n points with the smallest scatter and exhibits breakdown point of almost 50%. The FAST-MCD estimator is evaluated for a subset of $M < N$ points for m-variate dataset, for which the covariance matrix reaches the smallest determinant out of all possible subsets of size M. The associated location T and scale C are

$$\boldsymbol{T} = \frac{1}{M}\sum_{j=1}^{N} \boldsymbol{u}_{i_j} \tag{10.54}$$

$$\boldsymbol{S} = c_m \frac{1}{M}\sum_{j=1}^{M} \left(\boldsymbol{u}_{i_j} - \boldsymbol{T}\right)\left(\boldsymbol{u}_{i_j} - \boldsymbol{T}\right)^T \tag{10.55}$$

The subset size M can be chosen in different ways, but $M = \lfloor N + m + 1 \rfloor$ is often used.

The homogeneity analysis and discordance detection are pointless for the sample data used, as they should either come from a single complex process or, for the same loop, from different time periods. And the above does not apply to data X_i. In next sections the industrial case is analyzed and then the homogeneity analysis is visualized.

10.7.1 Concluding remarks on control homogeneity and sustainability

The aspects of data homogeneity, discordance and even sustainability take their roots from dissimilar research areas than control engineering. However, they can be relatively easily used and address issues, which remain open. The aspect of how to select control strategy for complex and multi-loop systems is one such example. Benchmarking and comparison of different loops is an alternative opportunity. Finally, the extremely important practical aspect of control loop and control system sustainability may be visualized and measured.

10.8 CAUSALITY ANALYSIS

The notion of causality has achieved much interest in recent years, both in its theoretical perspective and in industrial use. However, there is still an issue how to distinct between two separate problems [291]:

- "effects of causes" – EoC, which assesses the likely impacts of applied actions;
- or "causes of effects" – CoE, when we check whether or not a noted result is caused by an earlier exposure.

Though philosophers have split hairs since forever, statisticians have remained reluctant. Only recently they put attention to what exactly the "causality" means. Particularly influential have been terms of potential outcomes [291] and an introduction of graph-like visual representation [255]. The main issue is to develop inferences about the effects using data, generated using dedicated experiments or purely observational.

Several methods have been developed over the years. Similarly to the CPA, we may distinguish the ones that require plant model and the model-free ones. Actually, the methodology must be chosen carefully using known process behavior and the available data. We may find many approaches, like cross-correlation analysis, frequency domain techniques, information-theoretical methods, Bayesian nets. Some are successfully used in theoretical examples, but hardly applied to industrial cases. The most popular causality detection approaches are Cross-correlation (Cc), Granger Causality (GC), Partial Directed Coherence (PDC), and transfer entropy (TE).

The Cc adjusts two time series $x(t)$ and $y(t)$ in a way that they are lagged. Correlation factors between adjusted time series are evaluated, and we repeat the procedure for an assumed range of delays. The Cc advantage is due to its simplicity and low computational demand [214]. However, it only detects linear interactions for stationary data. In industrial problems, it may give inaccurate or opposite results [21, 360].

The underlying concept of Granger causality we may trace back to Wiener [353], who commented that, if we may improve the prediction of one data incorporating the knowledge of another one, then the second time series has influences causally the first one [48]. Granger formalized this idea using linear regression [124]. Granger causality is characterized by simple implementation and high efficiency. Despite that, it's limited due to the linearity assumption [39]. It assumes data, which are linear and stationary in a wide-sense [125]. Its performance depends on knowledge of the model structure and its proper identification. Despite its extensions, it should be used with high caution.

Partial directed coherence performs the analysis in a frequency domain. Energy transfer between pairs of data for each frequency reflects causality [205]. The method was developed as a frequency-based description of Granger causality [12]. It's especially efficient in case of oscillatory data [175]. Though it's frequency domain approach, it still incorporates the regression model. Computations using the model are sequential, so method's parallelization is limited. Moreover, regression may still introduce the bias limiting its applicability to industrial cases. Therefore, similarly to GC, the PDC should be used cautiously according to the situation.

Transfer entropy uses information interpretation of Wiener's causality, as it's a measure of information transfer between variables through the reduction of uncertainty [298]. The basic formulation behind the TE is as follows

$$T_{x \to y} = \sum_{y_{i+h}, \boldsymbol{Y}_i, \boldsymbol{X}_j} p(y_{i+h}, \boldsymbol{Y}_i, \boldsymbol{X}_j) \cdot \frac{p(y_{i+h}|\boldsymbol{Y}_i, \boldsymbol{X}_j)}{p(y_{i+h}|\boldsymbol{Y}_i)}, \tag{10.56}$$

where p the complete or conditional PDF function, $\boldsymbol{Y}_i = [y_i, y_{i-l}, \ldots, y_{i-(k-1)l}]$, $\boldsymbol{X}_j = [x_j, x_{j-l}, \ldots, x_{j-(k-1)l}]$ are time series of interest, l is a sampling interval, and h a prediction horizon.

It describes the difference between information about future x values observed in the concurrent past of x and y. The information about future behavior of x is obtained using past values of x. It delivers a reasonable formulation of the causality without any need for *a priori* assumptions, like delay [361]. The existence of bi-directional entropy is highly probable, so $T_{x \to y} = T_{x|y} - T_{y|x}$ is decisive delivering information about to quantity and direction, i.e. causality.

The practical implementation of Eq. 10.56 requires its simplification as in Eq. 10.57

$$T_{x \to y} = \sum_{y_i, y_{i-\tau_y}, x_{i-\tau_x}} p(y_i, y_{i-\tau_y}, x_{i-\tau_x}) \cdot \frac{p(y_i, y_{i-\tau_y}, x_{i-\tau_x}) p(y_{i-\tau_y})}{p(y_{i-\tau_y}, x_{i-\tau_x}) p(y_i, y_{i-\tau_y})}, \tag{10.57}$$

where p denotes conditional distribution function, while τ_x and τ_y are time delays in x and y. In its original formulation, the PDF uses Gaussian kernel density estimation. If data length is short, we set τ_x to 1 assuming that the maximum information auto-transfer occurs from the observation immediately prior to the target value of y.

Transfer entropy between two variables x and y can be calculated using the equation [298]

$$T_{x \to y} = \mathrm{TE}(x, y, \tau_x, \tau_y, N_p, C_{thumb}), \tag{10.58}$$

where x and y are one-dimensional data, τ_x and τ_y are time delays for x and y, N_p is a number of equally spaced points along each dimension and C_{thumb} is the coefficient adjusted by some rule of thumb [19]. In the original formulation, Gaussian function is used as a PDF. $T_{x \to y}$ indicates the information transfer between two variables x and y.

Transfer entropy uses dynamic and directional information. It is asymmetric and uses transition probabilities [338]. It does not assume in

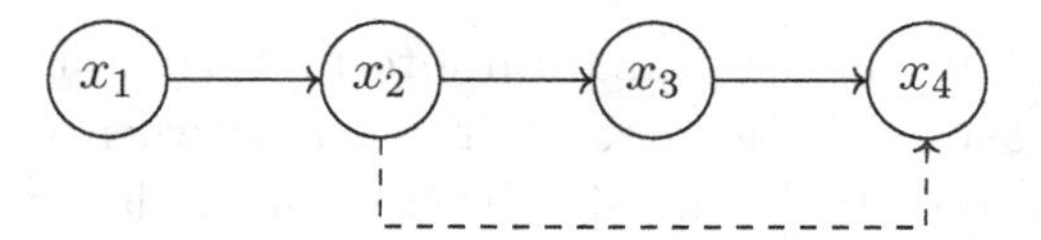

Figure 10.14: The sample layout of causality diagram

principle any particular interaction model. Thus, its sensitivity against all order correlations becomes an advantage in the analyses over model-based Cs, GC or PDC. Moreover, the method does not require any model. This is particularly important for unknown nonlinear interactions.

The TE is restricted to stationary data [102]. The approach requires PDFs being dependent on its proper definition. it may exhibit any non-Gaussian formulation. A few studies were carried out for other PDFs [166]. Fitting of the PDF to data does not have to improve the method performance [107]. Unfortunately, it results in non-negligible computational cost [108].

Nonetheless, the TE method constitutes a promising alternative in the causality assessment of real multi-loop systems. Each method can visualize identified relationships in form of a causality graph. Nodes represent considered time series (control errors in a CPA case), solid lines represent direct causal relation, while dashed ones indicate indirect relations. Fig. 10.14 presents such an exemplary diagram.

The causality analysis is pointless in case of sample data used in this book, as all datasets should originate from the same complex process, which is not the case.

10.8.1 Concluding remarks on causality analysis

The concept of causality, despite crucial in multi-variate data analysis, is somehow neglected in the research. It happens, despite the fact that correlation does not imply causality. Looking from the outside, one could conclude that scientists do not care about causality and correlation is sufficient. This is despite the existing physical correlations, which, however, does not bother anyone. Perhaps it is also due to the fact that complex multi-loop systems somehow do not find recognition among researchers [318] and are not addressed. In my personal opinion, the existing gap between the industry with its demands and the research with its sometimes random, artificial or self-imagined goals.

The purpose of this paper is to reverse this trend and draw attention to cause-effect relationships and their validity. This is the reason why causality testing is included, as relevant to statistical data analysis.

> Statistics is not only limited to probabilities, distributions, histograms, moments, quantiles,... There is much more. Analysis of time series, also those from control engineering uses other related methods. Aspect of the stationarity allows to detect trends, implying detrending or differencing. Oscillations and other periodic components may be decomposed and removed. Stochastic processes behind data can be identified with the use of regression and regressive models. As regression models capture the distribution peak area behavior, tail index facilitates measurement of tails.
>
> In contrary, the entropy allows to measure the information content in data. That direction might be followed by further data investigation and detection of long-memory dependencies, persistence, fractality and fractionality.
>
> In case of multiple time series, their benchmarking and interconnections the homogeneity analysis allows to find homogeneous or heterogeneous (discordant) datasets, while the causality analysis extends simplified concept of correlation allowing to perform root-cause assessment.
>
> Time series analysis, its interpretation and drawing conclusions is not a one-dimensional task. The challenge is full of different perspectives and the analyst's goal is to use an adequate and wide range of methods to make a correct and comprehensive control performance assessment.

GLOSSARY

DF: Dickey-Fuller test

ADF: Augmented Dickey-Fuller test

ADF-GLS: Augmented Dickey-Fuller – Generalized Least Squares

GLS: Generalized Least Squares

MAE: Mean Absolute Error

EMD: Empirical Mode Decomposition

IMF: Intrinsic Mode Functions

MS: Mode Splitting

EEMD: Ensemble Empirical Mode Decomposition

MEEMD: Median Ensemble Empirical Mode Decomposition

LS: Least Squares

ARMAX: Auto-Regressive Moving Average with auXiliary input

NARMAX: Nonlinear Auto-Regressive Moving Average with auXiliary input

GPR: Gaussian Process Regression

RR: Ridge Regression

RFR: Random Forest Regression

LTS: Least Trimmed Squares

LMS: Least Median of Squares

RLS: Recursive Least Squares

EVI: Extreme Value Index

MLE: Maximum Likelihood Estimator

MIE: Minimum Information Entropy

LRD: Long Range Dependence

DFA: Detrended Fluctuations Analysis

MF-DFA: Multi-Fractal Detrended Fluctuation Analysis

ARFIMA: Auto-Regressive Fractionally Integrated Moving Average

ARIMA: Auto-Regressive Integrated Moving Average

GPH: Geweke Porter-Hudak fractional order estimator

FAST-MCD: FAST Minimum Covariance Determinant

EoC: Effects of Causes

CoE: Causes of Effects

TE: Transfer Entropy

Cc: Cross-correlation

GC: Granger Causality

PDC: Partial Directed Coherence

III

Control performance assessment

CHAPTER 11

The feedback loop

FEEDBACK LOOP is an invention of James Watt and it constitutes the starting point of an industrial revolution and the backbone of modern industry. The feedback loop cannot exist without the plant. It requires the specific environment. And, what's crucial in the industry, it's never ideal. Its performance is affected by external and internal, known or unknown influences.

11.1 PROCESS CONTROL - WHAT MATTERS

Control engineering can be applied in different areas and industrial applications. Each of these domains has its specificity, which translates into the specific methodology and scope of use. Some issues are important in robotic applications, others in mechanics, and still others in chemistry, biology or economics. Yes, control engineering has many names. It has its place and time. It is implemented here and now, in a particular location, not on an imaginary, non-existent textbook plant.

This book is rather limited to *process control*. However, what is more important: *process* or *control*? What knowledge (and background education) is needed to properly design and build a control system? Is perfect knowledge of methods for designing and tuning control systems or technological process knowledge more important? Opinions are divided and two could discuss for a long time. Personally, I favor the opinion that process knowledge, being more difficult to understand, is nevertheless more important. I am in favor despite the fact that I myself am a control engineer by education and not a technologist.

All activities involving the implementation of any control system must harmoniously fit into plant operation. All elements, business,

DOI: 10.1201/9781003604709-11

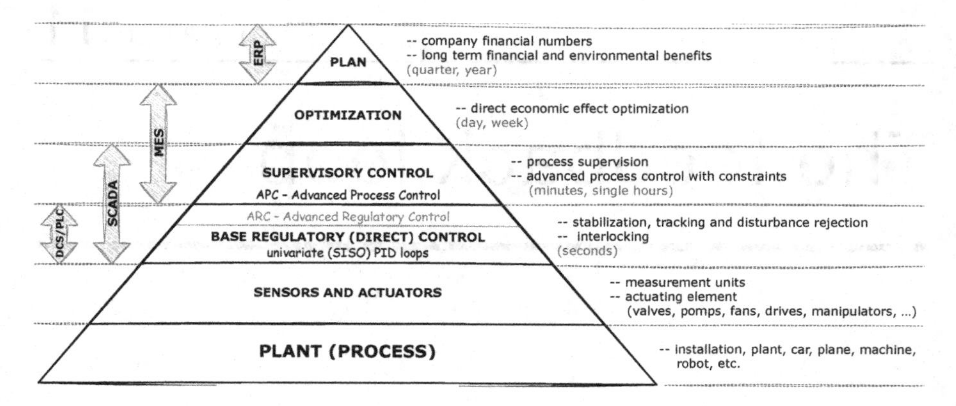

Figure 11.1: Control system hierarchy

technology, control and people, must collaborate. Control system can be interpreted and implemented on different levels of interests, on different time horizons and process domains. Most often, these relationships are presented in form of a pyramid. Fig. 11.1 shows a typical diagram, where the plant forms the necessary basis. Without the process, there is no control system. It may be a factory, an installation, an airplane, a robot, a machine tool or a room, in which we keep the temperature.

We must have the ability to communicate with the object two-way to be able to run any control actions. Getting information about a plant allows us to observe its condition and make actions. This is what measurements are for, while actuators allow you to influence the plant, i.e. implement decisions made by algorithms. Thus, sensors and actuators infrastructure layer constitute the base of the pyramid. Once the measuring and actuating infrastructure is in place, we can proceed to the installation of the direct control and safety systems. Safety system must be separated from the control system, so that any failure of the control system does not result in a risk of damage or destruction of the installation. Both systems must be fully independent and are powered independently.

Base regulatory control plays a fundamental role in a hierarchical layout. Its goal is simple. The control system takes information from the plant and, uses the a control law (algorithm) and generates the appropriate manipulated variable, which reaches the plant through actuators. This system works dynamically in a negative feedback loop, as shown in Fig. 11.2. Its purpose is to keep the plant output so that it is as close

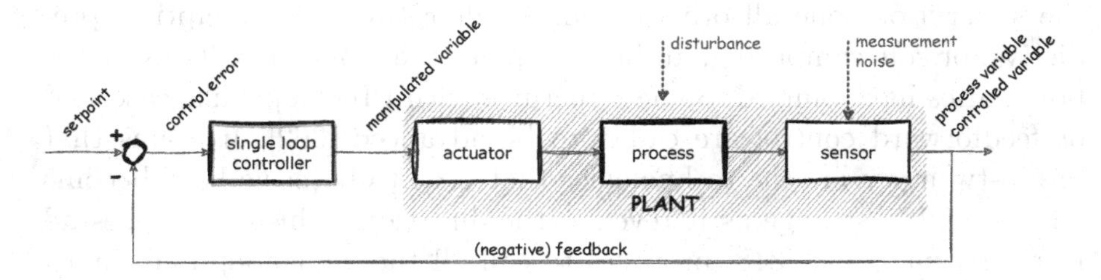

Figure 11.2: Univariate feedback loop

as possible to the setpoint value (reference, desired) regardless of any disturbances.

It is worth paying attention to an often overlooked aspect of the control hierarchy, i.e. **time**. The direct control layer operates with the shortest possible time horizons determined by the plant dynamics. Basic control systems most often use univariate (single-element – we measure one variable) control loops, analogous to the one shown in Fig. 11.2. Simple control system may include only one loop, while complex industrial processes consist of dozens or even hundreds of such loops (multi-loop systems). However, no matter the scale, its components remain the same. The design, use and development of feedback loops is the *raison d'être* of control engineering; Seemingly trivial, yet so complicated. One might wonder what accounts for this difficulty.

Systems consisting of the three lowest layers are sufficient to construct a proper control system. Practically for the first two hundred years of industrial control systems (counting from 1763, when James Watt's first steam engine was launched), the three-layer system has been enough.

The task of the supervisory layer is to select direct controllers in such a way that, in the long run, the system meets upper level goals associated with production efficiency. This is where economic or environmental goals are most often taken into account. To organize it properly, we must again recall the concept of time. The supervisory control layer has longer time horizons than the regulatory control. Sometimes supervisory control is called dynamic optimization or optimizing control systems. Most often, in such cases, we implement multidimensional and non-linear advanced control systems, called APC.

Unfortunately, these pictures, especially the distinction between the regulatory and supervisory layers is unclear. Forty or thirty years ago everything was clear. Regulatory control used PID-based structures, while the supervisory one all other advanced algorithms, like adaptive, predictive, or soft computing; all hidden under the name of APC. Now the boundaries have blurred. Some structures with PID loops, like cascaded or feedforward control are called to be advanced [362]. It seems that in ten-twenty years the technologies that used to be basic have become advanced. It's strange as it reverses the direction of history. Skogestad in his recent paper [318] proposes to call all the multi-loop and multi-component structures that still use PID algorithm but in complex interconnections an advanced regulatory control (ARC). This notation is used in further considerations.

Each higher system hierarchy extends operating horizons. We use static optimization solutions, very often multi-criteria, in the optimization layer. The static optimization task determines the goals included in the dynamic optimization performance index. In these systems, the *explicit* time does not occur, and the system assumes idealized steady state, assuming that any changes are so slow that we can ignore them. For the plant owner, m^3, kW or MJ are not important – what matters is the unified unit – money. He plans the maintenance policy and negotiates contracts with suppliers. His decisions regarding renovations or the use of gas, oil or electricity for heating affect the economic goals of the manager and the use of utilities by tenants.

The control system hierarchy presented above is analogous to hierarchical decision-making systems. A corporation is managed or a state is administered in a similar way. From the perspective of this book we focus on the single loop and its analysis. Therefore, the loop aspects of the negative feedback and disturbances are described below.

11.2 NEGATIVE FEEDBACK

Feedback, especially the desired negative one, has accompanied man since the dawn of time. Nature has put it in our body. Biology and medicine began to study this phenomenon only in the 1930s, starting with the work of Walter N. Cannon[41], who formulated the concept of homeostasis, i.e. the ability of the body to maintain relatively constant internal parameters.

The beginnings of using feedback in technical applications are lost in the darkness of history. Perhaps some solutions related to the use of water clocks or irrigation systems appeared earlier in ancient Egypt or Persia, but the historically confirmed record is held by the Greek technician Ktesibios (285–222 BC) and his water clock. Unfortunately, this revolutionary idea had no followers and humanity had to wait over two thousand years for further technical solutions. In the 18th century, two ideas appeared that practically revolutionized the development of industry and permanently introduced two solutions into technology: a controller with negative feedback and a servo-mechanism.

A servo was the first. Windmills use wind energy to power machines. Historically, they were used to provide energy for mills. However, this energy was difficult to maintain due to the variability of wind speed and direction. British blacksmith E. Lee patented a solution to this problem in 1745. He invented a mechanism that rotated windmill blades along their axis using a counterweight. Moreover, he proposed the use of additional blades placed behind the rotating blades, which positioned the entire windmill perpendicular to the wind direction and followed changes in its direction. This principle accompanies all servomechanisms.

The most famous solution using negative feedback is a centrifugal regulator. Similar invention, but without feedback, was proposed in the 17th century by the Dutch mathematician and physicist C. Huygen to maintain the force and distance between mill wheels. However, closing the loop was imminent. In 1705, English inventor T. Newcomen built the first atmospheric steam engine. It was improved by the young Scottish inventor James Watt in 1769. In 1788, in collaboration with J. Rennie, Watt introduced the main innovation that made it possible to automatically keep the speed of a machine – the centrifugal governor. Interestingly, a year earlier in 1787, T. Mead combined a centrifugal system with the windmill's winding and unwinding system to create a controller that maintained a constant rotational speed despite disturbances caused by varying wind [29].

Regardless of the above flagship inventions, there are more historical examples. Temperature control systems appeared in the alchemical furnace (C. Drebbel, 1624) or poultry incubators (J.J. Becher, 1680), although their industrial applications became established only with Bonnemain's solution (1777) for an incubator or a water-heating furnace. Float controllers that stabilized water level in toilets also appeared in the 18th century (W. Salmon, 1775), although boiler controllers gained much greater interest (J. Brindley, 1758; I.I. Polzunov, 1765; S.T. Wood, 1784). Pressure controls were first used with the invention of the safety valve (D. Papin, 1681), while general pressure controllers were applied to steam machines (R. Delap and M. Murray, 1799).

The steam engine, a symbol of the industrial revolution, came into existence and gained recognition only because the governor automatically allowed for the safe production of the required amount of energy of a given quality without the need for constant human supervision. However, it must be remembered that these solutions were mainly created thanks to the creative thinking and unlimited imagination of the designers. Their unrestrained imagination, entrepreneurship and willingness to beat the competition powered the development. Inventions were created mainly through a tedious process of trial and error. They were not supported by any mathematical theory.

G.B. Airy proposed in 1840 to use differential equations to analyze control problems. The first analytical solution to the stability of centrifugal controller was proposed in 1868 by J. Maxwell [233]. His work "On governors" is the first scientific paper dedicated to control theory. Further work on stability allowed for the independent algebraic formulations of E.J. Routh [289] in 1877 and A. Hurwitz [161] in 1895. A complete and elegant solution of the stability of nonlinear systems was proposed in Russia by A. Lyapunov [223] in 1892. Interestingly, the Western world became acquainted with his work only in the 1960s. At the end of the 19th century, the British engineer O. Heaviside developed operator calculus and introduced the concept of transfer function, which is now the basis of linear control theory. The 20th century brought further developments:

1910: gyroscope – stabilization and control (E.A. Sperry),

1922: PID controller (N. Minorsky) [240],

1932: stability criterion in the frequency domain (H. Nyquist) [249],

1934: feedback amplifier (H.S. Black) [32],

1934: artillery guidance – servo mechanism (H.L. Házen) [139],

1938: frequency characteristics of linear systems (H.W. Bode) [35],

1942: PID controller tuning (J.G. Ziegler and N.B. Nichols) [374],

1957: dynamic programming in discrete optimal control [25],

1960: minimum variance control (linear quadratic regulator) LQR [178],

1960: Kalman filter [180],

1977: IDCOM (IDentification and COMmand) predictive control [278],

1979: DMC (Dynamic Matrix Control) predictive control [60],

1987: GPC (Generalized Predictive Control) predictive control [50].

Finally, it is worth referring to an often overlooked and unnoticed aspect: disturbances and uncertainties. The control system is never fully isolated from its environment. F. Knight published in 1921 the theory of measurable and unmeasurable uncertainty. For measurable uncertainty, i.e. disturbance, we can determine the risk. We cannot assess or estimate unmeasurable uncertainty – it is a "pure and untainted". Uncertainties always occur. At most sometimes their impact is unnoticeable and negligible. We cannot ignore them or forget them. The worst mistake (apart from ignorance of course) is to confuse disturbance with uncertainty, and the consequences can be devastating.

Control system design must take into account the disturbance analysis. It is necessary to consider their existence, impact, the possible way of their measuring and the risk assessment. Disturbance analysis either allows to create an appropriately resistant and safely tuned system, or to introduce decoupling. Uncertainty in control systems can be viewed from two perspectives. On the one hand, we can interpret uncertainties as a kind of disruption, but we do not know when and how they will occur, in what direction they will change and what their impact will be. Uncertainty also exists in the context of measurements noise. Noise is generated by internal phenomena, most often modeled as stochastic processes (e.g. thermal phenomena in a measuring device), superimposed on the actual signal (information). Uncertainty occurs when we are unable to distinguish the actual signal from the noise. Disturbances and noises should be taken into account in the process of control system design and tuning. Their impact should be estimated and eliminated. We cannot measure uncertainty and therefore we can't predict it. Thus, we cannot use them directly in control system design.

As one can see, the control system design process uses more than just negative feedback. We must consider the entire control system. "A chain is only as strong as its weakest link, and life is after all a chain"[1]. Therefore, a control system will only be as good as its weakest link. There's no point in using sophisticated control algorithms if the control valve sticks and leaks - it's just a waste of time. Likewise, you cannot save on the measurement sensor if the rest of the ingredients are of very

[1]William James, *A chain is no stronger than its weakest link, and life is after all a chain.*

high quality. The control system must be homogeneous. Its individual elements must work together, because then the whole works best.

Generally, control system must fulfill two tasks: setpoint tracking and disturbance rejection. Sometimes one may be more important than the other. Sometimes setpoint does not change. Other times disturbances can be neglected. To achieve control goal we must select proper strategy, design adequate configuration, implement the system in plant infrastructure, tune it (set algorithms' parameters), commission the system and maintain it over time.

Two elements are important from the perspective of control performance assessment: the commissioning and the maintenance. They require the knowledge about control system quality. We need to know how good (or bad) the system is. This decision requires specific indexes. We desire algorithms and data to measure the performance. During commissioning we do it once, however afterward the system operates in normal certain conditions. They might be normal or not. The system may change and vary. Plant goals and operating regimes evolve in time. Equipment breaks. Generally, initial configuration is no longer valid. We need to find it out and react. Control system maintenance is a must as it allows to keep initial results sustainable. The evaluation of these algorithms and approaches is addressed in the following sections.

> In the considered narration control systems constitute the playing ground, in which statistics and tail-aware control performance assessment are the players. PID algorithm applied in the univariate control loop can be met in almost all process installations. It's the working horse of process control. The knowledge how and why it works is crucial any further analyzes.

GLOSSARY

PID: Proportional Integral Derivative

APC: Advanced Process Control

ARC: Advanced Regulatory Control

CHAPTER 12

Control loop performance assessment

CONTROL PERFORMANCE ASSESSMENT allows to asses and benchmark the quality of a control. In common understanding CPA aims at the system performance. However, if we return to the notion of a good control system, the performance (benefit) perspective seems to be very narrow. Actually, the system is often affected by external impacts: noises, disturbances and uncertainties. Nevertheless, it must maintain the quality. We call that feature robustness. Thus good controller must be efficient and robust. However, it still does not cover the entire picture.

We mustn't forget time. The control system, initially perfectly tuned and immune to noises and disturbances, should maintain its high quality in time. The environment changes, varying market imposes reconfiguration, new equipment brings improved functionality, the plant "lives". Moreover, the installation may break down, mechanically or due to the external reasons like cyber attacks. And it must recover as fast as possible. Therefore, our control system must be sustainable and resilient. Concluding, the good controller or the whole control system must:

- perform well,
- be robust,

DOI: 10.1201/9781003604709-12

- sustain its quality in time
- and be resilient.

This work aims at addressing all these issues. Everything wis easy if the industrial control systems works well, is properly designed, optimally tuned and maintains at the highest level over time. Unfortunately, it's almost never the case. Despite common awareness that good control brings money, industrial systems often do not operate effectively. The literature shows that as much as 60% of basic loops are poorly tuned and an even larger part (85%) has an inadequate structure [171]. The use of copy-paste templates is on a huge scale. To summarize, there are many reasons for poor control performance [20, 322]:

1. poor tuning due to the non-linearity, insufficient supervision, conservative approach, changes in operating regimes or field devices,
2. equipment problems and failures,
3. negative impact of disturbances plus with rare use of decoupling, not to mention the lack of uncertainty analysis attempts,
4. unmatched control structure,
5. human errors, such as accidental damage, lack of time, or insufficient skills/training/education.

Very often the situation is rescued in the simplest possible way using so-called good practices, safe or worst-case tuning. In particular, the worst-case tuning is the most common approach. Design for the worst case scenario makes perfect sense. After all, safety comes first. However, overusing it, or even using such an approach as a rule and not as a result of considering the properties of a given system, is a serious mistake. Why? Because tuning for the worst-case scenario, i.e. to minimize the probability of too aggressive control (fear of overshoot), leads to "sluggishness" of the controller away from optimal settings. This fact can be observed in most installations. The share of such slowly tuned control loops is dominant. Fear has big eyes.

Apart from rational arguments related to plant safety, there are three other reasons: outsourcing of I&C services, poor skills and fear. In the old days, control maintenance services worked directly at site. They knew the installation inside out, could react immediately and had time to look into the problem. Moreover, you could always ask and get advice from your roommate – sharing of expertise. With the winds of change, we have moved from one extreme to the other. Only costs and costs are important in the outsourcing model. The facility employs minimal staff, sometimes even just a single person (or even none). This model assumes that necessary support is called from an external company. The effect of such an extreme solution is to turn a blind eye to minor deficiencies, and calling for external support as a last resort. The service provider also needs to minimize his costs. There is rarely a dedicated person for a given plant and in the event of a call, actually free person travels, a person who does not necessarily know the plant thoroughly. The arriving engineer also minimizes his time and wants to solve the problem as quickly as possible, because he has to go to another plant soon. So he tunes the system for the worst-case scenario, because of time, because of money, because he won't be coming back any time soon. Therefore, for the so-called peace of mind, we tune loops as conservatively and safely as possible. We avoid not only unnecessary risk, but even any risk. And if it could have been better so what?

Control systems always need to be properly designed and tuned. It's the essence of a control job. The person making decisions regarding the choice of solution should have objective measures to assess control quality. A clear, repeatable and comparable methodology and so-called key performance indicators (KPIs) are expected. The issue of assessing the quality of control systems consists of two basic processes, as shown in Fig. 12.1. First, we need to decide whether the control system works properly or not. The analytical task enabling such an assessment requires information about its operation. We obtain it either as a result of conversations with plant personnel or through historical data review. The first step is to obtain data from plant data acquisition system or by installing dedicated data collection software.

Reliable plant data enable further analysis, i.e. calculation of selected (appropriate, required, possible to determine) indicators for assessing the operation of considered control systems. We can confirm whether a given system works as intended or requires improvement using obtained numbers. In the first case, the further development of the situation is simple. We are happy and wait for the next evaluation. Otherwise, we must take actions. We need to carry out diagnostics and find the cause of poor performance. We need to carry out installation diagnostics to see what failure has occurred. Next, we verify whether the control system as such, i.e. the hardware and software, work as designed. After that we

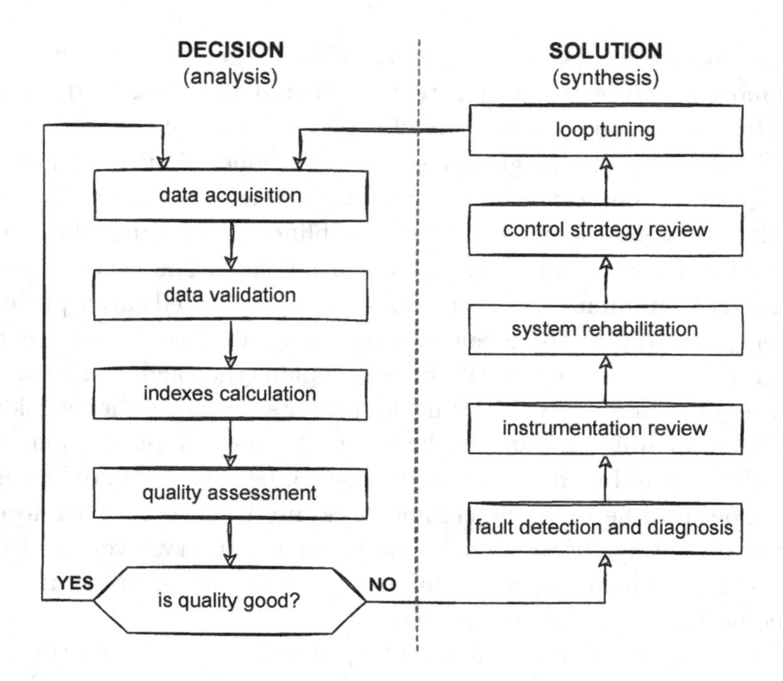

Figure 12.1: CPA functional procedure

should check control philosophy correctness, and if it's proper, we should verify the tuning.

As one can see, this process is a repeatable and ongoing activity. After all, we strive for perfection. This is an ideal situation. However, it is as it is. Control performance assessment is not performed often, if at all.

The control performance assessment story began with a simple, universal rating of a single input single output PID loop. The first reported adequate solution can be found as early as 1967 in the works of Åstrőm [269], where a benchmarking system using key variables' statistical properties (standard deviation) was applied to the pulp and paper plant. Over the next more than fifty years, solutions have evolved in various directions, providing the industry with more or less accurate tools, measures and methodologies. Fig. 12.2 presents classification of available methods.

Generally, we can distinguish two basic situations. Some methods require to conduct a dedicated plant experiment. These measures are based on dynamic close loop step responses: setpoint or disturbance. This group includes the two most popular indicators: *overshoot* and *settling time*. Together, they contain virtually all information for comprehensive

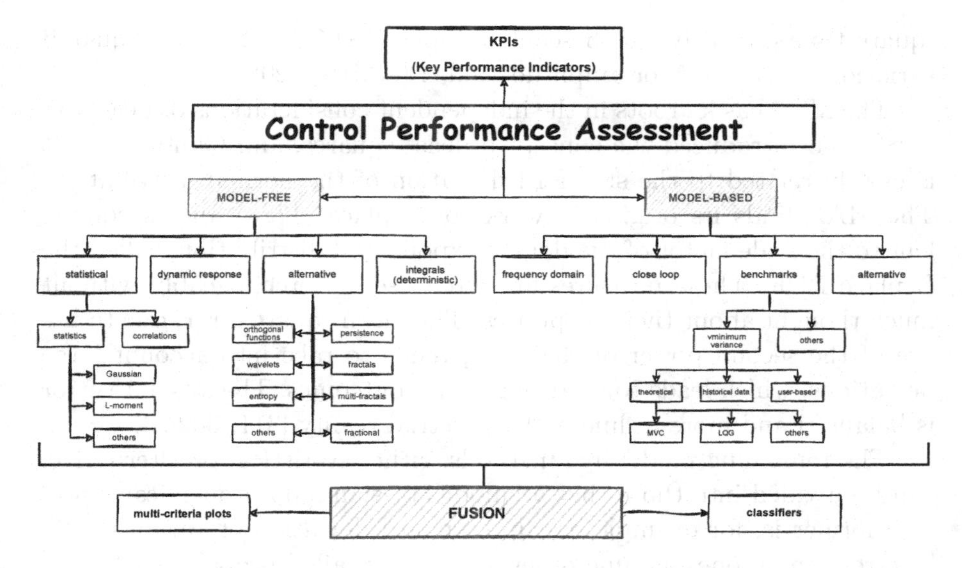

Figure 12.2: Classification of the CPA approaches

assessment: accuracy and speed. They are simple, known and liked. However, they have one shortcut – we need to have step response. The best way is to obtain it using dedicated experiment, however in industrial reality it's hardly achievable. Theoretically, we could also evaluate it from operational data, but the quality of such response is not very informative. Nonetheless, *overshoot* and *settling time* constitute ideal reference to assess and interpret other indexes. Among the other step response indicators, let's mention rise and peak times, peak value, decay ratio, steady-state error[320], Idle Index [142], Area Index, Output Index [339] or R-index [294].

The remaining classes of methods do not require any experiment and use normal operational data in the form of time series of basic variables describing control loop, such as the controlled variable or control error. At the same time, the second perspective divides the methods into those that require a modeling process and therefore *a priori* knowledge (*model-based*) and methods that do not use any plant knowledge (*model-free*).

The next classical choice is to use integral measures like, mean square error (MSE) and mean absolute error (MAE) [311] are the most frequently used. In addition to the above, there are less frequently used indexes, such as integral time absolute Value (ITAE) [371], integral of square time derivative of the control input (ISTC) [372], total squared variation (TSV) [365] or amplitude index (AMP) [320].

The MSE has its roots in the independent considerations of Legendre [208], who introduced the concept of least squares, and Gauss [117]. It is closely related to the standard deviation of the normal distribution. The MAE finds its origins in works of Laplace (1783) and is equivalent to the scale factor of the double-exponential distribution, called the Laplace PDF. These measures are often used interchangeably without much thought about their properties. The mean square error, due to the use of the second power of the error, tends to take into account large deviations, which leads to more aggressive outcomes. The absolute error is balanced and more in line with industrial reality [311, 302].

The remaining model-free methods, either statistical or alternative, can be divided into those that assume Gaussian loop properties or not. The former is, for example, mean, standard deviation or variance [49]. The remaining ones assume other, e.g. heavy-tailed probability density functions [373, 81], or do not make any assumptions, focusing on other properties of time series (persistence [79], fractal features [265], long-range memory [217] or entropy [368]).

Model-based methods, although there are many of them, have one disadvantage – they require initial *a priori* assumptions regarding the process (at least knowledge about its delay). These assumptions do not seem strong and theoretically they should be simple to implement. However, as often happens, theory goes hand in hand with practice only in theory. Correct identification of the delay is a practical challenge, not to mention stronger assumptions about the process model, such as the orders of the transfer function polynomials. The situation becomes even more complicated in the case of significant non-linearities and non-Gaussian properties of the control loop. Model-based methods should be treated with caution. I personally try to avoid them whenever I can.

The summary clearly shows that there is no single universal solution. Each one has its advantages and disadvantages. Andy user should know their specificity and apply them properly. Simultaneous use of various methods leads to a multi-criteria assessment definition [101].

When considering control quality assessment indicators, we cannot forget about industrial indicators, commonly known as KPIs. They are tailored to a given installation, habits, engineering culture, corporate policy and the strength of tradition. The simplest example is the loop operating time in automatic mode or the time when limits are exceeded in a specific unit of time. Such indicators are often used to settle the maintenance team or people responsible for automation. They can be easily calculated and presented in a spreadsheet. Moreover, they very often have a direct impact on the economics of the process, appealing to management and accounting staff.

In summary, human supervision is insufficient. Therefore, there is a growing demand for methods and solutions to support this process. They must meet many requirements to gain industry acceptance. In particular, they must work autonomously using data from everyday work without the need to carry out any specialized tests, and they must be formulated based on clearly defined and reliable measures (assessment indicators). On the market you can find many different computer programs or even entire systems that, after integration with object-oriented computer systems, continuously analyze and calculate the measures of the indicated control loops [88].

Interestingly, despite multiple CPA methods, there is no single measure of robustness, except frequency stability margins. Therefore, what to do? How to do? There is a whole area of robust control and it covers a large field of various algorithms, e.g. H_∞, H_2, control Lyapunov function, etc. Definition of robust control is extremely optimistic and says that "*robust control refers to the control of unknown plants with unknown dynamics subject to unknown disturbances*" [45]. A real panacea for all the world's ills. Unfortunately, in practice, at least in the process industry, this approach does not occur.

This work focuses on statistical tail-aware perspective of the CPA task. Therefore associated methods are further described below. The methods are used within the proposed multicriteria framework of the index ratio diagrams (IRD) and visualized in a simulation study

12.1 POINT OF REFERENCE

As it was summarized above the CPA task is a complex and multithreaded issue. We have many methods, many measures, many perspectives and many goals. The main element of the assessment is to validated dynamic loop properties. The very issue of assessing the quality of control comes down to evaluating two opposing properties: accuracy and speed of control. Thus, the evaluation should refer to the overshoot κ and the settling time T_{set}. Fig. 12.3 presents an exemplary step response, with indicated measures.

Steady-state error is evaluated as a difference between realized and desired process variable values in a steady state as time goes to

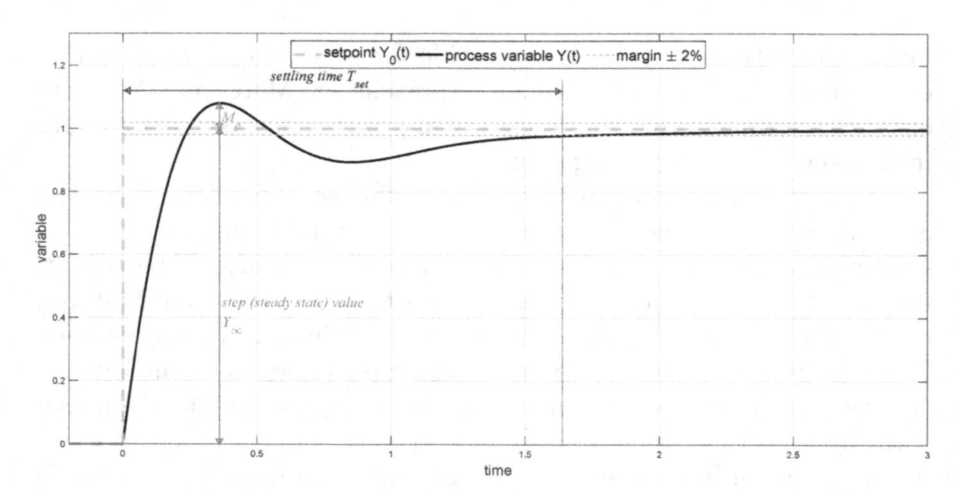

Figure 12.3: Exemplary dynamic step response of SISO control loop

infinity

$$\epsilon_\infty = Y_0 - Y_\infty. \tag{12.1}$$

We expect and demand it to be zero $\epsilon_\infty = 0$ (as in Fig. 12.3), but a non-zero value may occur. It may happen due to the lack of the integration in the feedback loop or because of nonlinearities or actuator saturation, which limit manipulated variable.

Overshoot κ measures control accuracy. We prefer to minimize the overshoot and in an ideal scenario we aim at $\kappa = 0$, as the higher it is, the worse tuning performance and accuracy are. Its large values denote *aggressive* control. We derive it using the following equation

$$\kappa = 100 \cdot \frac{M_p}{Y_\infty} \ [\%] \tag{12.2}$$

Settling time T_{set} is the second measure, which is often calculated with the overshoot. It measures how fast the system settles down within a given band around the steady state. Ideally, it should be also zero, which cannot happen, thus we aim at the shortest T_{set}. Its large values denote *sluggish* control.

The meanings of κ and T_{set} are contradictory. Achieving their simultaneous minimization cannot be met. Too low settling time we obtain at a cost of large overshoot and we have to sacrifice the κ to end up with low κ. Loop tuning is a kind of multi-criteria, actually two-criteria, art of compromise. These indexes share common practical feature. once we want to change them, like for instance to lower the settling time, we now what to do with the controller. Their meaning is **constructive** and **explanatory**.

Once both reference indexes are defined we may move forward toward more complex measures. Two KPI families are presented: integrals, because of their popularity and statistical as they may give deeper insight into loop tuning, unachievable with other indexes.

12.2 INTEGRALS – POWER OF TRADITION

SISO control loop is described by four variables:

- setpoint, denoted as $y_0(t)$,
- process variable (PV) or controlled variable (CV), denoted as $y(t)$,
- controller output or manipulated variable (CV), denoted as $m(t)$,
- control error $\epsilon(t) = y_0(t) - y(t)$.

Customarily, in the greatest number of cases, the control error signal is used as a loop variable, for which we evaluate measures. It is due to the fact, that it's expected to detrended with a zero mean [86]. Manipulated variable is taken into account, when the aspect of the energy spent on control needs to be addressed [91].

Traditionally, two indexes are used: the mean and absolute square errors. The MSE is calculated as a means of control error squared values

$$\text{MSE} = \frac{1}{N} \sum_{k=1}^{k=N} \epsilon(k)^2 = \frac{1}{N} \sum_{k=1}^{k=N} [y_0(k) - y(k)]^2 . \tag{12.3}$$

The MSE index highly penalizes errors with large values. They usually occur immediately after a disturbance and are observed in overshoots. It is shown [303] that tuning minimizing the MSE biases toward aggressive settings. It has marginal relation to economic control aspects [312], especially in disturbed cases[82]. We also remember that it is not robust against outliers. The reason behind its popularity lies in tradition, powered by its differentiability.

At this point, it should also be recalled that the MSE is equivalent to the scale factor (standard deviation or variance estimators) of normal distribution.

Mean absolute error we evaluate as a mean of control errors' absolute values over a given time period $k = 1, \dots, N$

$$\mathrm{MAE} = \frac{1}{N} \sum_{k=1}^{k=N} |\epsilon(k)| = \frac{1}{N} \sum_{k=1}^{k=N} |y_0(k) - y(k)| \,. \tag{12.4}$$

The MAE is less conservative than MSE and it is often used in controller tuning tasks. It exhibits the closest relation to process economics [312] and penalizes continued cycling. It well fits to any kind of operation data. It is equivalent to the scale factor b of Laplace double exponential distribution.

Above indexes are shown in their basic forms there exists their modifications, as in case of time-weighting [247] modifications, normalization [171] or fractional order [329] versions.

Integral measures have one important advantage: they are easy in evaluation and practically can be always calculated.

12.3 STATISTICAL MEASURES – OLD ANEW

Part II of this book describes background for the statistical data analysis. Selected factors are applied and validated in this part. The comparison is done with simulations with some supplementary material utilizing real industrial data. The selection should be done systematically by matching of control quality features and the properties of available statistical factors. As basic control quality features we may distinguish:

1. continuing and not disappearing offset between the process variable and setpoint, i.e. control error non-zero steady-state error,
2. aggressive and too fast overshooting loop behavior, addressed by high overshoot,
3. sluggish and slow convergence to the setpoint, addressed by long settling time,
4. loop nonlinearities,
5. continuing oscillations due to the actuator malfunctioning, low stability margins or integral (reset) wind-up [271, 63],
6. uncoupled disturbances and/or frequent outlying control error observations.

Steady-state error may be reflected by the offset measures of the control error signal. The simplest is the arithmetic mean μ or median $\overline{x}_{\text{med}}$, as the latter is robust to outliers. The use of more complicated and exotic measures (like offset factor of α-stable distribution or location M-estimators) is unjustified and practically meaningless. Though the median value is robust, and seems to better suit the task there is a sense to use both, as the comparison between them might indicate outliers.

Nonetheless, the shift measures and are not included, as they rather do not depend on the loop tuning, but its configuration. Moreover, they are not fitting into the designed simulation experiment, once they are important and included in the general assessment procedure.

Aggressive control might be addressed by two statistical properties: the scale or the persistence. Following this observation we might use different scale factors, starting from normal standard deviation. The research on statistical moments and distribution factors [99, 81, 82] points out several possible, already considered options: normal standard deviation σ, interquartile range IQR, median absolute deviation around median MADAM robust standard deviation M-estimator, for instance the one using logistic function $\hat{\sigma}_{\text{L}}$, scale factor of the α-stable distribution γ or scale L-moments : L-Cv or l_2.

Generally, it's useless to incorporate and analyze all of them, due to their equivalences and similarities. Using previous studies [88, 95] and industrial projects, a subjective choice is suggested:

– (M1): normal standard deviation σ,

– (M2): robust standard deviation logistic M-estimator $\hat{\sigma}_{\text{L}}$,

– (M3): L-scale moment l_2.

The selection of both normal standard deviation and its robust counterpart, is analogous as in case of mean and median; their comparison indicates outliers.

Moreover, two persistence measures are taken into account. Hurst exponent allows to indicate the aggressive/sluggish tuning as well. Therefore, its use is justified, apart from the fact that it also points out persistence, which reflects uncoupled disturbances. As research shows the shortest scale Hurst exponent estimator H_1 obtained with the R/S plot can be utilized [99], however the ARFIMA filter fractional order Geweke Porter-Hudak estimator d_{GPH} (M4) is compared during this research.

Non-linear loop behavior may be captured by distribution shape factors, especially the ones that reflect asymmetric behavior. Previous considerations introduced three possible estimators: excessive skewness γ_3, skewness factor β of α-stable distribution and L-skewness γ_3 (M5). The last one is selected as it is robust and limited to finite set $(-1, +1)$.

Following the idea of the L-moment ratio diagrams, also the L-kurtosis γ_4 (M6) is considered as it allows to indicate signal normal properties, which, as assumed, reflect proper tuning and decoupled disturbances.

Finally, two more measures are considered: Huismann tail index estimator $\hat{\xi}$ (M7) and the rational entropy H^{RE} (M8). One could imagine also the use of other statistical factors, like for instance stability index of α-stable distribution or coefficients of variation, but previously mentioned research has not identified significant advantages of these measures and thus they are not included.

Above selected eight pre-selected measures are incorporated into multi-criteria framework of the index ratio diagrams to assess loop dynamical properties. At this point, it should be noted that IRD charts do not exhaust the whole issue of studying control quality, but only allow assessing its dynamic aspects. A comprehensive and elaborate procedure will be presented in Chapter 16.

12.4 INDEX RATIO DIAGRAMS

Index ratio diagram is an extension of the idea of moment ratio diagrams. we show two indexes, one versus the other one. The simplest approach would be to plot overshoot versus the settling time to present the Pareto front of possible optimal solutions according to the combination of these two indexes. This fundamental in its meaning diagram constitutes the basis for the proposed idea, as it shows accuracy versus time behavior.

Actually, a given pair of selected measures in Cartesian coordinates cannot be random. It matters what we put as abscissa and as ordinate. It would be rather useless to show robust standard deviation versus l_2 as both reflect similar features.

The formulation of the IRD plots enables and introduction of the reference or a kind of the third dimension of the diagram. If we exchange points with circles, we may shade them according to the other third index. We may do it to emphasize the relation of the diagram to loop features or extend the assessment with another feature.

The analysis starts with the IRD showing reference plot with the overshoot versus settling time. It is followed with diagrams between scale parameters σ, $\hat{\sigma}_L$ and l_2 as abscissa and other five as ordinate. Next, the L-skewness will be related to L-kurtosis , as standard LMRD(τ_3, tau_4) and related to persistence measure, Huisman tail index estimator $\hat{\xi}$ and the rational entropy H^{RE}. Similarly, the L-kurtosis will be related to those indexes. Finally, the relationship between $\hat{\xi}$, H^{RE} and d_{GPH} will be assessed.

Actually, IRD diagram offers visual information, as we still do not obtain any single control performance indicator. Control engineer might choose that the shortest distance from the ideal point, like the origin $[x_0; y_0] = [0; 0]$ indicates the best tuning. We would wish to get this metrics independent on real units of time and overshoot, so it is scaled. We evaluate the following IRD distance index $d_{\text{IRD(x,y)}}$ obtained for scaled x and y

$$d_{\text{IRD(x,y)}} = \frac{1}{\sqrt{2}}\sqrt{(x - x_0)^2 + (y - y_0)^2}. \tag{12.5}$$

The analysis is conducted using the simulations. It will show how the IRDs work, how should we interpret them and which ones are the most interesting from the practical point of view.

12.5 SIMULATION STUDY

The Matlab simulation environment is designed to reflect classical raw single-element PID control loop, which is the most common in process industry. It does not include any supplements, such as filters, feedforward elements, nonlinear blocks, etc. It is done intentionally, as such a loop (unfortunately) is the most frequently observed in industry. The loop is driven by setpoint signal $y_0(t)$ and is affected by the additive disturbance $d(t)$, which is introduced before the process and the measurement noise $z(t)$, which is added to the process variable y(t). Control error signal $\epsilon(t) = y_0(t) - y(t)$ is used as the loop assessment variable. Fig. 12.4 graphically presents adopted simulation environment.

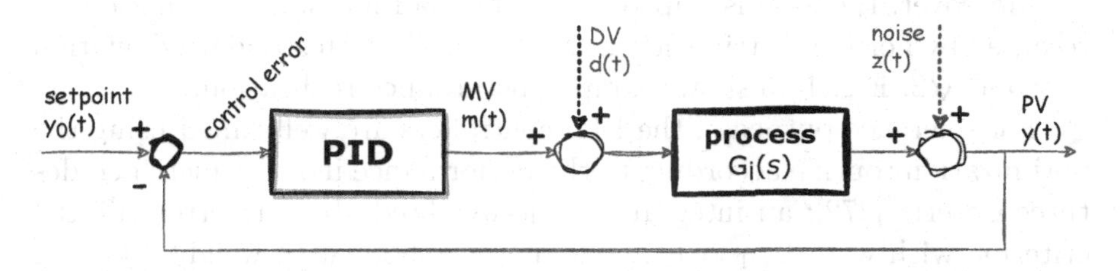

Figure 12.4: Simulation environment with a single-element PID control

The univariate PID controller in a parallel form is applied as the control algorithm

$$G_{\mathrm{PID}}(s) = k_p \left(1 + \frac{1}{T_i s} + T_d s\right). \tag{12.6}$$

The analysis is conducted for three exemplary transfer function, which were proposed as the plant benchmarks in process industry [273]:

- system with multiple equal poles

$$G_1(s) = \frac{1}{(s+1)^4}, \tag{12.7}$$

- first-order system with a dead time

$$G_2(s) = \frac{1}{(0.2s+1)^2} e^{-s}, \tag{12.8}$$

- system with two modes: fast and slow

$$G_3(s) = \frac{1}{(s+1)(0.04s^2+0.04s+1)}. \tag{12.9}$$

Above transfer functions represent a wide scope of cases that are observed in process industry. They are simulated in the continuous time mode and data is sampled with time $T_p = 0.1$ [s]. Setpoint signal is constant and set to zero $y_0(t) = 0$. The intention for this assumption is to minimize possible loop asymmetric effects, which are not aimed at that experiment. As the setpoint does not change, the loops are generally driven by the additive disturbance $d(t)$. The signal is generated as filtered by the first-order inertia summation of the rectangle wave and the symmetric SαS noise with coefficients: $\alpha = 1.95$, $\gamma = 2.0$ and $\beta = \delta = 0$. Fig. 12.5 visualizes disturbance generation scheme.

Moreover, the loop is impeded by Gaussian measurement noise represented by normal distribution $z(t) \sim N(0, \sigma^2)$ with standard deviation $\sigma = 0.5 \cdot \sqrt{2}$. Fig. 12.6 shows sample disturbance realization.

For reference purposes, the PID controllers are well-tuned using the optimization routine according to the performance index, which includes three criteria [272]: an integral of time-weighted absolute error (ITAE) criterion with weight $\beta_1 = 1$, maximum overshoot with weight $\beta_2 = 10$, and sensitivity weighted with $\beta_3 = 20$. Resulting well-tuned coefficients of the PID controllers are presented in Table 12.1.

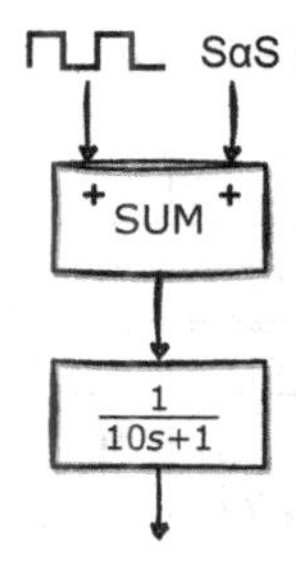

Figure 12.5: Disturbance signal $d(t)$ generation mechanism

The analysis uses large set of different PID parameters, to compare performance assessment at various controllers tuning. Table 12.2 shows the ranges and change decrements for parameters of the controllers. As

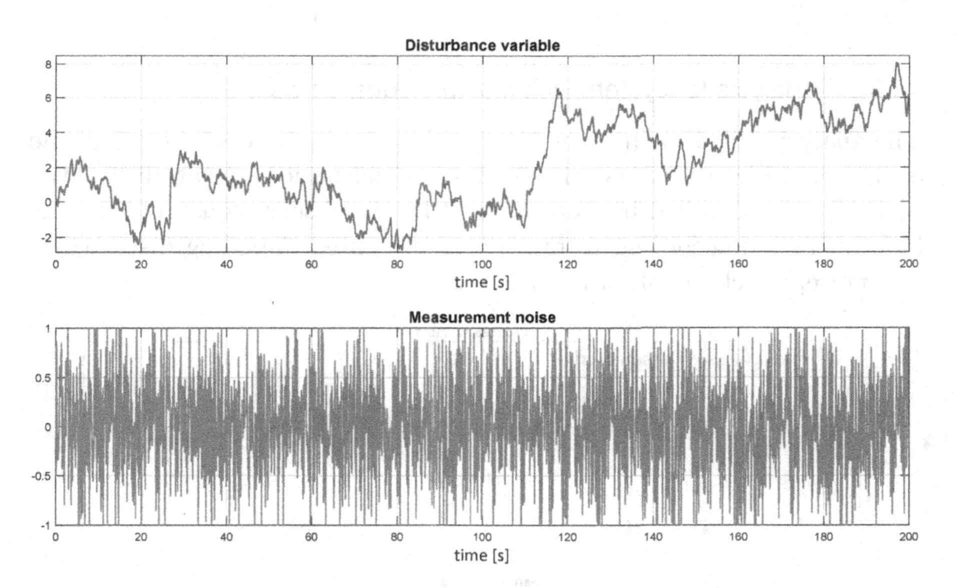

Figure 12.6: Sample representation of $d(t)$ and $z(t)$

Table 12.1: Properties of well-tuned PID controllers

	k_p	T_i	T_d	κ	T_{set}
$G_1(s)$	1.0503	2.9977	0.9293	0.091%	5.849
$G_2(s)$	0.2653	0.6066	0.2121	0.000%	5.594
$G_3(s)$	0.1330	0.2585	0.0808	2.199%	7.427

Table 12.2: Parameters ranges used during simulation experiments

plant	k_p			T_i			T_d
	min	decrement	max	min	decrement	max	const
$G_1(s)$	0.05	0.25	2.05	0.2	1.0	10.2	0.9293
$G_2(s)$	0.2	0.2	1.6	0.1	0.5	5.1	0.2121
$G_3(s)$	0.02	0.1	1.02	0.05	0.25	2.3	0.0808

the result, we obtain the grids of 99 cases in case of $G_1(s)$, 88 for $G_2(s)$ and 110 settings for $G_3(s)$.

Each set of parameters is simulated *5* times with different realization of disturbances and noises. Resulting measures are averaged, to exclude statistical biases that could be introduced by specific disturbance realizations. The same experiment is repeated for all plants.

12.5.1 Analysis for system with multiple equal poles

The analysis starts with plant $G_1(s)$, which is characterized by multiple equal poles. Fig. 12.7 shows time series data for basic loop variables of the plant $G_1(s)$ controlled by well-tuned PID. We observe constant setpoint loop operation according to previously presented impeding disturbance and noise, which are shown in Fig. 12.5.

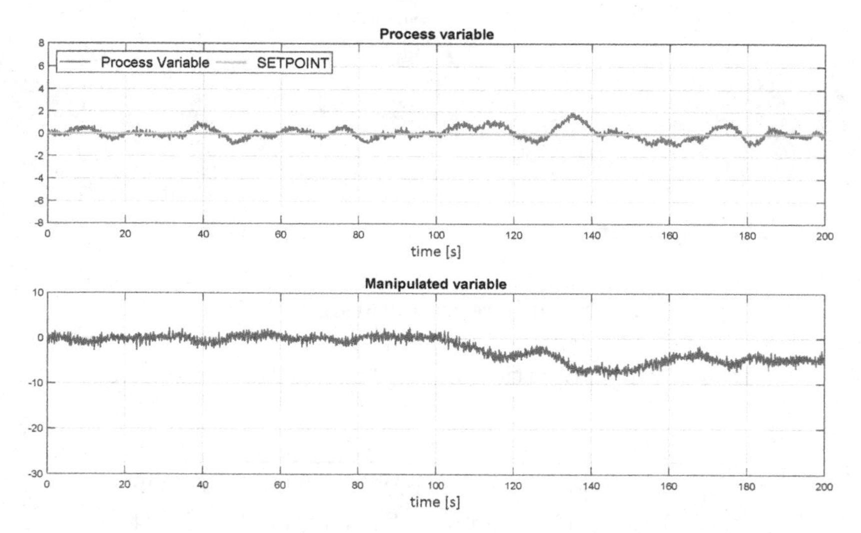

Figure 12.7: Time series for loop $G_1(s)$ with well-tuned controller

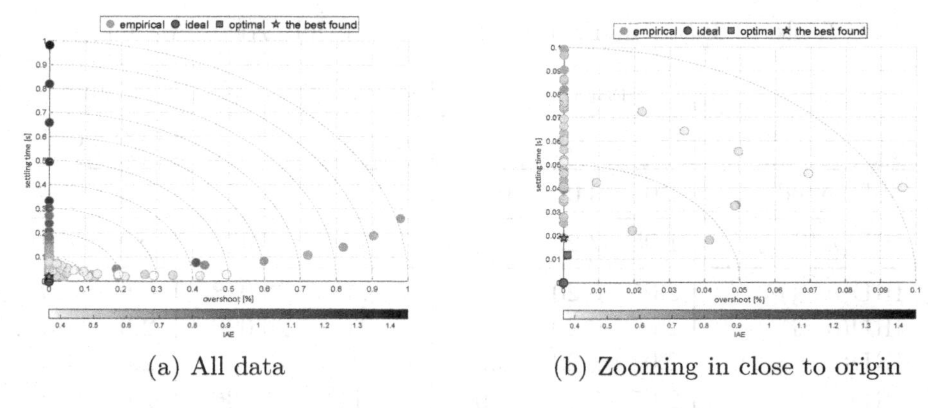

(a) All data

(b) Zooming in close to origin

Figure 12.8: IRD(κ,$\mathrm{T_{set}}$) diagram for $G_1(s)$ plant

The analysis starts with the reference diagram, combining basic dynamic step response indexes. Fig. 12.8 presents the reference IRD(κ,$\mathrm{T_{set}}$) diagram. It relates overshoot κ to the settling time $\mathrm{T_{set}}$.

We observe that points associated with various tuning are arranged in a kind of a curve, shaping the Pareto front. Optimized well-tuned controller is denoted with the red square, while the green star represents the best-found one. Circles in the diagram are shaded according to the IAE integral measure.

The best controller indicated by the diagram has the following settings: $k_p = 0.55$, $T_i = 2.20$ and $T_d = 0.9293$, for which $\kappa = 0.0\%$, $\mathrm{T_{set}} = 5.84$ sec. and MAE $= 1.212$. Generally, such a drawing allows only a simplified incomplete visualization and still we do not have any single value. Once we do the normalization, we might measure the distance from an ideal, but unachievable point $[x_0; y_0] = [0; 0]$. Scaling makes us independent of the values. The plot uses the following scaling factors: $x_{max} = 80$ and $y_{max} = 500$. They lead to the measure $d_{\mathrm{IRD}(\kappa,\mathrm{T_{set}})} = 0.013$.

As we are considering eight statistical measures, which additionally might combine with integral ones, the analysis consists of many different plots. In order to facilitate the interpretation of the results after the calculation, the obtained numbers are collected in two Tables: 12.3 and12.4. The first one summarizes relations with scale estimators, while the second other diagrams. The most interesting and representative plots are then shown and discussed in detail.

Review of the results from Table 12.3 shows a recurring pattern. Generally, detection tends to select relatively aggressive settings with

Table 12.3: Summary of IRD assessment for $G_1(s)$ – part I

	parameter			step response KPI		scaling	
	k_p	T_i	T_d	κ [%]	T_{set}	OX	OY
well-tuned	1.05	3.00	0.93	0.09	5.85	—	—
IRD(κ,T_{set})	1.05	3.20	0.9293	0.01	9.47	80	500
IRD(σ,τ_3)	2.05	6.20	0.93	7.694	20.078	2.0	1.0
IRD(σ,τ_4)	2.05	6.20	0.93	7.694	20.078	2.0	1.0
IRD(σ,d_{GPH})	2.05	2.20	0.93	39.522	13.737	2.0	1.5
IRD(σ,$\hat{\xi}$)	2.05	5.20	0.93	10.745	15.236	2.0	20.0
IRD(σ,H^{RE})	2.05	5.20	0.93	10.745	15.236	2.0	2.0
IRD($\hat{\sigma}_L$,τ_3)	2.05	3.20	0.93	23.385	12.058	2.0	1.0
IRD($\hat{\sigma}_L$,τ_4)	2.05	3.20	0.93	23.385	12.058	2.0	1.0
IRD($\hat{\sigma}_L$,d_{GPH})	2.05	2.20	0.93	39.522	13.737	2.0	1.5
IRD($\hat{\sigma}_L$,$\hat{\xi}$)	2.05	5.20	0.93	10.745	15.236	2.0	20.0
IRD($\hat{\sigma}_L$,H^{RE})	2.05	5.20	0.93	10.745	15.236	2.0	2.0
IRD(L-l_2,τ_3)	2.05	3.20	0.93	23.385	12.058	1.0	1.0
IRD(L-l_2,τ_4)	2.05	3.20	0.93	23.385	12.058	1.0	1.0
IRD(L-l_2,d_{GPH})	2.05	2.20	0.93	39.522	13.737	1.0	1.5
IRD(L-l_2,$\hat{\xi}$)	2.05	5.20	0.93	10.745	15.236	1.0	20.0
IRD(L-l_2,H^{RE})	2.05	5.20	0.93	10.745	15.236	1.0	2.0

large gain value, which in all case is the same $k_p = 2.05$. Relatively the best, however still quite different from the well-tuned controller are diagrams connecting scale estimators with tail index or rational entropy.

Therefore, only the diagrams for the L-scale l_2 are shown. The relationship with the shape L-skewness factor IRD(L-l_2,τ_3) is shown in Fig. 12.9. There are two plots included. One of them is related to overshoot, while the second one is shaded according to the settling time. The best setting, i.e. the one closest to the ideal point denoted as a blue circle is highlighted with a green star. The red square shows the well-tuned controller.

The ideal point in case of all the selected factors generally equals zero except for L-kurtosis , for which this value is $\tau_4 = 0.1226$, what relates to the normal distribution; see IRD(L-l_2,τ_4) diagram in Fig. 12.10.

Table 12.4: Summary of IRD assessment for $G_1(s)$ – part II

	parameter			step response KPI		scaling	
	k_p	T_i	T_d	κ [%]	T_{set}	OX	OY
well-tuned	1.05	3.00	0.93	0.09	5.85	—	—
IRD(κ,T_{set})	1.05	3.20	0.9293	0.01	9.47	80	500
IRD(τ_3,τ_4)	0.30	6.20	0.93	0.000	87.391	1.0	1.0
IRD(τ_3,d_{GPH})	1.80	1.20	0.93	78.232	130.007	1.0	1.5
IRD(τ_3,$\hat{\xi}$)	2.05	5.20	0.93	10.745	15.236	1.0	20.0
IRD(τ_3,H^{RE})	2.05	5.20	0.93	10.745	15.236	1.0	2.0
IRD(τ_4,d_{GPH})	1.80	1.20	0.93	78.232	130.007	1.0	1.5
IRD(τ_4,$\hat{\xi}$)	2.05	5.20	0.93	10.745	15.236	1.0	20.0
IRD(τ_4,H^{RE})	2.05	5.20	0.93	10.745	15.236	1.0	2.0
IRD(d_{GPH},$\hat{\xi}$)	1.80	1.20	0.93	78.232	130.007	1.5	20.0
IRD(d_{GPH},H^{RE})	2.05	2.20	0.93	39.522	13.737	1.5	2.0
IRD($\hat{\xi}$,H^{RE})	2.05	5.20	0.93	10.745	15.236	20.0	2.0
IRD(MAE,L-l_2)	2.05	3.20	0.93	23.385	12.058	1.5	5.0
IRD(MAE,τ_3)	2.05	3.20	0.93	23.385	12.058	1.5	1.0
IRD(MAE,τ_4)	2.05	3.20	0.93	23.385	12.058	1.5	1.0
IRD(MAE, d_{GPH})	2.05	2.20	0.93	39.522	13.737	1.5	1.1
IRD(MAE,$\hat{\xi}$)	2.05	5.20	0.93	10.745	15.236	1.5	20.0
IRD(MAE,H^{RE})	2.05	5.20	0.93	10.745	15.236	1.5	2.0

Control error time series for all considered controllers reflect persistent operation as the ARFIMA filter fractional order d_{GPH} is always positive, what is visualized in Fig. 12.11 with IRD(L-l_2,d_{GPH}) diagram.

The other diagrams show IRD(L-l_2,$\hat{\xi}$) relationship in Fig. 12.12 and IRD(L-l_2,H^{RE}) in Fig. 12.13.

The next group of diagrams is summarized in Table 12.3. These relationships use statistical shape factors, persistence and information measures, and for the sake of curiosity the integral MAE measure. In this cases we observe much higher variability in obtained results. We observe that the same relatively the best controller indicated by the scale combined with the tail index and rational entropy is also indicated in all diagrams that include these two factors. It repeats in case of L-skewness, L-kurtosis, fractional order and mean absolute error.

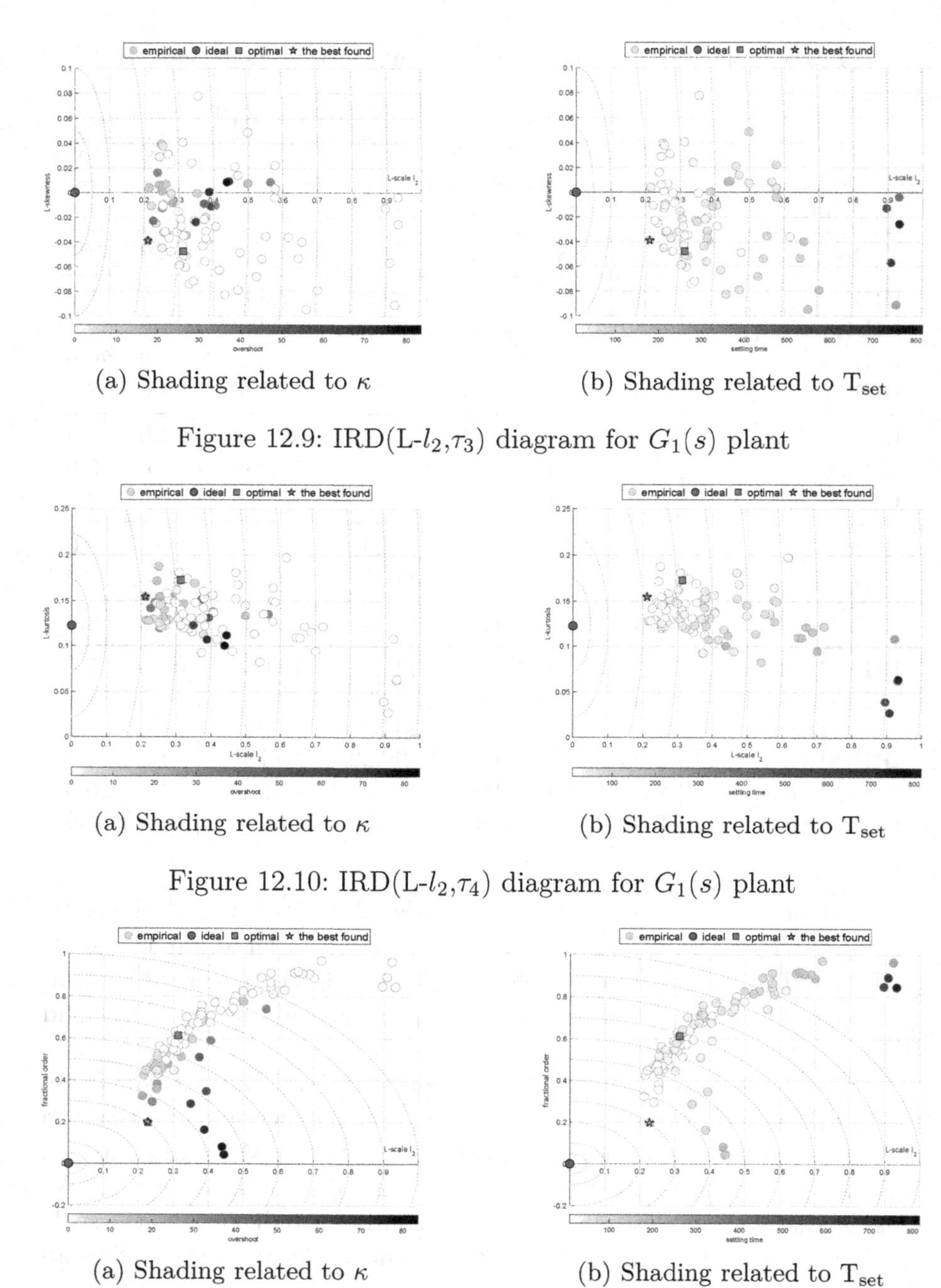

(a) Shading related to κ

(b) Shading related to T_{set}

Figure 12.9: IRD(L-l_2,τ_3) diagram for $G_1(s)$ plant

(a) Shading related to κ

(b) Shading related to T_{set}

Figure 12.10: IRD(L-l_2,τ_4) diagram for $G_1(s)$ plant

(a) Shading related to κ

(b) Shading related to T_{set}

Figure 12.11: IRD(L-l_2,d_{GPH}) diagram for $G_1(s)$ plant

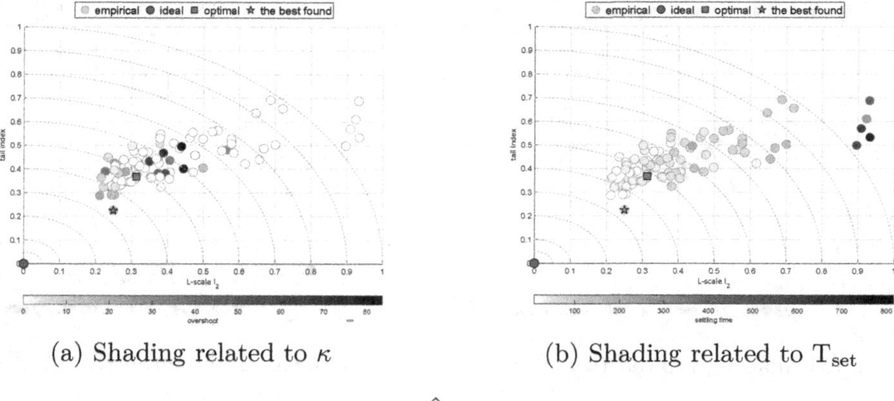

(a) Shading related to κ (b) Shading related to T_{set}

Figure 12.12: IRD(L-l_2,$\hat{\xi}$) diagram for $G_1(s)$ plant

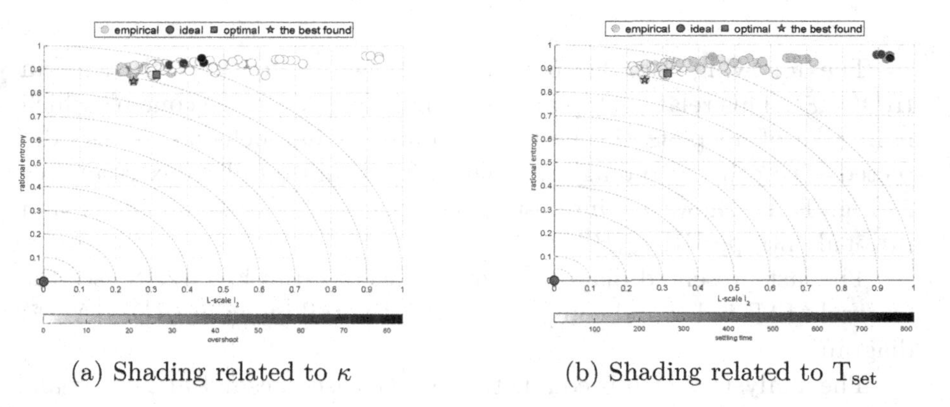

(a) Shading related to κ (b) Shading related to T_{set}

Figure 12.13: IRD(L-l_2,H^{RE}) diagram for $G_1(s)$ plant

On the opposite side lie completely unacceptable settings indicated by diagrams that include ARFIMA filter fractional order, i.e. IRD(τ_3,d_{GPH}), IRD(τ_4,d_{GPH}) and IRD($\hat{\xi}$,H^{RE}). The IRD(τ_3,τ_4) points out the PID controller that has no overshoot but highly sluggish performance.

Five diagrams are shown for the visualization purposes. Fig. 12.14 shows the IRD(τ_3,τ_4) plot which is exact L-moment ratio diagram indexL-moment ratio diagram LMRD(τ_3,τ_4). It points ot the tuning, which is the closest to the normal distribution LMRD point $(0, 0.1226)$. In this case it exhibits a zero overshoot, but seriously sluggish controller.

The plot that compares L-kurtosis with ARFIMA filter fractional order IRD(τ_3,d_{GPH}) is shown in next Fig. 12.15. It points out highly aggressive and unacceptable control.

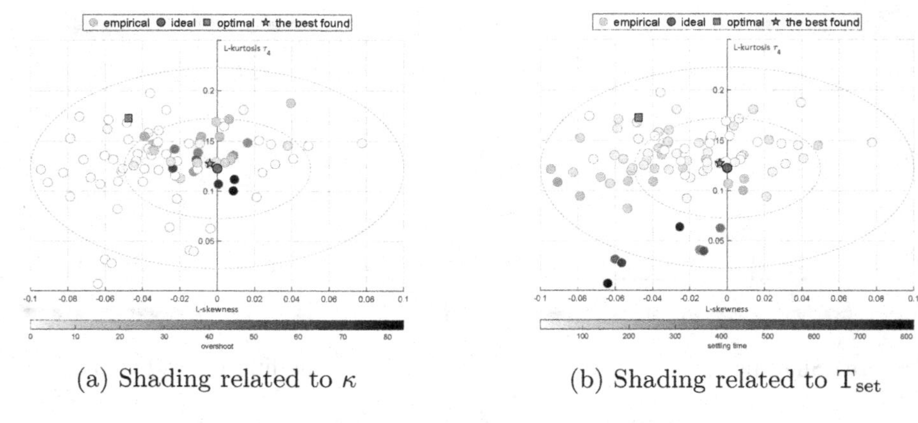

(a) Shading related to κ

(b) Shading related to T_{set}

Figure 12.14: IRD(τ_3,τ_4) diagram for $G_1(s)$ plant

The following Fig. 12.16 relates L-skewness to the tail index in IRD(τ_3,$\hat{\xi}$). This relationship allows to indicate acceptable control, which is in case of all plots the best one found. Similar behavior is also detected with other two diagrams, which incorporate tail index estimator $\hat{\xi}$. Fig. 12.17 shows an interesting combination of the tail index and rational entropy IRD($\hat{\xi}$,H^{RE}).

The last presented diagram, which is sketched in Fig. 12.18 uses the integral MAE index and compares it with the tail index in IRD(MAE,$\hat{\xi}$) diagram.

The analysis, which is conducted for one case does not allow to make any credible conclusions, therefore similar experiments are repeated for other two plants.

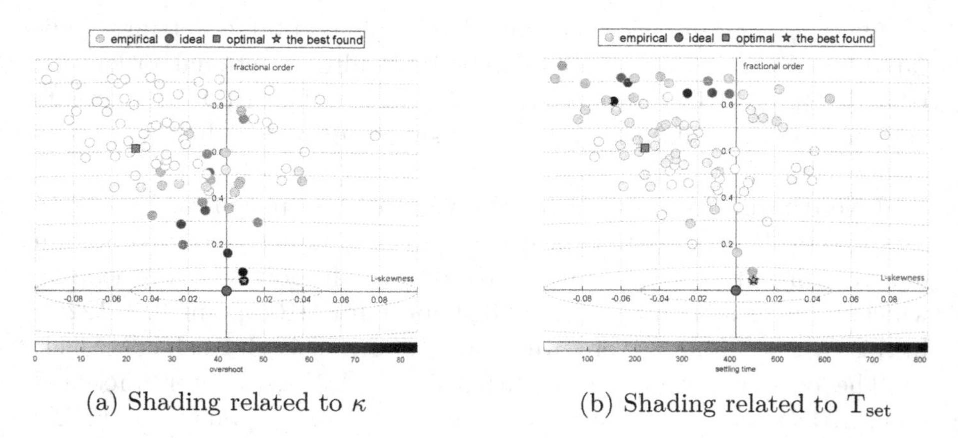

(a) Shading related to κ

(b) Shading related to T_{set}

Figure 12.15: IRD(τ_3,d_{GPH}) diagram for $G_1(s)$ plant

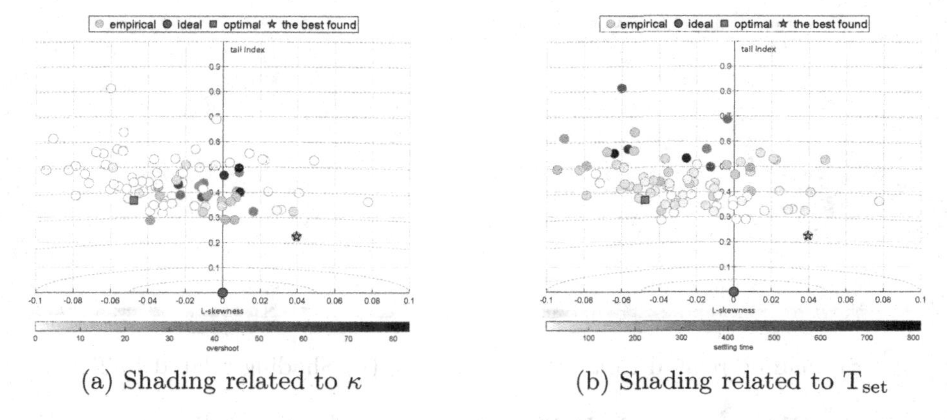

(a) Shading related to κ

(b) Shading related to $\mathrm{T_{set}}$

Figure 12.16: IRD(τ_3,$\hat{\xi}$) diagram for $G_1(s)$ plant

12.5.2 Analysis for first-order system with a dead time

The analysis continues with $G_2(s)$ transfer function, which incorporates delayed first-order plant. Fig. 12.19 shows time series data for basic loop variables of the plant $G_2(s)$ controlled by well-tuned PID . We consider constant setpoint loop operation as previously. The analysis starts with the diagram that shows relationship between basic step response indexes IRD(κ,$\mathrm{T_{set}}$) in Fig. 12.20.

We as previously observed a kind of the Pareto front. Optimized well-tuned controller is denoted with the red square, while the green star represents the best-found one. Circles in the diagram are shaded according to the IAE integral measure.

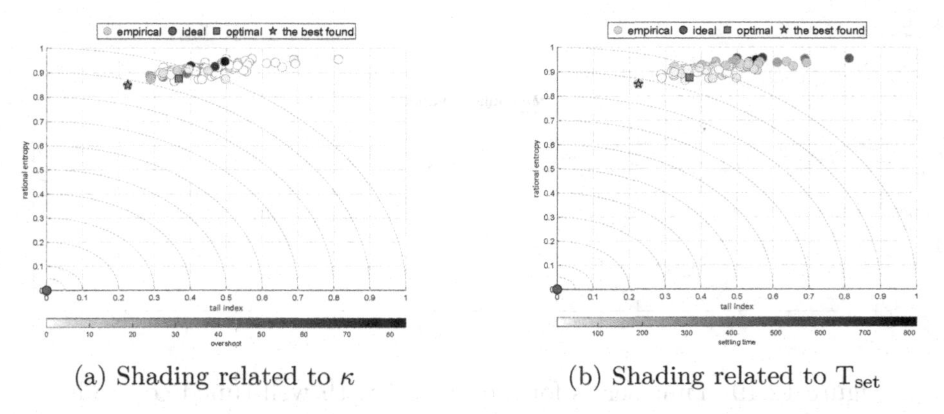

(a) Shading related to κ

(b) Shading related to $\mathrm{T_{set}}$

Figure 12.17: IRD($\hat{\xi}$,H^{RE}) diagram for $G_1(s)$ plant

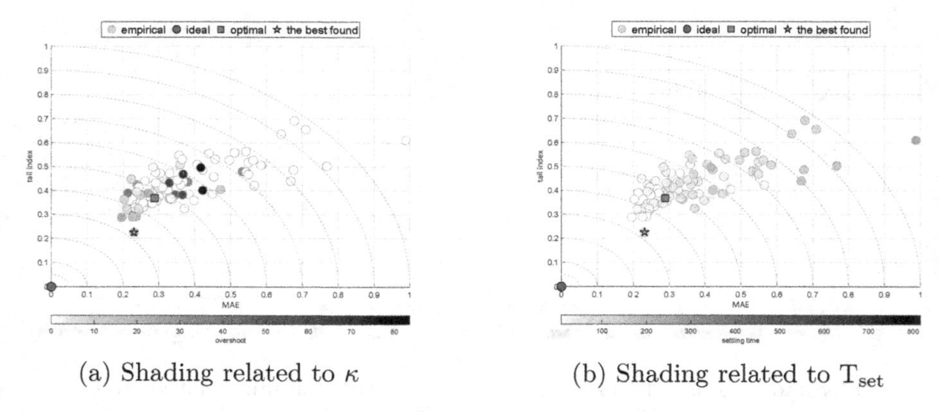

(a) Shading related to κ (b) Shading related to T_{set}

Figure 12.18: IRD(MAE,$\hat{\xi}$) diagram for $G_1(s)$ plant

The best controller indicated by the diagram obtained following settings: $k_p = 0.60$, $T_i = 1.10$ and $T_d = 0.2121$, for which $\kappa = 0.00\%$, $T_{set} = 5.84$ sec. and MAE $= 0.664$. Generally, this drawing allows only a simplified incomplete visualization and still we do not have any single value. Once we do the normalization, we might measure the distance from an ideal, but unachievable point $[x_0; y_0] = [0; 0]$. Scaling ($x_{max} = 80$ and

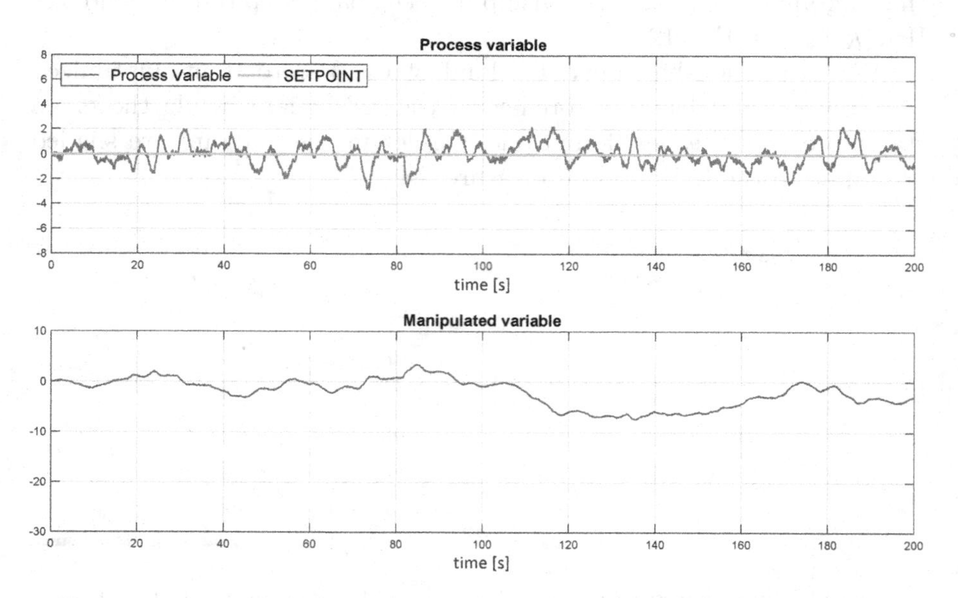

Figure 12.19: Time series for loop $G_2(s)$ with well-tuned controller

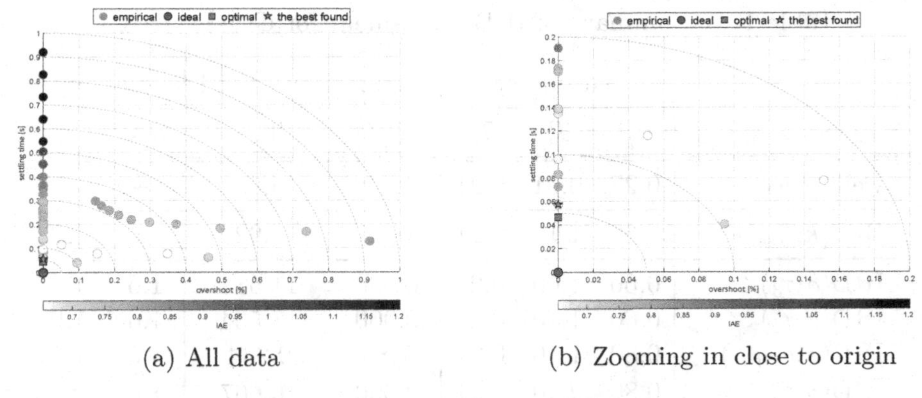

(a) All data (b) Zooming in close to origin

Figure 12.20: IRD(κ,T$_{\text{set}}$) diagram for $G_2(s)$ plant

$y_{max} = 120$) makes detection independent on the values and leads to the measure $d_{\text{IRD}(\kappa,\text{T}_{\text{set}})} = 0.041$.

The results as previously are summarized in Tables 12.5 and 12.6. The analysis of the results obtained confirms previous observations, despite the fact that, in contrast to the previous multiple-pole plant, we see much greater variation in results. The object described by the $G_1(s)$ transfer function is slow and it is difficult to obtain indications that allow greater distinction at the level of overshoot, which in most cases indicates values of zero or close to zero.

The $G_2(s)$ transfer function, due to the embedded delay, allows for a much greater variation in the received control performance which, however, makes it easier to evaluate and allows indications of PID controllers close to optimal or simply correctly tuned.

Review of the results summarized in Table 12.5 shows similar recurring pattern. We see that diagrams relating scale estimators with tail index give the best results, which are the same as shown by the reference IRD(κ,T$_{\text{set}}$) diagram. Rational entropy in case of robust scale estimator delivers the same results, while in case of other scale estimators gives higher overshoot (more aggressive control). Some other diagrams also point out zero-overshoot solutions, but with higher settling time.

An overview of the subsequent results confirms previous observations. The utilization of tail index and rational entropy in the analysis leads to the appropriate controller tuning. Also the observation about the worst indicated control is confirmed as in case of IRD(τ_3,d_{GPH}) and IRD(τ_4,d_{GPH}) and the most sluggish tuning found by IRD(τ_3,τ_4). Visualization of the results is performed with a diminished number of plots

Table 12.5: Summary of IRD assessment for $G_2(s)$ – part I

	parameter			step response KPI		scaling	
	k_p	T_i	T_d	κ [%]	T_{set}	OX	OY
well-tuned	0.27	0.61	0.21	0.00	5.59	—	—
IRD(κ,T_{set})	0.60	1.10	0.21	0.00	6.92	80	120
IRD(σ,τ_3)	0.60	1.60	0.21	0.000	11.534	1.6	1.0
IRD(σ,τ_4)	0.60	1.60	0.21	0.000	11.534	1.6	1.0
IRD(σ,d_{GPH})	0.80	1.10	0.21	27.852	9.234	1.6	1.5
IRD(σ,$\hat{\xi}$)	0.80	2.60	0.21	0.000	16.667	1.6	10.0
IRD(σ,H^{RE})	0.80	1.60	0.21	12.092	9.302	1.6	2.0
IRD($\hat{\sigma}_L$,τ_3)	0.80	1.10	0.21	27.852	9.234	1.6	1.0
IRD($\hat{\sigma}_L$,τ_4)	0.80	1.10	0.21	27.852	9.234	1.6	1.0
IRD($\hat{\sigma}_L$,d_{GPH})	0.80	1.10	0.21	27.852	9.234	1.6	1.5
IRD($\hat{\sigma}_L$,$\hat{\xi}$)	0.60	1.10	0.21	0.000	6.921	1.6	10.0
IRD($\hat{\sigma}_L$,H^{RE})	0.60	1.10	0.21	0.000	6.921	1.6	2.0
IRD(L-l_2,τ_3)	0.60	1.60	0.21	0.000	11.534	1.0	1.0
IRD(L-l_2,τ_4)	0.60	1.60	0.21	0.000	11.534	1.0	1.0
IRD(L-l_2,d_{GPH})	0.80	1.10	0.21	27.852	9.234	1.0	1.5
IRD(L-l_2,$\hat{\xi}$)	0.60	1.10	0.21	0.000	6.921	1.0	10.0
IRD(L-l_2,H^{RE})	0.80	1.60	0.21	12.092	9.302	1.0	2.0

comparing to the previous $G_1(s)$ plant. Two diagrams that incorporate scaling are presented in Figs. 12.21 and 12.22. They show IRD($\hat{\sigma}_L$,d_{GPH}) and IRD($\hat{\sigma}_L$,$\hat{\xi}$), respectively. Both plots indicate a very good PID tuning.

Two other plots present the LMRD relation, i.e. IRD(τ_3,τ_4) in Fig. 12.23. As previously the LMRD diagram indicates very sluggish and no overshoot controller performance, while Fig. 12.24 with IRD(τ_3,d_{GPH}) diagram exemplifies the worst tuning indication case. Two final diagrams, i.e. IRD(d_{GPH},$\hat{\xi}$) and IRD(d_{GPH},H^{RE}), which are shown in Figs. 12.25 and 12.26 present appropriate tuning indications. They are still achieved with use of the tail index and rational entropy.

At the end of this section it's worth to notice that the use of normal standard deviation estimator σ gives worse (inconsistent) results than its robust equivalents: M-estimator with logistic function $\hat{\sigma}_L$ or L-l_2.

Table 12.6: Summary of IRD assessment for $G_2(s)$ – part II

	parameter			step response KPI		scaling	
	k_p	T_i	T_d	κ [%]	$\mathrm{T_{set}}$	OX	OY
well-tuned	0.27	0.61	0.21	0.00	6.92	—	—
IRD(κ,$\mathrm{T_{set}}$)	0.60	1.10	0.21	0.00	6.92	80	120
IRD(τ_3,τ_4)	0.40	5.10	0.21	0.000	60.572	1.0	1.0
IRD(τ_3,d_{GPH})	1.00	1.10	0.21	58.979	20.226	1.0	1.5
IRD(τ_3,$\hat{\xi}$)	0.80	2.60	0.21	0.000	16.667	1.0	10.0
IRD(τ_3,H^{RE})	0.80	3.10	0.21	0.000	20.477	1.0	2.0
IRD(τ_4,d_{GPH})	1.00	1.10	0.21	58.979	20.226	1.0	1.5
IRD(τ_4,$\hat{\xi}$)	0.80	2.60	0.21	0.000	16.667	1.0	10.0
IRD(τ_4,H^{RE})	0.80	3.10	0.21	0.000	20.477	1.0	2.0
IRD(d_{GPH},$\hat{\xi}$)	0.60	1.10	0.21	0.000	6.921	1.5	10.0
IRD(d_{GPH},H^{RE})	0.60	1.10	0.21	0.000	6.921	1.5	2.0
IRD($\hat{\xi}$,H^{RE})	0.80	3.10	0.21	0.000	20.477	10.0	2.0
IRD(MAE,L-l_2)	0.80	1.10	0.21	27.852	9.234	1.5	5.0
IRD(MAE,τ_3)	0.80	1.10	0.21	27.852	9.234	1.5	1.0
IRD(MAE,τ_4)	0.80	1.10	0.21	27.852	9.234	1.5	1.0
IRD(MAE, d_{GPH})	0.80	1.10	0.21	27.852	9.234	1.5	1.1
IRD(MAE,$\hat{\xi}$)	0.60	1.10	0.21	0.000	6.921	1.5	10.0
IRD(MAE,H^{RE})	0.80	1.60	0.21	12.092	9.302	1.5	2.0

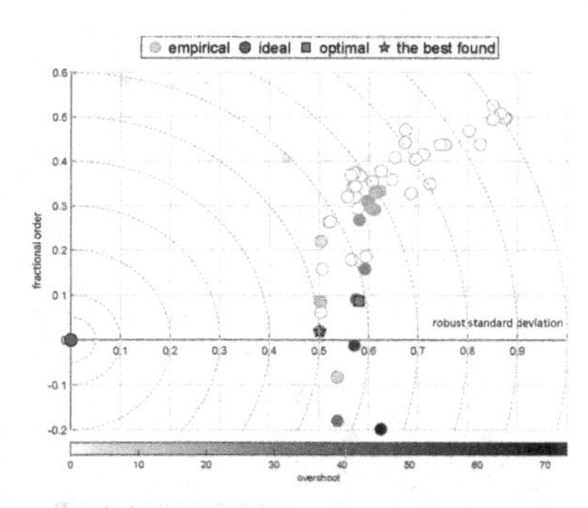

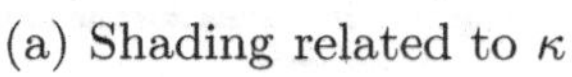

(a) Shading related to κ

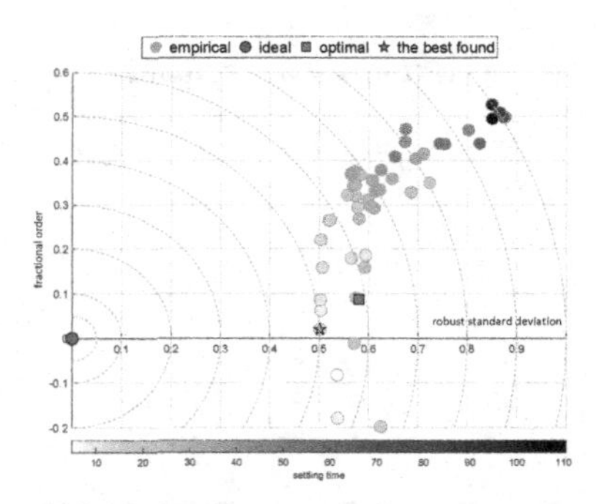

(b) Shading related to $\mathrm{T_{set}}$

Figure 12.21: IRD($\hat{\sigma}_{\mathrm{L}}$,$d_{\mathrm{GPH}}$) diagram for $G_2(s)$ plant

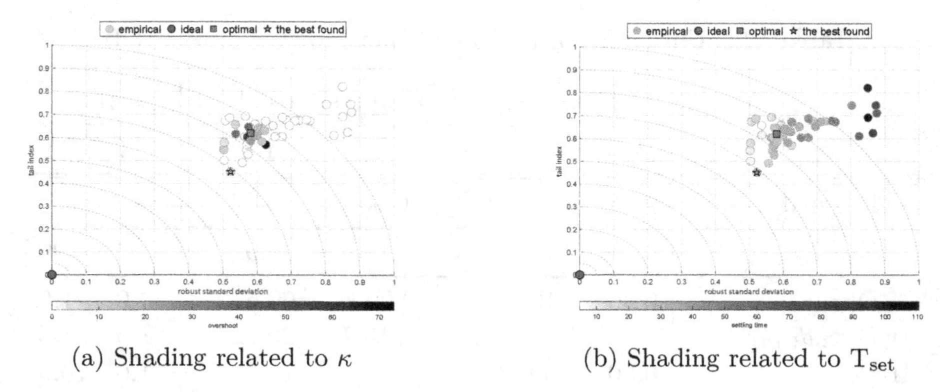

(a) Shading related to κ

(b) Shading related to $\mathrm{T_{set}}$

Figure 12.22: IRD($\hat{\sigma}_{\mathrm{L}}$,$\hat{\xi}$) diagram for $G_2(s)$ plant

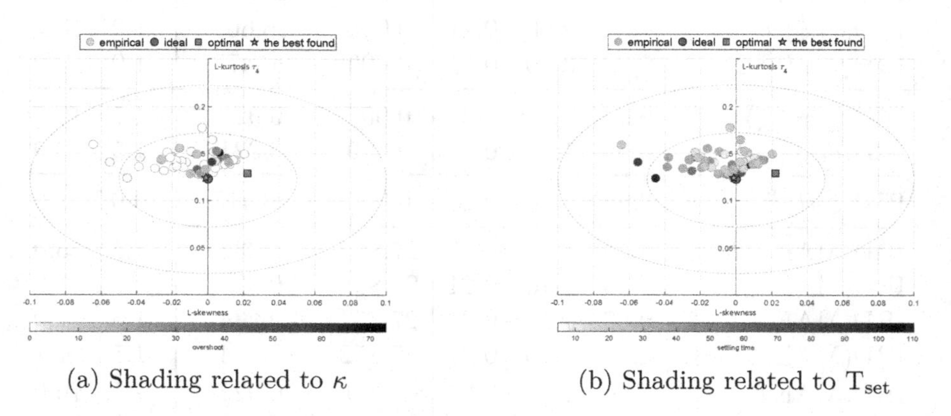

(a) Shading related to κ

(b) Shading related to $\mathrm{T_{set}}$

Figure 12.23: IRD(τ_3,τ_4) diagram for $G_2(s)$ plant

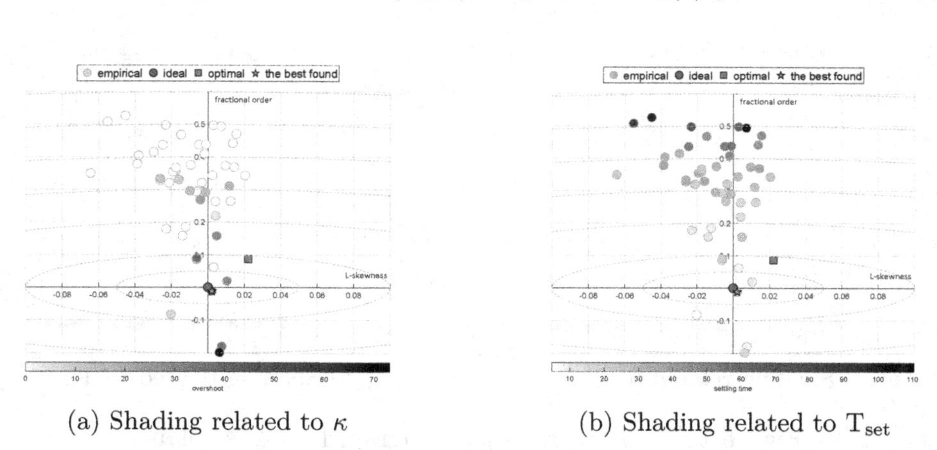

(a) Shading related to κ

(b) Shading related to $\mathrm{T_{set}}$

Figure 12.24: IRD(τ_3,d_{GPH}) diagram for $G_2(s)$ plant

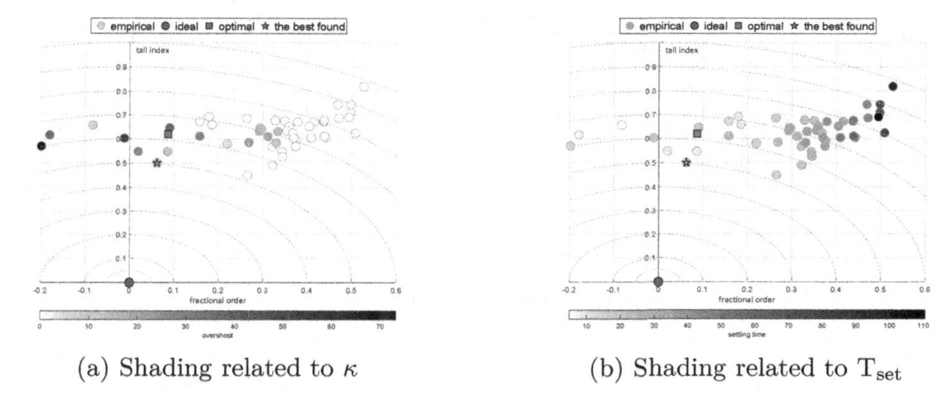

(a) Shading related to κ (b) Shading related to $\mathrm{T_{set}}$

Figure 12.25: IRD(d_{GPH},$\hat{\xi}$) diagram for $G_2(s)$ plant

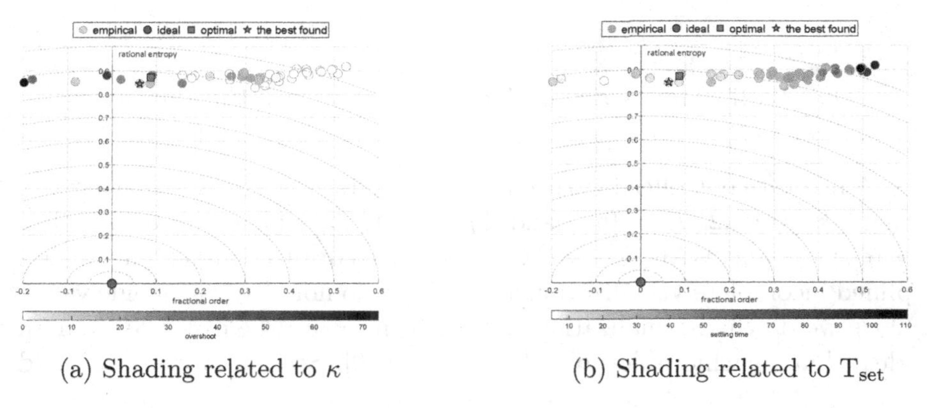

(a) Shading related to κ (b) Shading related to $\mathrm{T_{set}}$

Figure 12.26: IRD(d_{GPH},H^{RE}) diagram for $G_2(s)$ plant

12.5.3 Analysis for system with two modes: fast and slow

Simulations finish with $G_3(s)$ transfer function, which uses a plant with fast and slow modes. Fig. 12.27 shows time series data for basic loop variables of the plant $G_2(s)$ controlled by well-tuned PID . We consider constant setpoint loop operation as previously. The analysis starts with the diagram that shows relationship between basic step response indexes IRD(κ,$\mathrm{T_{set}}$) in Fig. 12.28.

As previously, we observe a kind of the Pareto front. Optimized well-tuned controller is denoted with the red square, while the green star represents the best found one. Circles in the diagram are shaded according to the IAE integral measure.

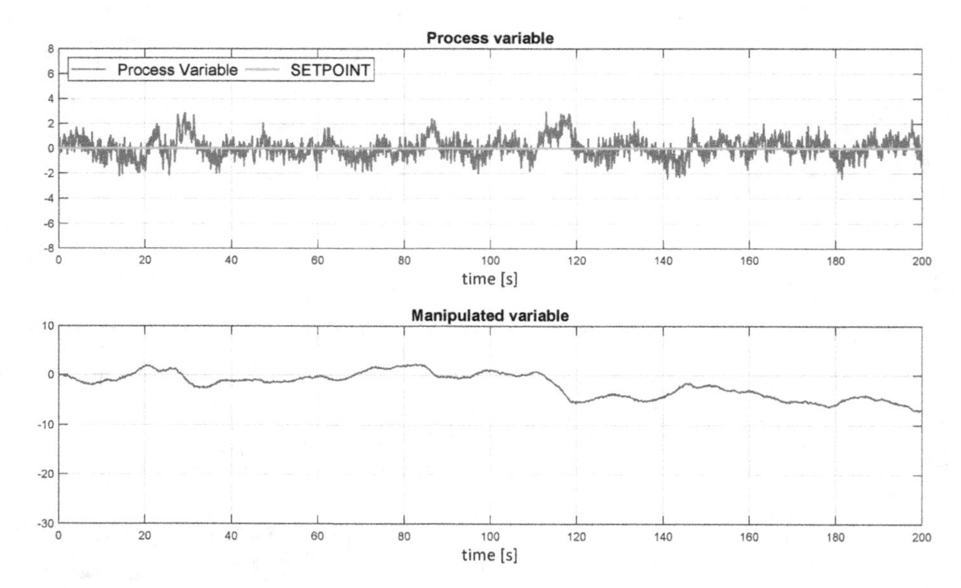

Figure 12.27: Time series for loop $G_3(s)$ with well-tuned controller

The best controller indicated by the diagram obtained following settings: $k_p = 0.42$, $T_i = 0.80$ and $T_d = 0.08$, for which $\kappa = 0.30\%$, $\text{T}_{\text{set}} = 6.75$ sec. and MAE = 0.631. Generally, this drawing allows only a simplified incomplete visualization and still we do not have any single value. Once we do the normalization, we might measure the distance from an ideal, but unachievable point $[x_0; y_0] = [0; 0]$. Scaling ($x_{max} = 80$ and

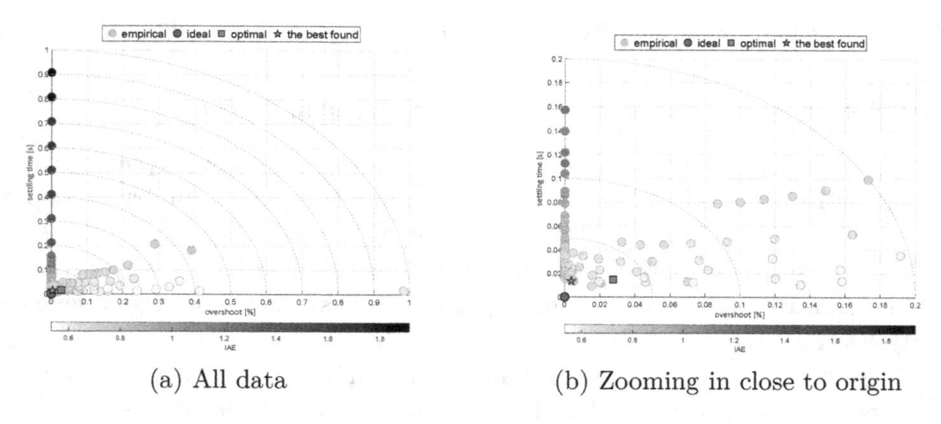

(a) All data

(b) Zooming in close to origin

Figure 12.28: IRD(κ,T_{set}) diagram for $G_3(s)$ plant

Table 12.7: Summary of IRD assessment for $G_3(s)$ – part I

	parameter			step response KPI		scaling	
	k_p	T_i	T_d	κ [%]	T_{set}	OX	OY
well-tuned	0.12	0.28	0.08	2.20	7.43	—	—
IRD(κ,T_{set})	0.42	0.80	0.08	0.30	6.75	80	500
IRD(σ,τ_3)	0.62	0.30	0.08	23.435	9.903	2.1	1.0
IRD(σ,τ_4)	0.62	0.30	0.08	23.435	9.903	2.1	1.0
IRD(σ,d_{GPH})	0.72	0.55	0.08	12.587	11.577	2.1	1.5
IRD(σ,$\hat{\xi}$)	0.72	0.80	0.08	6.536	11.403	2.1	10.0
IRD(σ,H^{RE})	0.12	0.05	0.08	33.092	6.872	2.1	2.0
IRD($\hat{\sigma}_L$,τ_3)	0.62	0.30	0.08	23.435	9.903	2.2	1.0
IRD($\hat{\sigma}_L$,τ_4)	0.62	0.30	0.08	23.435	9.903	2.2	1.0
IRD($\hat{\sigma}_L$,d_{GPH})	0.72	0.55	0.08	12.587	11.577	2.2	1.5
IRD($\hat{\sigma}_L$,$\hat{\xi}$)	0.72	0.80	0.08	6.536	11.403	2.2	10.0
IRD($\hat{\sigma}_L$,H^{RE})	0.12	0.05	0.08	33.092	6.872	2.2	2.0
IRD(L-l_2,τ_3)	0.62	0.30	0.08	23.435	9.903	1.2	1.0
IRD(L-l_2,τ_4)	0.62	0.30	0.08	23.435	9.903	1.2	1.0
IRD(L-l_2,d_{GPH})	0.72	0.55	0.08	12.587	11.577	1.2	1.5
IRD(L-l_2,$\hat{\xi}$)	0.72	0.80	0.08	6.536	11.403	1.2	10.0
IRD(L-l_2,H^{RE})	0.12	0.05	0.08	33.092	6.872	1.2	2.0

$y_{max} = 500$) makes detection independent of the values and leads to the measure $d_{IRD(\kappa,T_{set})} = 0.010$.

The results are summarized in Tables 12.7 and 12.8. The analysis of the results obtained confirms previous observations, with one exception as in this case only the tail index estimator allow quite proper detection.

Review of Table 12.7 shows similarities, but it's impossible to point out the well-tuned controller or the one indicated in the IRD(κ,T_{set}) plot. Only diagrams that use tail index estimator or its combination with rational entropy allow for reliable assessment. An interesting solution is given in case of combination of the fractional order estimator with the tail index, rational entropy or L-kurtosis. Still the LMRD diagram IRD(τ_3,τ_4) shows a very sluggish controller setup. Graphical presentation is limited to five characteristic diagrams. Fig. 12.29 shows the IRD(L-l_2,$\hat{\xi}$) diagram, which relates scale estimator with tail index. Fig. 12.30 shows the LMRD, i.e. IRD(τ_3,τ_4) plot with a very sluggish control. An

Table 12.8: Summary of IRD assessment for $G_3(s)$ – part II

	parameter			step response KPI		scaling	
	k_p	T_i	T_d	κ [%]	T_{set}	OX	OY
well-tuned	0.12	0.28	0.08	2.20	7.43	—	—
IRD(κ,T_{set})	0.42	0.80	0.08	0.30	6.75	80	500
IRD(τ_3,τ_4)	1.02	0.55	0.08	23.106	103.607	1.0	1.0
IRD(τ_3,d_{GPH})	1.02	0.80	0.08	17.207	59.686	1.0	1.5
IRD(τ_3,$\hat{\xi}$)	0.72	0.80	0.08	6.536	11.403	1.0	10.0
IRD(τ_3,H^{RE})	0.82	1.05	0.08	6.133	15.635	1.0	2.0
IRD(τ_4,d_{GPH})	0.62	0.55	0.08	9.605	7.907	1.0	1.5
IRD(τ_4,$\hat{\xi}$)	0.72	0.80	0.08	6.536	11.403	1.0	10.0
IRD(τ_4,H^{RE})	0.82	1.05	0.08	6.133	15.635	1.0	2.0
IRD(d_{GPH},$\hat{\xi}$)	0.72	0.80	0.08	6.536	11.403	1.5	10.0
IRD(d_{GPH},H^{RE})	0.72	0.80	0.08	6.536	11.403	1.5	2.0
IRD($\hat{\xi}$,H^{RE})	0.72	0.80	0.08	6.536	11.403	10.0	2.0
IRD(MAE,L-l_2)	0.62	0.30	0.08	23.435	9.903	1.2	5.0
IRD(MAE,τ_3)	0.62	0.30	0.08	23.435	9.903	1.2	1.0
IRD(MAE,τ_4)	0.62	0.30	0.08	23.435	9.903	1.2	1.0
IRD(MAE, d_{GPH})	0.72	0.55	0.08	12.587	11.577	1.2	1.1
IRD(MAE,$\hat{\xi}$)	0.42	0.30	0.08	12.664	6.806	1.2	20.0
IRD(MAE,H^{RE})	0.12	0.05	0.08	33.092	6.872	1.2	2.0

interesting IRD(τ_4, d_{GPH}) plot is shown in Fig. 12.31 with fast control. Two last figures present the IRD(d_{GPH},$\hat{\xi}$) diagram in Fig. 12.32 and the IRD($\hat{\xi}$,H^{RE}) diagram in Fig. 12.33.

12.6 CONCLUDING REMARKS ON CPA

Above simulations consider quite artificial configuration, which does not occur in real work. We never have possibility to measure various performance for a single loop. No one will allow it, nor can anyone afford it. The purpose of the above exercise is quite different.

First, it was to compare the various possible measures taken from statistics, especially those not previously used, i.e. robust and using knowledge of the tails of distributions. Secondly, it was to indicate the

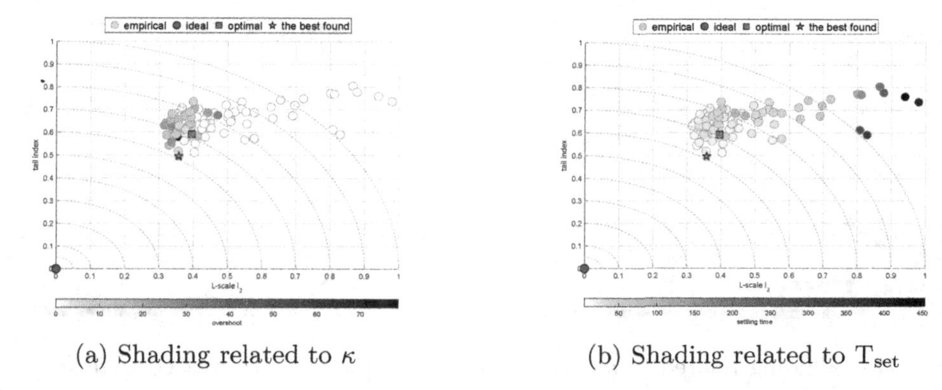

(a) Shading related to κ (b) Shading related to $\mathrm{T_{set}}$

Figure 12.29: IRD(L-l_2,$\hat{\xi}$) diagram for $G_3(s)$ plant

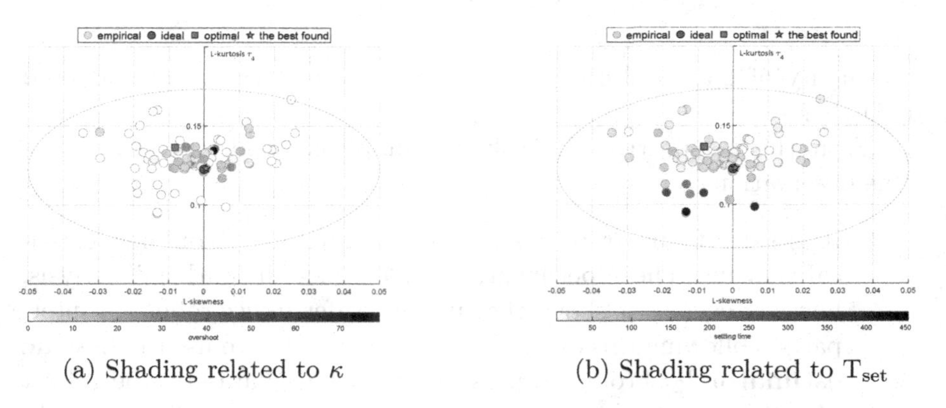

(a) Shading related to κ (b) Shading related to $\mathrm{T_{set}}$

Figure 12.30: IRD(τ_3,τ_4) diagram for $G_3(s)$ plant

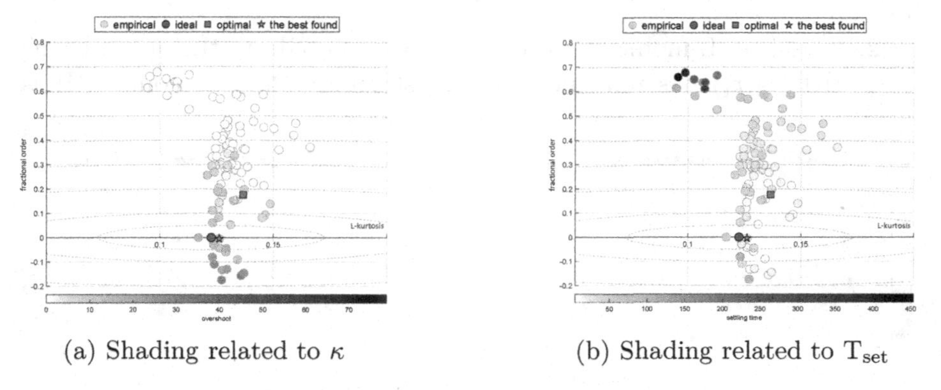

(a) Shading related to κ (b) Shading related to $\mathrm{T_{set}}$

Figure 12.31: IRD(τ_4, d_{GPH}) diagram for $G_3(s)$ plant

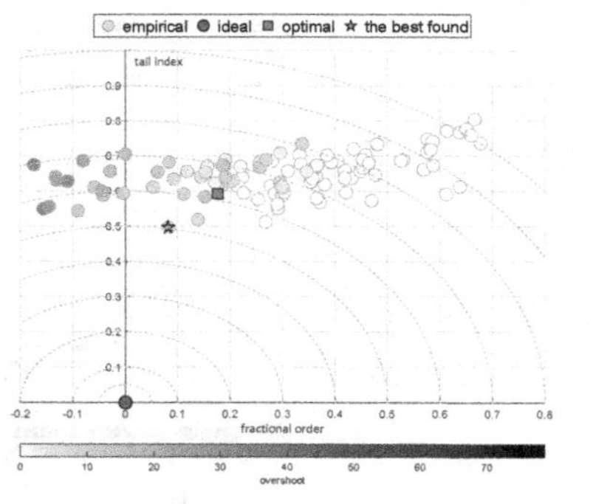

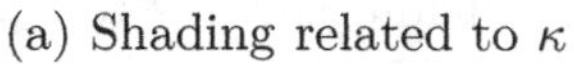

(a) Shading related to κ

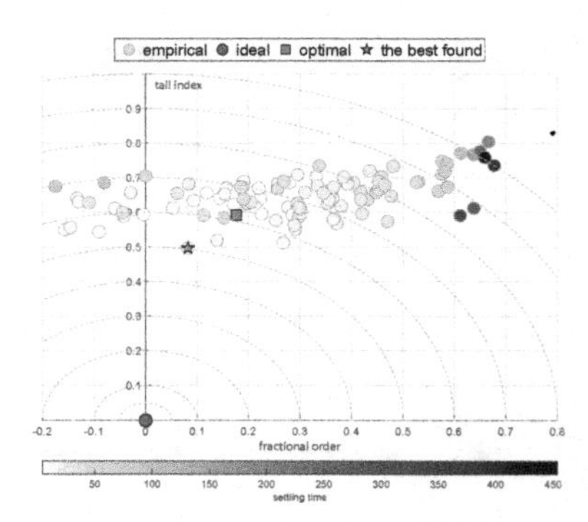

(b) Shading related to $\mathrm{T_{set}}$

Figure 12.32: IRD(d_{GPH},$\hat{\xi}$) diagram for $G_3(s)$ plant

rationality of multi-criteria presentation of results using the proposed IRD charts.

From this perspective, the task is accomplished and it provides valuable observations:

1. It is safer to use robust statistical estimators, than their classical versions. The experiment, did not utilize the outliers' analysis, however even so it shows that it's just safer to use robust counterparts. Following this observation. it is suggested to use the **L-scale estimator $\boldsymbol{\lambda_2}$**. From the perspective of long datasets, the robust M-estimators are similarly good, but in case of small samples L-moments give higher reliability.

2. Classical L-moment ratio diagram denoted in the analysis as IRD(τ_3,τ_4) has tendency to suggest highly sluggish control. It is

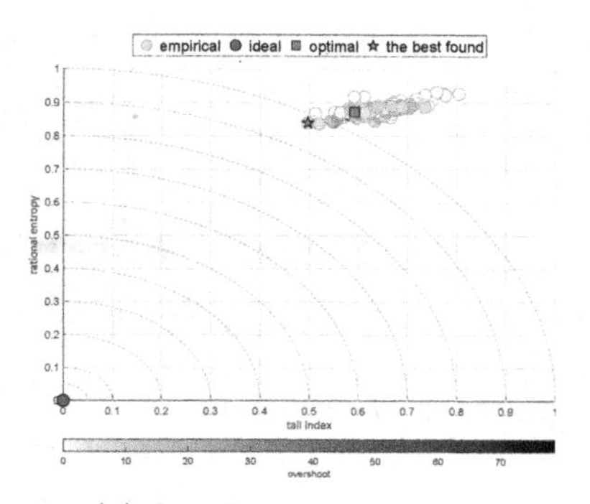

(a) Shading related to κ

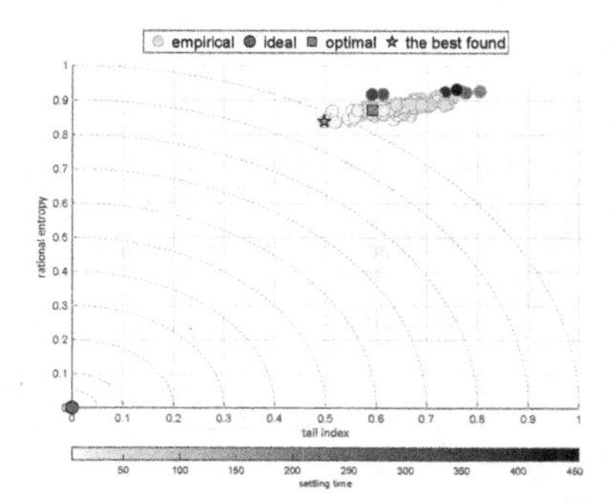

(b) Shading related to $\mathrm{T_{set}}$

Figure 12.33: IRD($\hat{\xi}$,H^{RE}) diagram for $G_3(s)$ plant

probably due to the fact that such controller filters or undesirable data producing in this way control error closer to the normally distributed data. This observation suggests also another hypothesis. The control error of well-tuned controller does not have to purely Gaussian, i.e. described by point $(o, 0.1226)$ in the LMRD plane. We may assume that the good control does not exhibit normal time series properties. It is inline with previous works showing that majority of control data, even those properly tuned is not Gaussian in industry [77, 79]. There is uneasy question what that value should be, but unfortunately it probably depends on the plant. That hypothesis means that good control is not addressed by L-kurtosis $\tau_4 = 0.1226$, while for the sake of symmetric behavior the L-skewness should be equal to zero $\tau_3 = 0.0$.

3. The analysis shows to favorite measures, which behave in a correct, repeatable and consistent manner. These are **tail index estimator** $\boldsymbol{\hat{\xi}}$ and **rational entropy** $\boldsymbol{H^{\mathrm{RE}}}$. They help in distinguishing relatively good control, with minimized overshoot and not too high settling time.

4. The Geweke Porter-Hudak estimator of the ARFIMA filter **fractional order** $\boldsymbol{d_{\mathrm{GPH}}}$ offers one advantage of distinguishing sluggish/aggressive control, but should be used with cautiousness. This index in some cases, especially with small variability between step response indexes might be biased, therefore it is suggested to be used separately, not in the IRD framework.

5. It is useful to use the symmetricity measure, like **L-skewness** $\boldsymbol{\tau_3}$, once the nonlinearities are expected.

6. It is suggested to use the IRD plots that include above proposed indexes, in particular:

   - IRD(L-l_2,τ_3),
   - IRD(L-l_2,$\hat{\xi}$),
   - IRD(L-l_2,H^{RE}),
   - IRD(τ_3,$\hat{\xi}$),
   - IRD(τ_3,H^{RE}),
   - IRD($\hat{\xi}$,H^{RE}),
   - IRD(d_{GPH},$\hat{\xi}$),
   - IRD(d_{GPH},H^{RE}).

7. Final comment seems to be not promising, but is very important. Once we have normal assessment procedure we get one point in the IRD plane. Such a point, except the planes using fractional order d_{GPH}, are not wery informative. The diagrams start be play an important role once we get more points, like for instance situations *before* and *after*, work in different operating regimes, or a comparison between several different loops. Incorporation of a time and data from variable operating periods leads to the sustainability analysis, which is discussed in the following chapter.

It should be noted that the above analysis is neither complete nor perfect. It is not complete, because it shows only a part of the tasks carried out during actual industrial feasibility study projects – it addresses possible measures for evaluating the performance of the control system and how the information can be presented against two/three criteria. It does not consider signal preprocessing, stationarity analysis, persistence, decomposition of signals into basic components, regression analysis, nor homogeneity, neither causality.

It is not perfect, because simulations, even the most refined, will not take into account the richness of reality despite all efforts.

Control performance assessment task plays a very important, however frequently undervalued, role i control engineering. We do not know what to do without an information whether the loop operates good or bad. Reality is not perfect. Many loops work far away from their proper settings. An engineer requires respective assessment procedures and performance measures. As we do have plenty of them, the choice is not clear.

We propose to use robust measures that are not biased by outlying observations. They need to incorporate knowledge about non-Gaussian behavior. They must be tail-aware.

We rarely have one single loop in an industrial plant. Control system constitutes of many loops and thus a global view of the entire installation is needed. We need to be able to perform a root-cause analysis and find bottlenecks. Control system rehabilitation should be organized and must start from the improvement of the loops, which are limiting an entire installation.

Finally, the CPA task is only a part of an overall procedure, which should not only measure the loop variables, but also perform time series analysis and preprocessing, and take into account time.

GLOSSARY

I&C: Instrumentation and Control

KPI: Key Performance Indicator

SCADA: System Control Alarming and Data Acquisition

SISO: Single Input Single Output

MSE: Mean Square Error

MAE: Mean Absolute Error

ITAV: Integral Time Absolute Value

ISTC: Integral of Square Time derivative of the Control input

TSV: Total Squared Variation

AMP: Amplitude index

IRD: Index Ratio Diagram

PV: Process Variable

CV: Controlled Variable

MV: Manipulated Variable

SαS: Symmetric α-Stable

ITAE: Integral of Time-weighted Absolute Error

LMRD: L-Moment Ratio Diagram

CHAPTER 13

Time matters – control sustainability

NOWADAYS, the concept of sustainability is being said by everyone and in all cases. It plays an important role in the development of our society, especially once environmental issues are regarded [106, 317]. It is defined as a term for accomplishing the needs of the present without compromising the ability of future generations to meet their own needs. We use this notion in the engineering context as well, though from different perspectives.

Respective regulations and mitigation strategies are investigated especially in the management and a sustainable development area [209]. The research also addresses the subject, how certain control approach impacts the environment [66, 122, 209]. These are very high-level considerations, while the idea can be also applied at low levels, i.e. at the level of control algorithms. The well target is to design and implement proper control strategy, which will meet process and environmental sustainability goals.

Simultaneously these already achieved goals have to be maintained for a long time (longer the better) despite the presence of all uncertainties and the impact of interference. It is also the same concept of the sustainability, but implemented at the very low, control algorithm level. The performance must be kept for a long time at least at the same values as just after the commissioning. We cannot afford its too fast degradation.

Control system sustainability is crucial from the practical perspective though seldom addressed in the research. Time matters. The installation varies, its performance changes in time. This effect is a daily practice for

DOI: 10.1201/9781003604709-13

the field automation engineer. If we have to implement a control system or tune it, we don't want to have to quickly return to the site to make tuning. The control system being designed must adapt itself to changing operating conditions. There are two ways to accomplish this task. In the case of multi-loop control systems, we try to achieve this by using robust algorithms. The PID controller is well suited for this, which is why it continues to enjoy unflagging popularity.

The second approach is to use adaptive algorithms, and here we enter the field of advanced process control (APC). Especially in the latter case, the sustainability of the performance is critical. This is due to the much higher cost of such a solution, the significantly increased implementation time, and what has recently become increasingly apparent the limited number of people who can do it.

A sustainable control system that does not require constant supervision from this perspective is superior even to more perfect, optimal, solutions, which, however, need constant observation. In the APC market, there have been companies that offered magical solutions that worked great when the supervising engineer was on site and quickly lost results when he was not. Interestingly, analogous situations are now resurrecting with magical, solve-all-the-problems black boxes with artificial intelligence.

Trust matters, both in people and control solutions.

This work aims at proposing a solution that might help in that task. It has been already validated in the industrial conditions and gives promising results [98]. The idea is to transfer the term of the homogeneity from life sciences and hydrology [153] into control engineering. Its original formulation uses the L-moment ratio diagrams (LMRD), however it might be naturally incorporated into the index ratio diagram (IRD) framework.

Homogeneity assessment approaches are successfully used in regional frequency analysis in life sciences, especially in hydrology. In contrary, similar applications are not existent in control engineering research context. The lack of control applications does not mean that it's infeasible or impossible. Statistical moments, like a mean, standard deviation, variance, skewness and kurtosis or other statistical concepts are commonly utilized during the CPA activities [170, 364, 10, 97]. Robust L-moments might be considered as a promising alternative, because they are not affected by outliers and can be used in small sample time series.

The possibility to present a given data properties as a single point in the LMRD plane [259] allows further applications. Among others, we may investigate empirical stochastic process and compare it with known theoretical distributions. In case of many time series data originating from different sources, we may compare them, assess their homogeneity or identify discordant data. We may solve this task with different techniques. Practitioners may even use simple visual data inspection.

This approach aims at different use of the LMRD concept. The idea is to include time into the analysis. Time brings forward new opportunity – new degree of freedom. Visualization of time-varying statistical properties enables to see and measure their evolution. In a steady and stationary situation statistical properties remain constant and the point, which corresponds to these data does not change its position. The variable sustains its statistical properties. If the variable describes control loop, it means that the loop sustains its properties (performance) as well. In such a way, representation of a point, its position and eventual movement in the LMRD plane represents loop performance evolution. Therefore it measures its sustainability, as we may observe the direction in which it travels. Such a technology may extend currently available best practice methodologies [313].

The movement of a point in two-dimensional LMRD plane fulfills simple visualization expectations of the sustainability phenomenon. However, we may get even more. We can develop a new quantitative measure, a number representing the point (data) travel range. If the point does not leave the area defined by this range, we may say that the data sustains its properties – the loop sustains its performance. If the point is found outside of that area, we may signal that some improvement actions are required. At least we inform that somebody should take a closer look at the loop.

Natural tool to realize that task is to apply a homogeneity test. However, the outcome of the test might not be practical enough. It will inform that the loop is homogeneous over time (good – it sustains its performance) or heterogeneous (bad – its performance degrades). However, in the second case the reason remains unknown and we do not proper actions to be done. Discordance analysis gives suitable answers.

When none discordant is found (all time series are inside the homogeneous region of properties) considered loop sustains its performance. Once any discordant is detected, we get information which one it is and how far it goes beyond the acceptable sustainability area. Discordance analysis allows to trace loop performance, perform root-cause analysis and find a solution.

To evaluate the homogeneity (sustainability) area we may also introduce the concept of the concentration area. We propose a dedicated measure. We start with the evaluation of an area center. The calculation algorithm should be robust and therefore we do not use simple center of area (mean) approach. We propose to use two-dimensional geometric median (GeoMed) [104] $\overline{x}_{\mathrm{GM}}$, which originates from GIS applications. It is formulated as a value of the argument x_0 from where the sum of all points x_i Euclidean distances is minimized

$$\overline{x}_{\mathrm{GM}} = \underset{x_0 \in \mathbb{R}^m}{\arg\min} \sum_{i=1}^{N} \|x_i - x_0\|_2, \tag{13.1}$$

where N denotes number of points.

We calculate the GeoMed center using Weiszfeld's algorithm [349]. Once we know the position of GeoMed data center, the respective concentration point range of can be obtained. We propose to use two-dimensional modification of the MADAM measure. Thus, we obtain concentration range r_{cc} and respective maximum distance r_{max}. They enable to compare different points (loop variables realizations) in a robust way

$$r_{\mathrm{cc}} = 1.5 \cdot \underset{i}{\mathrm{median}} \left| x_i - \overline{x}_{\mathrm{GM}} \right|, \tag{13.2}$$

$$r_{\mathrm{max}} = \max_{i} \left| x_i - \overline{x}_{\mathrm{GM}} \right|. \tag{13.3}$$

The sustainability assessment should be performed in a repeatable and easy-to-follow way. It might be done in two ways: on the continuous basis using dedicated software integrated with the plant control system or as a single-shot custom project. Fig. 13.1 presents the diagram, how it might be done. The procedure distinguishes between a dedicated project and on-line operation.

Two steps of the proposed procedure must be explained as they are crucial for its successful operation. First of all, we have to guarantee that we will compare apples with apples. The time series for selected periods must fulfill two requirements. They must reflect enough distance in time to address time sustainability. From the control system perspective probably monthly basis would be enough, though weekly timeline will work as well. Longer periods might delay control personnel reaction and the according to rehabilitation plan.

The second issue is more important. The data must be comparable and consistent. i.e. within each selected period (week, month) they must adhere to the same (similar) operating regimes (no startups/shutdowns, the same installation load, product composition, ambient conditions, etc). As the system might have different dynamics according to the

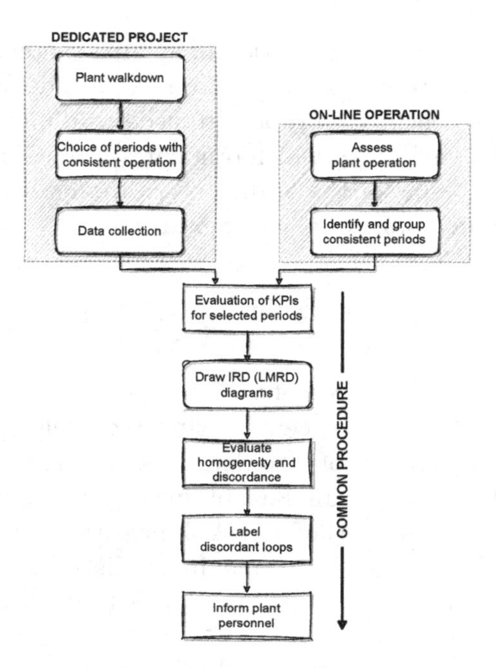

Figure 13.1: Control sustainability assessment procedure

load changes, the wrong data selection can bias the assessment. Moreover, the data must be clean of any external effects (manual interventions,data loses, equipment failures, etc.) and data collection system artifacts. These time series must also be long enough to fulfill proper evaluation of selected indexes. The phase of period selection for data collection must be performed with extreme caution, as it matters [248].

The next important issue is connected with the proper selection of the loop performance indexes and the resulting type of the IRD diagram. Selected measures must be comparable, respective to the problem and possible to be evaluated according to the available time series properties (sampling, dataset length, oscillatory and noise components). The utilization of the source LMRD chart, i.e. the IRD(τ_3,τ_4) is quite obvious. But other options suggested in Chapter 12 should be also accounted for.

This chapter finishes with the presentation of the procedure. The simulation study for the sustainability analysis is possible, although without a reliable practical reference. It will be presented in Chapter 17, which concludes the book with industrial data.

GLOSSARY

CPA: Control Performance Assessment

APC: Advanced Process Control

LMRD: L-Moment Ratio Diagram

IRD: Index Ratio Diagram

GeoMed: Geometric Median

GIS: Geographic Information System

MADAM: Median Absolute Deviation Around Median

CHAPTER 14

Single-element, multi-element and multivariate controls

MULTIVARIATE control system performance analysis is just another cup of tea. By design, this book is intended to focus on univariate control. The reason is simple. In the process industry, multidimensional control systems are extremely rare. And virtually all of them involve advanced process control solutions that use model predictive control strategy. Therefore, this chapter might not exist at all or its content should be hidden inside other sections.

It seems that there exists a mismatch in the literature about what is multivariate control, as it is frequently confused. One should distinguish between multi-loop configurations (multi-element controls) that consists of several single-element loops, which are in general using PID-like control algorithm. Cascaded and ratio control, feedforward disturbance decoupling, three-element structure constitutes simple examples of such controls. Being exact, this is by no means a multivariate control, despite the fact that multiple variables are controlled. By multivariable control we should consider control algorithms that have multiple inputs and/ot multiple outputs.

Following above introduction, the multi-loop control is just a solution consisting of several single-loop elements. From that perspective nothing prevents from considering elementary loops independently using the univariate framework. The literature shows a lot of research addressing specifically multi-element configurations, like cascaded control

DOI: 10.1201/9781003604709-14

[194, 295, 47, 369, 174, 266], feedforward [72, 366, 365], three-element control (drum control example) [370, 254]. There also other, clearly multi-loop reports, like [155, 129, 358, 48, 348, 274, 68]. Probably there is a lot more of them and these are only selected works. They use different basic CPA approaches and do their joint visualization or combination.

Coming back to the beginning of this chapter, the real multivariate control is just a different story. It is mostly connected with the industrial APC applications that mostly use the MPC technology. Such works also already exist, for instance [116, 296, 207, 348, 100, 103]. Collected review of these approaches can be found in [87].

The assessment of the real multivariate control is both easy and hard. Generally, multivariable strategy has a model and some internal performance index. It's not close magical black box. The loss in performance of a multidimensional control algorithm most often takes its cause from model mismatch or loss of adequacy of the internal performance index. Fortunately, the multi-criteria control quality assessment of such an algorithm exists in this internal cost function. Since we have access to both, the situation is relatively simple. By comparing the changes in the value of the performance index against its previous values for analogous situations, we automatically get information that something is wrong. If, on the other hand, it turns out that the model output does not match reality then we immediately have a diagnosis.

This is that easy. The harder part is getting inside the controller structure and make proper interpretation. Far more difficult is the appropriate modification in the internal cost function in relation to changes in the environment, not to mention the most difficult part – model adaptation. It depends on the exact structure of the APC solution and as such is not within the scope of this study.

However, if we put together various pieces of the research presented in the book the multi-element control system assessment is just at hand. Actually, we have two promising concepts: Index Ratio Diagrams and causality analysis. Both exactly suit the task.

Causality analysis traces the root-cause path of the multivariate system [108]. Comparing causality graph with the single-element CPA we may assess all controllers separately and trace which one of them is a potential source of loss of performance.

The IRDs offer an alternative solution. As in one plane we can compare several different loops we may check their homogeneity, and observe their evolution. Integration with the causality graphs would only help.

Multi-criteria property of the IRD charts offers another one option. If we have two-element control, like it happens in case of the cascaded control, we may compare each loop performance measure, for instance the tail index estimator, as each dimension of the plot. Moreover, we may make the shading according to the another element and we might be able to assess three-element control in a single diagram.

> Therefore, the task of the multi-loop control system performance assessment is just putting together known elements. Knowledge of the entire process configuration only helps, as the analysis should or even must rely on this information. The role of good engineer is to be open for different research contexts and no existing possibilities. Once a challenge appears, it would be just a selection of proper tools. This is one of the goals of that book – to deliver an tool-box of open statistical measures and methods.

GLOSSARY

APC: Advanced Process Control

MPC: Model Predictive Control

IRD: Index Ratio Diagram

IV

Case studies

CHAPTER 15

Systematic time series analysis procedure

TIME SERIES analysis procedure summarizes one path of this book – how to interpret properties for a single variable. The goal is to evaluate potential features of a given variable using its historical dataset collected from the site data acquisition system. We expect to obtain information about signal properties and their components, as a result of the time series assessment procedure. This information allows us to assess the process being a source of studied signal. We may assess the following time series features:

- stationarity,
- normality,
- persistence.

Moreover, as the results we may obtain several models and parameters composing the analyzed data:

- outliers,
- long-term trend,
- oscillatory components,
- regression model,
- tail index,
- residual noise signal distribution with its factors.

DOI: 10.1201/9781003604709-15

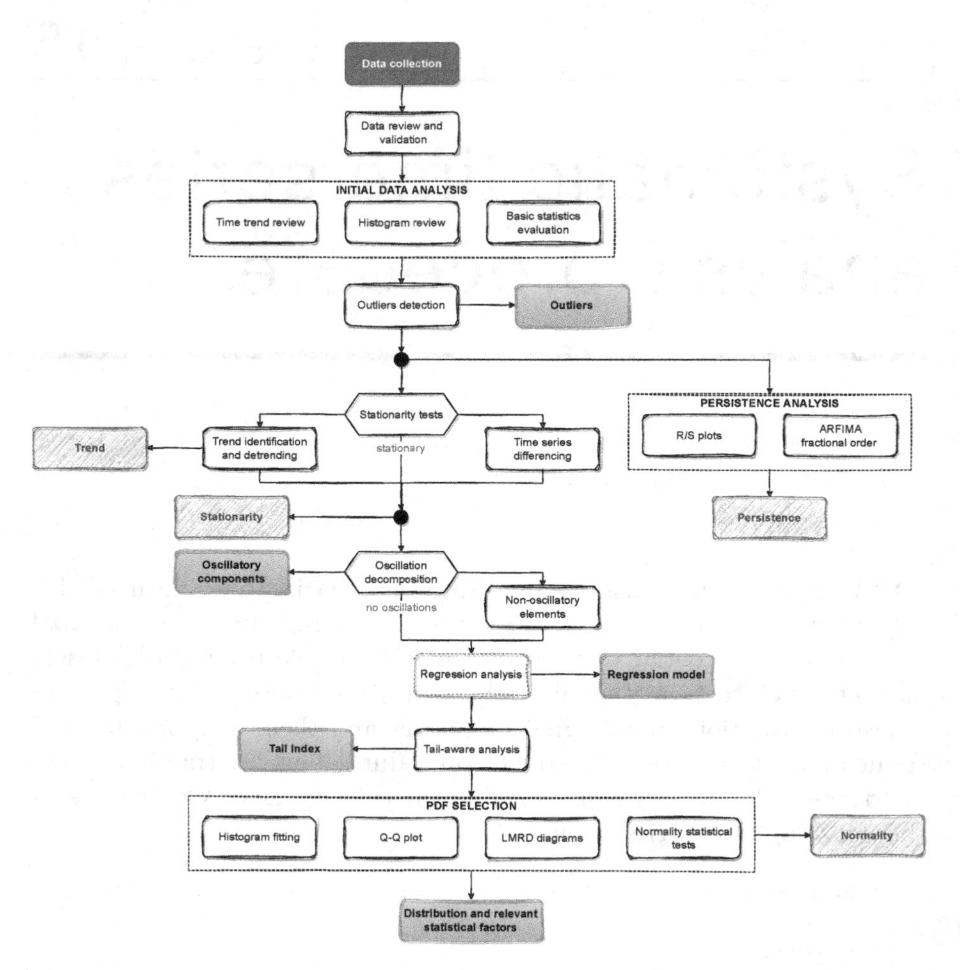

Figure 15.1: Schematic representation of the control engineering-oriented time series analysis

Fig. 15.1 presents graphical representation of the proposed time series analysis procedure. Please note, that this proposal is dedicated toward the assessment of control engineering data and it is not considered universal. It should be also regarded as a kind of an example for a framework, as specific project conditions might require its modification.

The basic elements and steps of the proposed procedure are described in the following paragraphs.

Data collection phase seems to be trivial from the academic perspective. Everybody expects that we only need to go to the site, request data and we get it, just like this. Unfortunately, the reality is not just like that. The objective is to collect true time series data of the selected variables. Although it looks straightforward, the reality presents unexpected challenges [84]. We observe two groups of issues that we may meet during of, i.e. technological limitations and I&C constraints.

Industrial processes frequently behave non-stationary, they are affected by various influences that deteriorate normal operation. We must be aware of that during data collection. We need to choose proper and comparable time periods, when the plant operates in normal operating regimes, and definitely not in emergency. What is obvious, the installation cannot be impacted by human interventions [248].

These influences can come from two types of sources: external and internal. Changes in technology, installation reconfiguration, variations in product quality, fluctuations in plant operating regimes done by site personnel or varying weather conditions have external origin. Failures in equipment, its modifications, maintenance or calibration, human (mostly done by operators) interventions, like manual operator's actions are considered to be internal. Knowing the influences and risks to normal operations we may properly design data collection scenarios.

We also have to remember about possible limitations originating from data collection software (DCS/PLC/SCADA/MES). Their internal database systems frequently have limited storage volumes and thus use custom data compression mechanisms, like averaging, collection deadbands, interpolations, etc.

The acquisition of such data may significantly influence its character, in contrary to its raw values. Moreover, we often observe empty observations (no data) in datasets, strange artifact values, step-wise or linear saw-tooth profile (effect of data interpolation). It causes that the historian software must be reviewed and eventually reconfigured, before the start of any data-collection process.

Even if we properly set up data acquisition software, we need conduct **data review and validation**. Once the data are collected. we have to find "no data" fields, somehow fill them (interpolation, data cutting and records removal), check data synchronization, especially if we collect data from different sources. This process is tedious, however crucial, as any errors done at that moment may waste all the later efforts.

The **outlier detection** process is highlighted, due to its relevance and a set of specific methods. We use dedicated algorithms and detect outlying observations or entire time series subsets. We remember that those outliers might be just the errors or rare realizations of the real process so high cautiousness is required [93].

We may remove those records or exchange them with interpolated data, in case of erroneous observation. Frequent outliers might suggest further usage of robust estimation/regression techniques and show the premise of potential occurrence of tails in probability distributions.

Outlier detection and labeling are followed by the **stationarity tests** done using presented algorithms. As we remember, four options are possible. The data can be stationary and we may just pass to the next phase of signal decomposition. In case of detected trend stationarity we have to remove the embedded trend. Therefore, we get the trend itself and detrended, already stationary data. We have to remember that the trend may have periodic, oscillatory character, and then we remove oscillations during the time series decomposition process.

In case of incremental stationarity we need to perform data differencing and follow the procedure. The question remains, how to proceed in case of non-stationary data. Personally, I suggest to stay with that and follow the procedure. However, we must remember about potential bias in consecutive interpretations.

We may proceed with the **persistence analysis** parallel to the stationarity assessment. We may do it using the canonical long-range dependence R/S plots, by estimation of the ARFIMA filter fractional order or by any other Hurst exponent estimation. As a result we get conclusion if the time series is pure independent, persistent or average reverting.

We do the signal **oscillation decomposition** after stationarity assessment. The time series is decomposed to eventual oscillations of different frequencies and the residuum. Resulting residual signal can be directly considered as the noise, for which we perform further evaluation of statistical properties or we conduct **regression analysis**.

Identification of regression models may be conducted using various algorithm, but we have to remember to apply previously acquired knowledge about eventual outliers. In such a case we should utilize suitable robust regression technique [71].

Afterwards, we may focus on purely statistical analysis. At first, we can assess histogram tails with **tail-aware analysis** and evaluate the tail index. Simultaneously, we may proceed with the appropriate **PDF selection** procedure. We may follow different methods, such a distribution to histogram fitting, quantile plots, L-moment ratio diagrams or just simple normality testing to validate the hypothesis about signal Gaussian procedure. As the result, we end up with properly selected probabilistic density function and a set of its statistical factors of shift, scale and shape. The L-moments can also be evaluated especially in case of poor fitting of known theoretical PDF or in case of short dataset size. Small sample statistics moment estimators should be taken into account in the latter case.

As stated in the beginning of this section, this procedure is the Author's proposal. Though validated during several industrial projects, specific situations may imply its modifications. Rigid adherence to procedure is not always effective, especially in unusual situations. *Ad hoc* modifications dependent on the specifics encountered are sometimes simply necessary.

This procedure applies to the assessment of single, univariate time series. The multivariate case can be assessed by successive applications of the above procedure. Once data origin from the same, comparable site, the assessment might be extended by causality, homogeneity and discordance analysis.

Decomposition embedded in the procedure allows to apply causality and homogeneity analysis to different components separately [109], what gives more degrees of freedom during assessment.

GLOSSARY

I&C: Infrastructure and Control

DCS: Distributed Control System

PLC: Programmable Logic Controller

SCADA: Supervisory Control And Data Acquisition

MES: Manufacturing Executions System

PDF: Probabilistic Density Function

CHAPTER 16

Systematic control system assessment procedure

CONTROL SYSTEM combines information about all its components, like control algorithms, filters, feedforward elements, nonlinear blocks, etc. The subject has been addressed in many papers and books. However, this issue is closely linked to industrial practice and cannot be considered in isolation from installation on-site conditions.

Blind calculation of measures without deeper analysis and practical reflections is erroneous and misleading [82]. Generic CPA procedures have already been proposed. This one, shows an expanded version of an earlier solution presented in [85]. The procedure uses experience from previous works/projects and results for the single time series analysis procedure presented in Fig. 16.1.

The procedure consists of three main parts. The first one is site-based and practitioners usually call it a **plant walkdown**. It consists of several activities that are performed during the site visit. Here we owe a comment to the reader. After the pandemic period and the outburst of remote activities, there is a growing temptation to perform all possible activities on line. The same happens with the initial getting plant information for the CPA assessment project.

However, some activities cannot be done remotely. Only the site visit enables getting the touch and feeling of the plant. You have to see the installation, you have to assess yourself equipment conditions, touch the control system, talks to plant key personnel. As a key personnel we

DOI: 10.1201/9781003604709-16

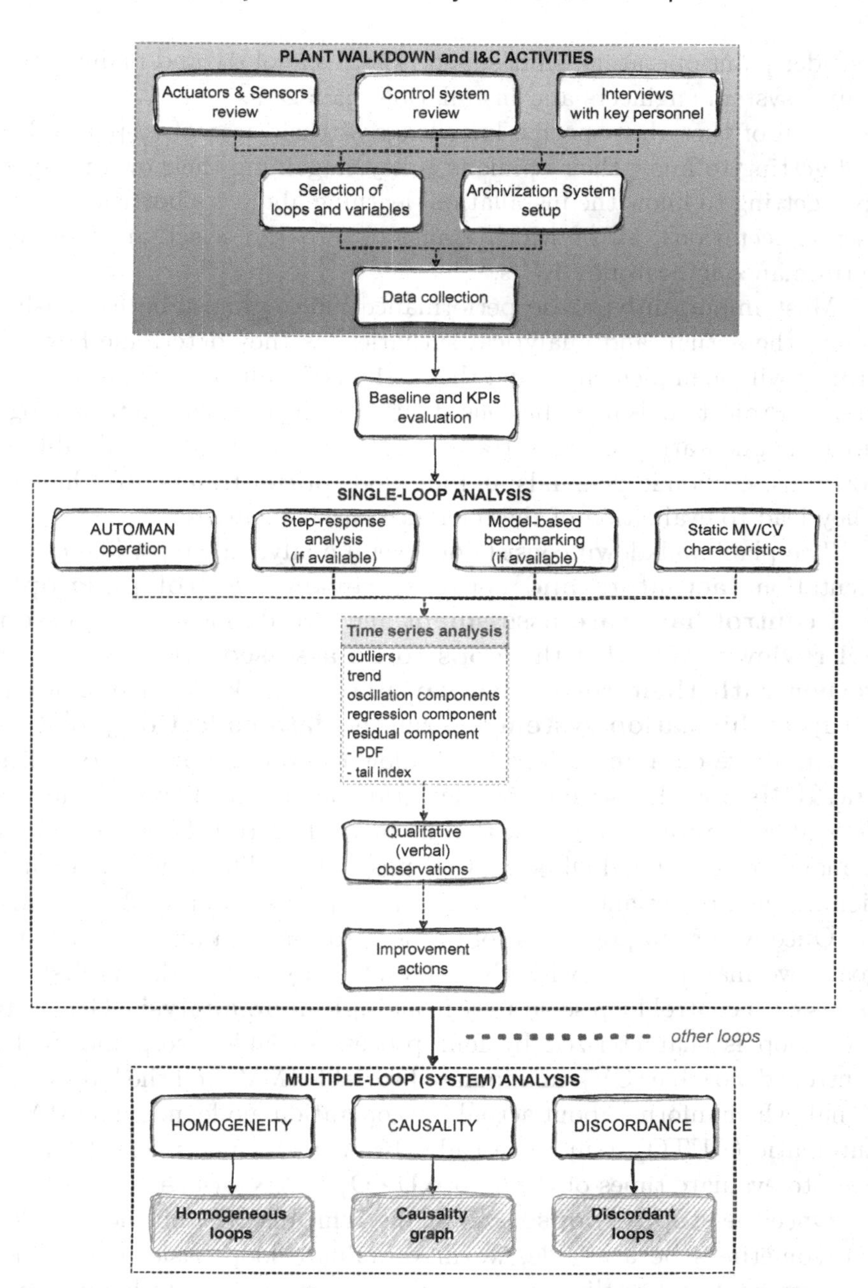

Figure 16.1: Schematic representation of the industrial control performance assessment analytical procedure

consider plant operators, shift supervisors, technology and maintenance teams, system engineers and installation managers.

Each of them has specific knowledge of their area of responsibility, and getting to know their opinions is key to gaining their own perspective, getting to know the installation, learning about its bottlenecks and user expectations, all of which translate into the selection of project performance indicators – KPIs.

Most importantly, these performance indexes must be established before the actual, and analytical activities, as they determine how the project will be implemented and the method of evaluating the results obtained. Project goals must be consistent with the installation technology, site instrumentation, control strategies, I&C infrastructure, realizable algorithms, economic demands and all kinds of constraints, also human. They lead to evaluation of the reference baseline curves.

The plant walkdown consists of several analytical steps, like instrumentation **(actuators and sensors) review**, **control philosophy and control hardware assessment** and already mentioned **personnel reviews**. After that the **loops to be assessed are selected together with their respective variables**. This knowledge allows to **setup archivization system** and run the **data collection** process.

All above data and information allow to evaluate project **baseline and KPIs**, together with their calculation procedure. This step must be done at early project stages, at that moment the latest. Forgetting about it may lead to later doubts, mutual misgivings, difficulties in commissioning the project and, in the worst case, quarrels and legal problems.

Once we are in possession of proper process data and we know the goals, we may proceed with the relevant analytical work. At first, we assess each control loop separately in a **single-loop analysis**. Generally, each loop is characterized by four process variables: setpoint (STP), controlled variable (CV), manipulated variable (MV) and the loop status signal, which informs about actual loop operation mode: manual (MAN), automatic (AUTO), remote cascade (RCAS) or supervised (SPV). We need to evaluate times of the loop **AUTO/MAN operation**.

Once the **step response** is available, which is rare, but nevertheless can sometimes be available, we may evaluate loop basic indexes, i.e. the overshoot and settling time. Similar opportunity might happen once the process model is available. In such a case we may proceed with the **model-based benchmarking**, for instance the one that uses minimum variance approach [56].

Another step, which is not described in this work, however important [84], is the evaluation of loop static curves in form of the X-Y drawings. These are simple scatter plots showing CV versus MV variable. These diagrams allow to identify the kind of eventual loop nonlinearity.

The main element of the assessment is the **single-loop analysis**, which is described in detail in previous Chapter 15. The analysis is applied to the loop control error and allows to evaluate loop properties:

- stationarity,
- normality,
- persistence.

and certain groups of loop-describing factors, like:

- outliers,
- long-term trend,
- oscillatory components,
- regression model,
- tail index,
- residual noise signal distribution with its factors.

The analysis and interpretation of these factors allow to make loop **qualitative observations**, which are the correct result of the analysis, allowing subsequent decision-making on control systems. Some of the observations are sketched below:

- High time percentage of MAN mode indicates that the operator has a low level of confidence in the loop and its control performance, leading to its switching to manual operation.
- Steady-state error we observe in the dynamic step response or by the shift factor of fitted distribution.
- Frequent outliers mean uncoupled disturbances or errors in the data communication or malfunctioning of the data archivization system.
- Sluggish or aggressiveness can be assessed using step response indexes or Hurst exponent,

- Asymmetric operation may be identified using statistical analysis with shape skewness factors, or a histogram.
- Nonlinearity in loop operation (in process or actuator) we observe in MV/CV charts or in asymmetric operation.
- Frequent loop constraints violation we may observe in skewed loop properties.
- Uncoupled disturbances we observe in system persistence and in distribution heavy tails – large tail index.
- Slowly decaying oscillation that occur just after disturbance incident indicate integration wind-up in the PI/PID controller.
- Long-term correlations cause persistent operation and heavy tails.
- Properly designed and tuned control we observe in the normality of loop variable statistics.
- Significant loop oscillations may indicate poor tuning, relay operation, wrong operating point or actuator malfunctioning (valve stiction).
- Multi-fractal properties indicate the presence of frequent human actions (MAN mode switching, operator biases, etc.) in control operation.

Each of the above observations is constructive and leads to subsequent **improvement suggestions**, like:

- Low operators confidence means the need for review/change of the control philosophy or the required maintenance to sensor or actuating units.
- Steady-state error requests adding of the integral action inside the feedback loop or verification of the operating points against saturation, or the need to add linearization.
- Frequent saturated operation may be addressed by adding setpoint feedforward filtering.
- Uncoupled disturbances mean the need for disturbance decoupling with feedforward.

- Communication errors or data historian malfunctioning require the review of the control system infrastructure.
- Sluggish control requires making less aggressive controller tuning, while the aggressive control performance need contradictory actions.
- Actuator nonlinearity can be addressed with the controller output linearization unit,
- Frequent oscillation of small gains may be addressed by feedback or control error filtering.
- Integration wind-up demands for anti-windup mechanism in the control loop.
- Process nonlinearity can be solved by proper controller gain scheduling or static MV signal linearization.,
- The measurable disturbances can be decoupled by properly designed feedforward unit.
- We can take care of unmeasurable disturbances with control error or feedback signal tuning or by robust controller design/tuning,
- Undamped loop oscillations can be solved with more sluggish tuning or filtering.

Above procedure is evidently nor universal, neither comprehensive. Process uniqueness demands flexibility and creativity from an assessment engineer. Comprehensive control performance assessment is an engineering art that cannot be fully dehumanized and any software packages should be considered as the decision-making support.

GLOSSARY

I&C: Infrastructure and Control

STP: SeTPoint

CV: Controlled Variable

MV: Manipulated Variable

MAN: MANual mode

AUTO: AUTOmatic mode

RCAS: Remote CAScade mode

SPV: SuPerVised mode

CHAPTER 17

Industrial case studies

INDUSTRIAL case studies allow theoretical considerations to collide with industrial reality. Below example presents how a given control loop may be assessed in practice. Let's assume that we have univariate SISO loop. This loop, as any other in multi-loop system, is described by a `tag name` (some systems name it an `item`) and text description. There exists a standard for tag naming, which is called KKS (Kraftwerk Kennzeichnen System / Identification Systems for Power Plants). The standard was intended to name equipment in Power Plants and Refineries under a single code set. It was designed by a committee convened in Germany in 1970 (consisting of engineers, suppliers, and regulators of the energy sector). Its organization is based on Function, Type, and Location in accordance with ISO, IEC, and DIN 40719 Part2 (IEC750) standards.

It allows quick finding of a type and approximate location of equipment from the P&ID drawing. Moreover, it enables unambiguous nomenclature of control system tags within the DCS system. Any control loop is described by four main variables, for which we may collect relevant time series:

- setpoint,
- controlled variable,
- manipulated variable,
- loop control mode.

We perform the analysis using collected datasets for above variables. The results, according to the methods described in this book, can be summarized and presented in form of the **Loop CPA Charts**. Exemplary ones, prepared for real industrial data are sketched below.

DOI: 10.1201/9781003604709-17

Loop CPA Chart

Loop tag name: KKSnum
Loop description: Pressure
Number of data samples: 40751
Time in AUTO mode: 100.0%
Setpoint statistics:

- Min = 4700.04; Max = 5016.20
- Mean = 4871.28; Median = 4853.27
- Std.Dev. = 66.84; M-Std.Dev. = 70.92
- MAD = 47.83; IQR = 85.38
- Skewness = -0.22; Kurtosis = 2.32
- L-l2 = 37.56; L-Skewness = -0.22; L-Kurtosis = 2.32

Controlled variable statistics:

- Min = 4698.85; Max = 5063.24
- Mean = 4871.25; Median = 4854.48
- Std.Dev. = 66.93; M-Std.Dev. = 70.95
- MAD = 48.78; IQR = 86.02
- Skewness = -0.22; Kurtosis = 2.33
- L-l2 = 37.69; L-Skewness = -0.22; L-Kurtosis = 2.33

Manipulated variable statistics:

- Min = 84.03; Max = 94.79
- Mean = 90.30; Median = 90.54
- Std.Dev. = 2.09; M-Std.Dev. = 2.52
- MAD = 1.89; IQR = 3.79
- Skewness = -0.21; Kurtosis = 1.82
- L-l2 = 1.20; L-Skewness = -0.21; L-Kurtosis = 1.82

Control error statistics:

- Min = -71.25; Max = 96.37
- Mean = 0.03; Median = 0.02
- Std.Dev. = 5.09; M-Std.Dev. = 4.38
- MAD = 2.92; IQR = 5.85
- Skewness = 0.28; Kurtosis = 18.86
- L-l2 = 2.69; L-Skewness = 0.28; L-Kurtosis = 18.86
- Time series is not normally distributed (Lilliefors)
- α-stable PDF: $\alpha = 1.85$; $\gamma = 3.07$; $\beta = 0.10$; $\delta = 0.07$
- tail index = 0.32
- Time series is stationary
- GPH test: d= 0.017

Loop time trends:

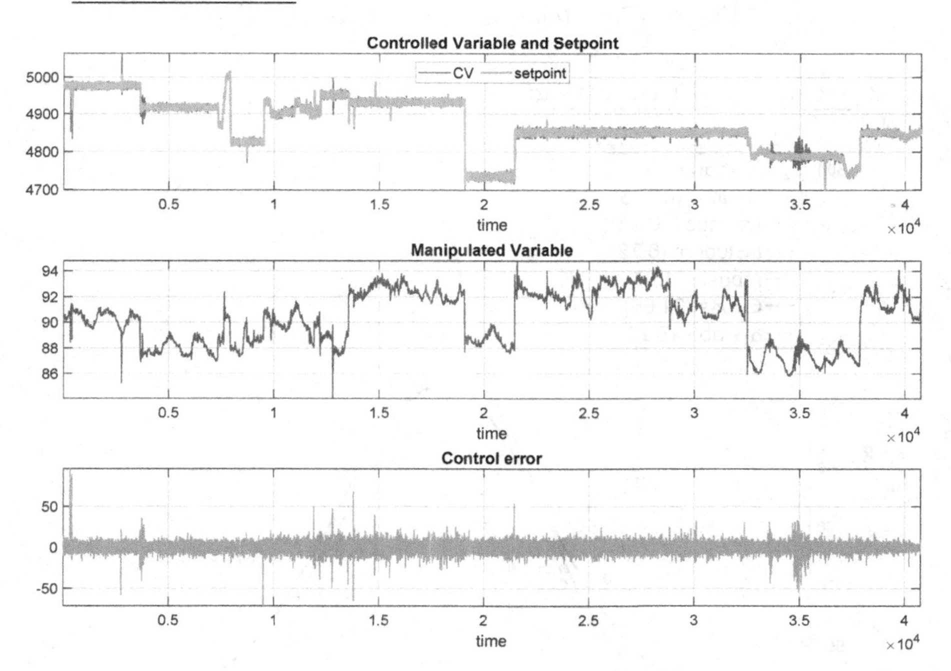

Figure 17.1: Loop time trends

Static characteristics X-Y:

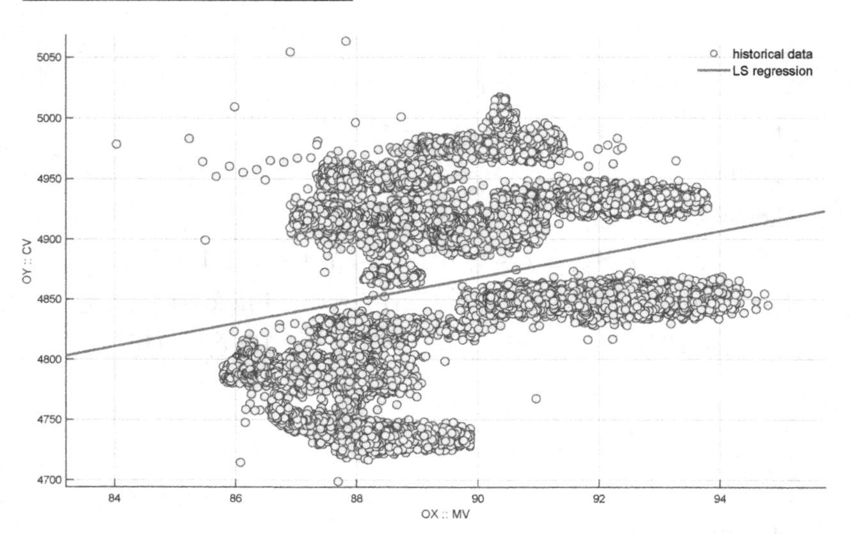

Figure 17.2: Relationship X-Y for the loop

Histogram and PDF fitting:

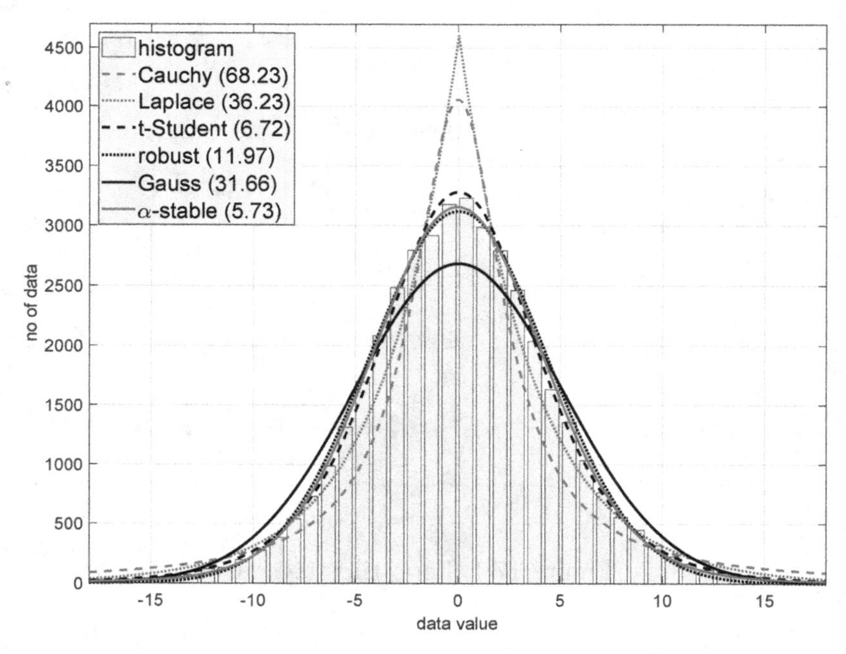

Figure 17.3: Histogram and PDF fitting

Q-Q plot:

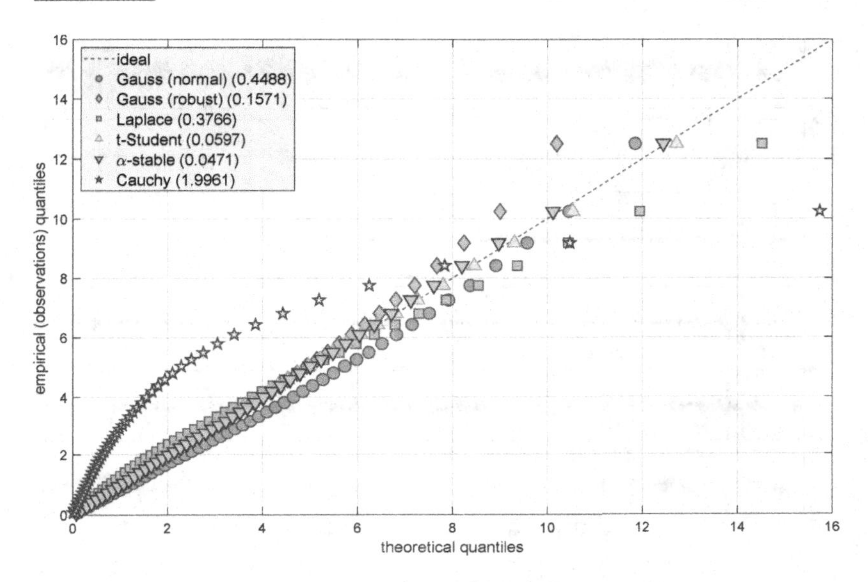

Figure 17.4: Q-Q plot

Histogram and boxplot:

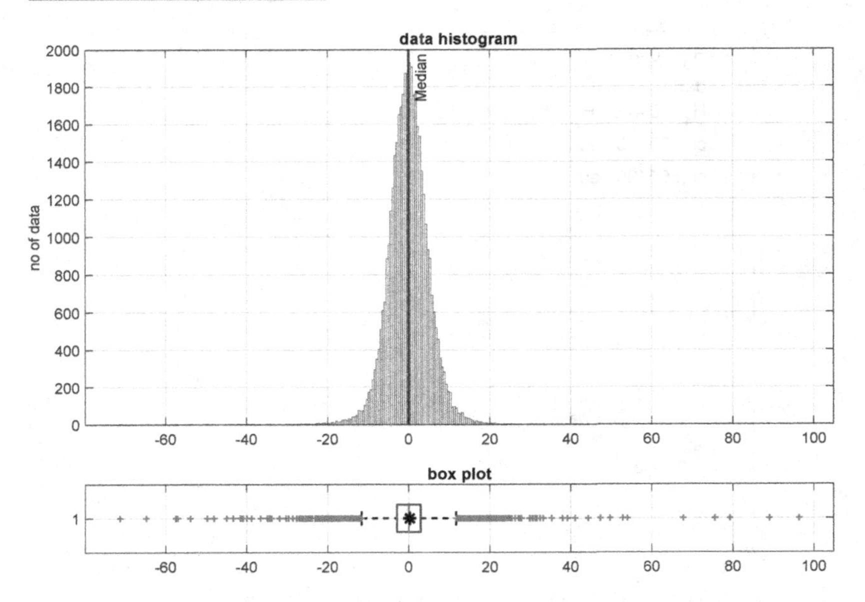

Figure 17.5: Histogram and boxplot

MEEMD decomposition:

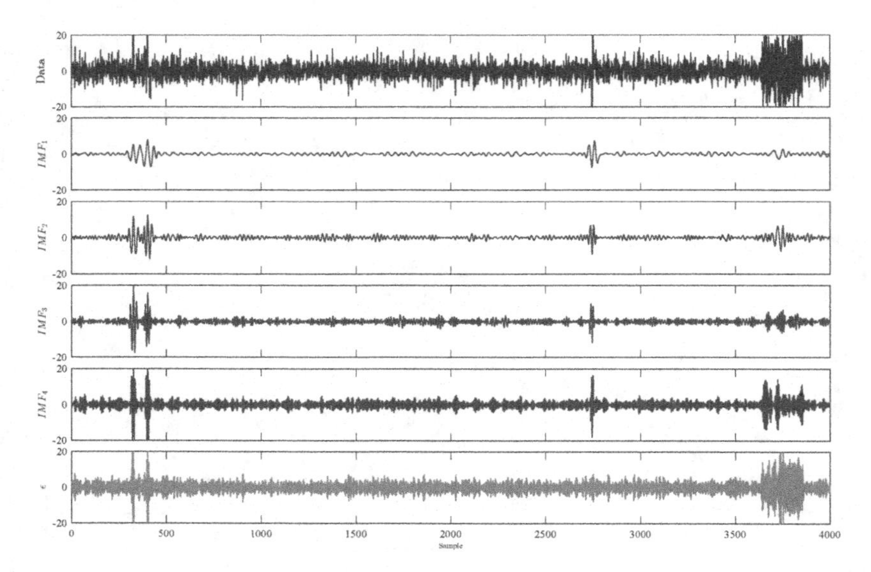

Figure 17.6: MEEMD decomposition

Rescaled range R/S diagram:

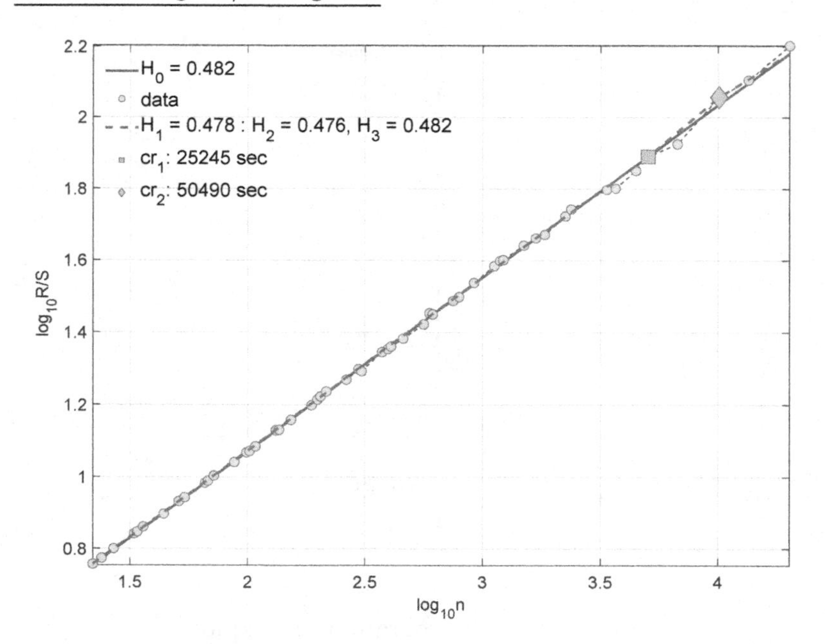

Figure 17.7: Rescaled range R/S diagram

Rescaled range R/S analysis:
- single Hurst exponent: H0 = 0.482
- triple exponents: H1 = 0.478; H2 = 0.476; H3 = 0.482
- crossover: cr1 = 25245.000; cr2 = 50490.000

LMRD diagram:

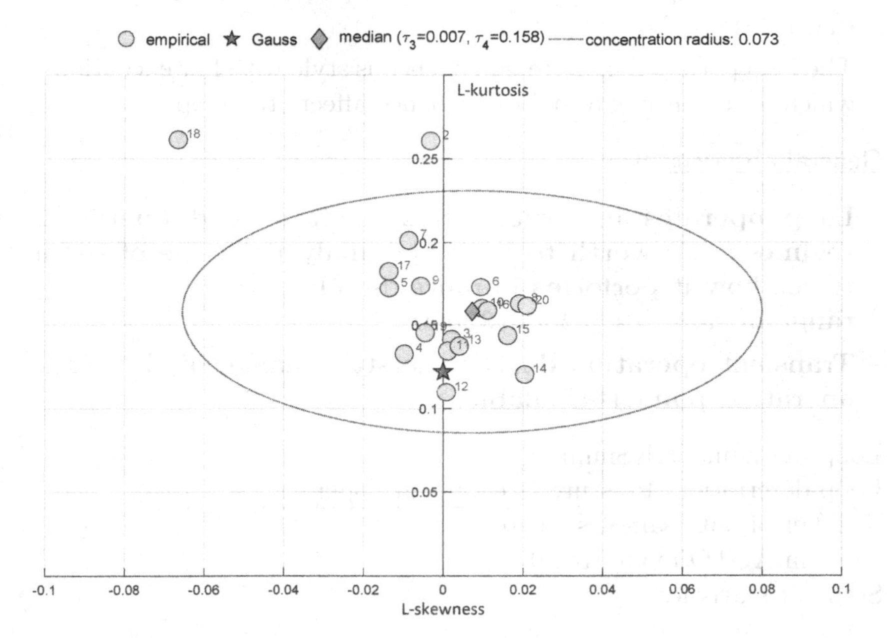

Figure 17.8: LMRD diagram

<u>Loop observations:</u>

- Loop continuously operates in AUTO mode.
- Steady-state error equals to zero.
- Control error is not normally distributed.
- Control error exhibits small skewness toward positive values.
- Stability exponent suggests independent properties (with slight persistence), which is confirmed by the Hurst exponent values and the fractional order.
- Loop is stationary.
- Non-Gaussian loop properties are confirmed by the histogram with a PDF fitting and by a Q-Q plot.
- The static MV-CV relationship indicates loop nonlinearity of different kinds. It means that data exhibit different operating regimes, which is confirmed in time trends.

- Box plot denotes relatively frequent outliers, which as we observe in time trends, occur in three moments, probably caused by external disturbances.
- There are visible loop oscillation. Further causality analysis for oscillations requires system-wide assessment and review of the actuator.
- The loop performs in quite homogeneous style with few exceptions, which occur once external disturbance affects the loop.

General observation:

- **Loop operates in various steady states and operating regimes - it's worth to limit the analysis to one of them to see how it performs in more detail: range of samples $(22000, 32000)$.**
- **Transient operation during the step change of the plant operating point is available**

Loop tag name: KKSnum
Loop description: Pressure (steady state operation)
Number of data samples: 10001
Time in AUTO mode: 100.0%
Setpoint statistics:

- Min = 4831.31; Max = 4883.50
- Mean = 4851.03; Median = 4850.97
- Std.Dev. = 3.55; M-Std.Dev. = 3.10
- MAD = 2.04; IQR = 4.07
- Skewness = 0.10; Kurtosis = 4.72
- L-l2 = 1.95; L-Skewness = 0.10; L-Kurtosis = 4.72

Controlled variable statistics:

- Min = 4816.49; Max = 4873.39
- Mean = 4850.97; Median = 4851.01
- Std.Dev. = 4.87; M-Std.Dev. = 4.76
- MAD = 3.22; IQR = 6.42
- Skewness = -0.09; Kurtosis = 3.60
- L-l2 = 2.73; L-Skewness = -0.09; L-Kurtosis = 3.60

Manipulated variable statistics:

- Min = 89.48; Max = 94.33

- Mean = 92.13; Median = 92.20
- Std.Dev. = 1.01; M-Std.Dev. = 1.13
- MAD = 0.85; IQR = 1.70
- Skewness = -0.20; Kurtosis = 2.04
- L-l2 = 0.58; L-Skewness = -0.20; L-Kurtosis = 2.04

Control error statistics:

- Min = -21.49; Max = 32.41
- Mean = 0.05; Median = 0.04
- Std.Dev. = 4.91; M-Std.Dev. = 4.79
- MAD = 3.23; IQR = 6.45
- Skewness = 0.08; Kurtosis = 3.67
- L-l2 = 2.74; L-Skewness = 0.08; L-Kurtosis = 3.67
- Time series is not normally distributed (Lilliefors)
- α-stable PDF: $\alpha = 1.97$; $\gamma = 3.37$; $\beta = 0.37$; $\delta = 0.06$
- tail index = 0.32
- Time series is stationary
- GPH test: d=-0.025

Loop time trends:

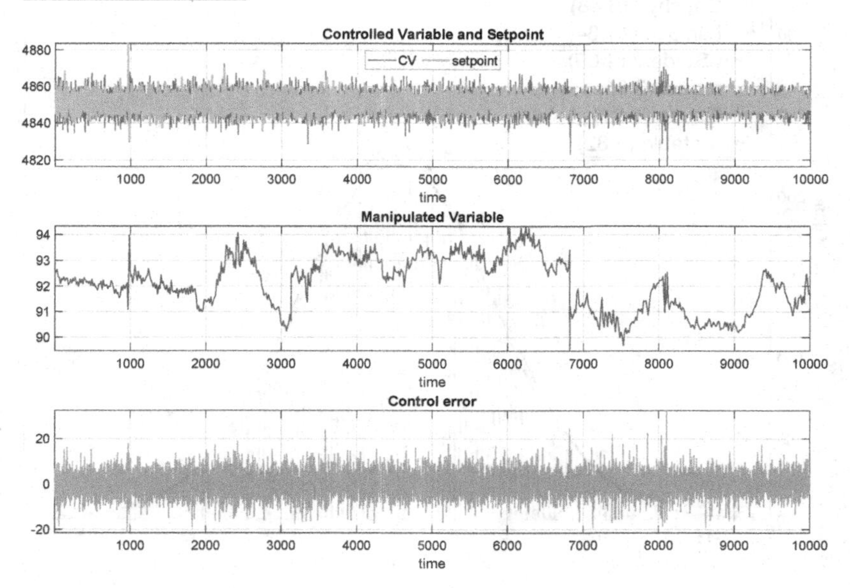

Figure 17.9: Loop time trends – steady state operation

Static characteristics X-Y:

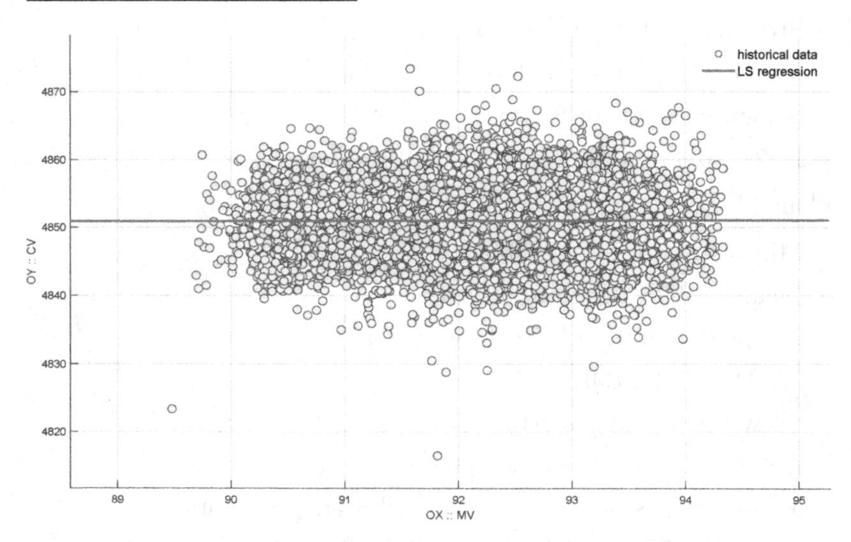

Figure 17.10: Relationship X-Y for loop – steady state operation

Histogram and PDF fitting:

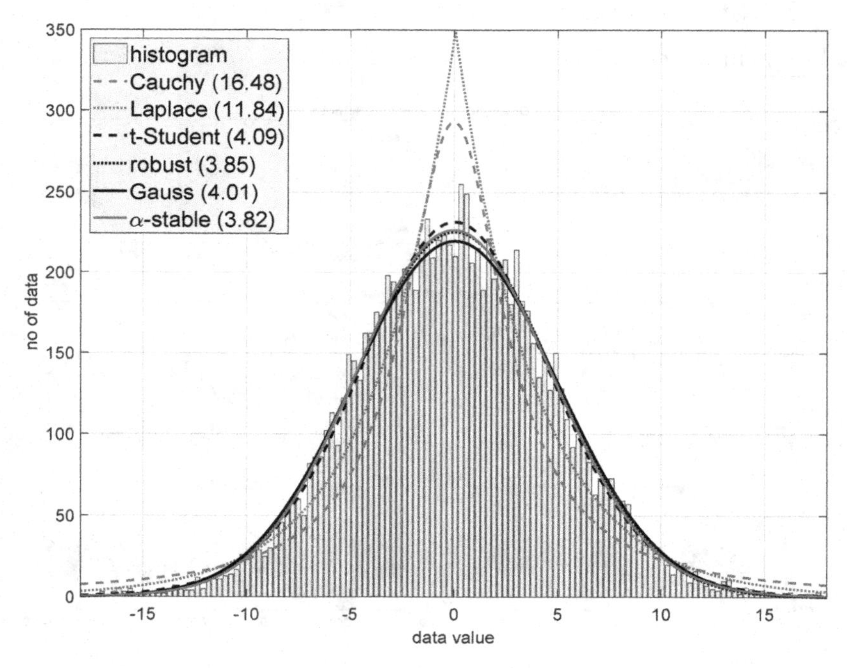

Figure 17.11: Histogram and PDF fitting – steady state operation

Q-Q plot:

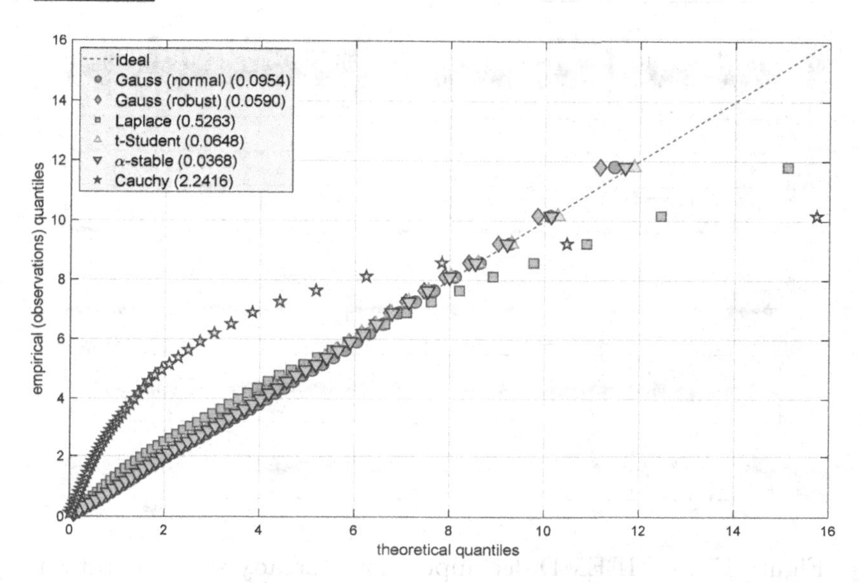

Figure 17.12: Q-Q plot – steady state operation

Histogram and boxplot:

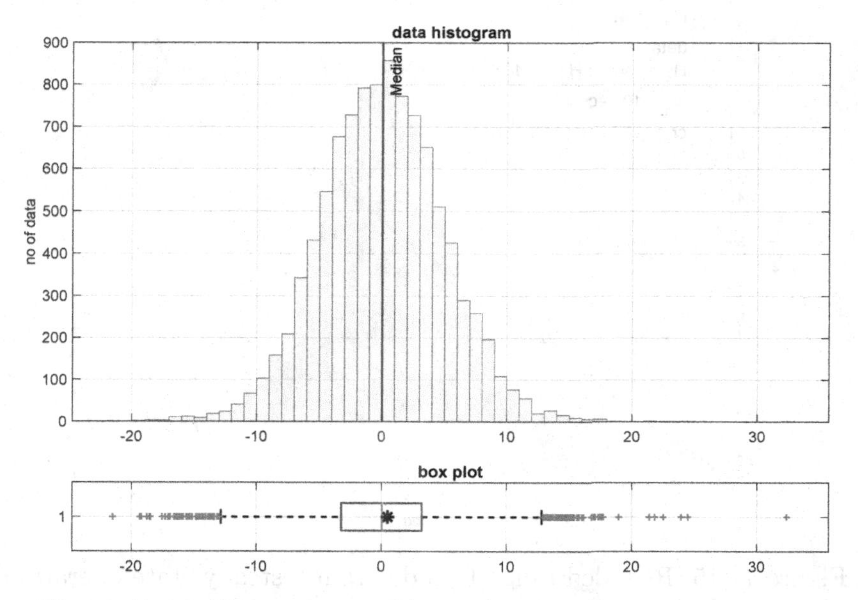

Figure 17.13: Histogram and boxplot – steady state operation

MEEMD decomposition:

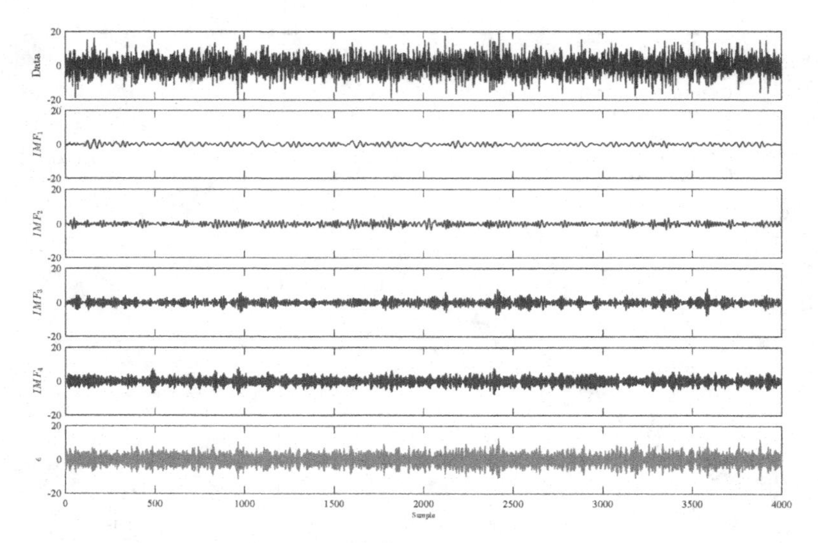

Figure 17.14: MEEMD decomposition – steady state operation

Rescaled range R/S diagram:

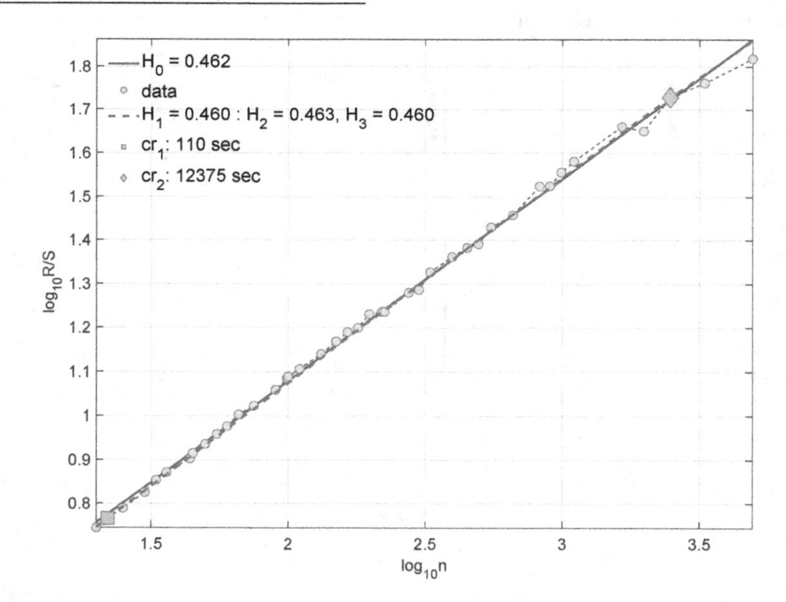

Figure 17.15: Rescaled range R/S diagram – steady state operation

Rescaled range R/S analysis:
- single Hurst exponent: H0 = 0.462
- triple exponents: H1 = 0.460; H2 = 0.463; H3 = 0.460
- crossover: cr1 = 110.0; cr2 = 12375.0

LMRD diagram:

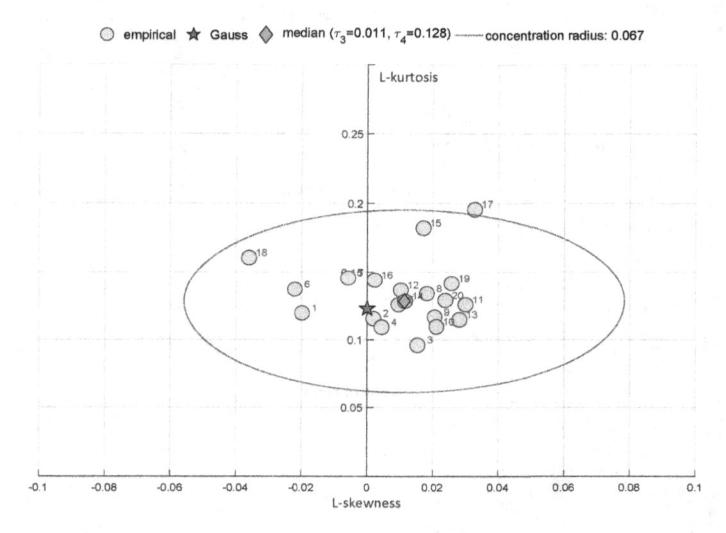

Figure 17.16: LMRD diagram – steady state operation

Loop steady state observations:

- Control error is not normally distributed according to the normality test, though stability exponent, Hurst exponent and ARFIMA filter fractional order indicate independent operation.
- Control error is symmetrical and stationary.
- The static MV-CV relationship confirms single operating point.
- Box plot denotes smaller number of outliers, though still occurring.
- There are visible small loop oscillation. Further causality analysis for oscillations requires system-wide assessment and review of the actuator.
- The loop performs in a homogeneous style with few exceptions.

Loop tag name: KKSnum
Loop description: Pressure (transient operation)

Transient operation is visualized with a time trend showing the change of setpoint from around 5000 to 4740. We see very good control with minimal overshoot. It confirms previous observations on control quality.

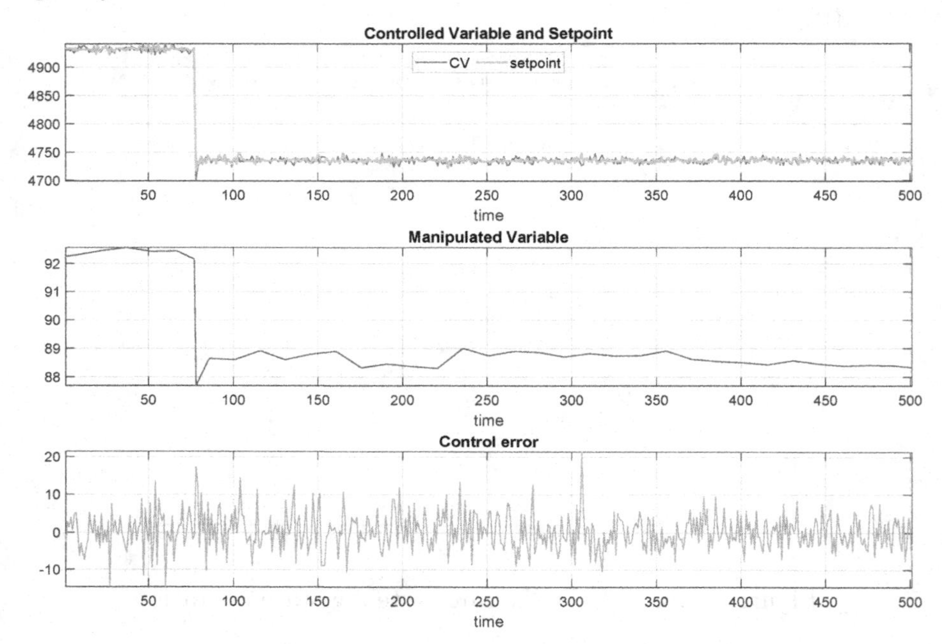

Figure 17.17: Loop time trends – transient operation

<u>Loop maintenance suggestions:</u>

- Review possible nonlinearity in the actuating unit, however the control seems to be good.
- It seems that all major disturbances are uncoupled.
- Noise filtering of the control error or the feedback measurement (controlled variable) might be helpful.
- The option of the controller gain scheduling should be evaluated.

Control Performance Assessment is a non-standard activity. It requires engineering flexibility of the responsible team. A loop does not equal a loop, and each installation differs, despite design similarities. The results depend on the engineer knowledge, experience and smart utilization of available tools. Incorporation of the tail-aware statistics allows to step outside the comfort zone of Gaussian assumptions.

This chapter includes single-loop assessment. Multi-loop cases pose different issues that extend univariate analysis. Good examples of such multi-variate assessment can be found in the previously published industrial studies for the homogeneity [98] and causality [108].

GLOSSARY

KKS: Kraftwerk Kennzeichnen System

P&ID: Piping and Instrumentation Diagram

DCS: Distributed Control System

Bibliography

[1] P. Abry and D. Veitch. Wavelet analysis of long-range-dependent traffic. *IEEE Transactions on Information Theory*, 44(1):2–15, 1998.

[2] F.A. Adam, A. Kurnia, I.G.P. Purnaba, I.W. Mangku, and A.M. Soleh. Modeling the amount of insurance claim using Gamma linear mixed model with AR (1) random effect. *Journal of Physics: Conference Series*, 1863(1):012027, 2021.

[3] C.C. Aggarwal and C.K. Reddy. *Data Clustering. Algorithms and Applications.* CRC Press, Taylor & Francis Group, Boca Raton, London, New York, 2014.

[4] C.C. Aggarwal and S. Sathe. *Outlier Ensembles: An Introduction.* Springer, Cham, Switzerland, 2017.

[5] M.I. Ahmad, C.D. Sinclair, and A. Werritty. Log-logistic flood frequency analysis. *Journal of Hydrology*, 98(3):205–224, 1988.

[6] A. Alfons, M. Templ, and P. Filzmoser. Robust estimation of economic indicators from survey samples based on pareto tail modelling. *Journal of the Royal Statistical Society Series C*, 62(2):271–286, 2013.

[7] S.A. Amburn, A.S.I.D. Lang, and M.A. Buonaiuto. Precipitation forecasting with Gamma distribution models for gridded precipitation events in eastern Oklahoma and northwestern Arkansas. *Weather and Forecasting*, 30(2):349–367, 2015.

[8] T.W. Anderson and D.A. Darling. A test of goodness-of-fit. *Journal of the American Statistical Association*, 49:765–769, 1954.

[9] T. Antal, M. Droz, G. Győrgyi, and Z. Rácz. 1/f noise and extreme value statistics. *Physical Review Letters*, 87(24):240601, 2001.

[10] R.K. Arjomandi. Statistical methods for control loop performance assessment. In *Proc. of 2011 International Conference on Communications, Computing and Control Applications*, pages 1–6, 2011.

[11] P. Axensten. Cauchy cdf, pdf, inverse cdf, parameter fit and random generator. `http://www.mathworks.com/matlabcentral/fileexchange/11749-cauchy/`, 2006.

[12] L.A. Baccala and K. Sameshima. Partial directed coherence: a new concept in neural structure determination. *Biological Cybernetics*, 84:463–474, 2001.

[13] S. Banerjee, T. Chattopadhyay, and U. Garain. A wide learning approach for interpretable feature recommendation for 1-d sensor data in iot analytics. *International Journal of Automation and Computing*, 16:800–811, 2019.

[14] A-L. Barabási, P. Szépfalusy, and T. Vicsek. Multifractal spectra of multi-affine functions. *Physica A: Statistical Mechanics and its Applications*, 178(1):17–28, 1991.

[15] J. M. Bardet and P. Bertrand. Identification of the multiscale fractional Brownian motion with biomechanical applications. *Journal of Time Series Analysis*, 28(1):1–52, 2007.

[16] V.D. Barnett and T. Lewis. *Outliers in Statistical Data*. John Wiley & Sons, Chichester, UK, 3 edition, 1994.

[17] K. Basiaga and T. Szkutnik. The application of generalized pareto distribution and copula functions in the issue of operational risk. *Econometrics. Ekonometria. Advances in Applied Data Analytics*, 1(39):133–143, 2013.

[18] D.M. Bates and D. G. Watts. *Nonlinear Regression Analysis and Its Applications*. Wiley Series in Probability and Statistics. John Wiley & Sons, New York, NY, 1988.

[19] M. Bauer, J.W. Cox, M.H. Caveness, J.J. Downs, and N.F. Thornhill. Finding the direction of disturbance propagation in a chemical process using transfer entropy. *IEEE Transactions on Control Systems Technology*, 15(1):12–21, 2007.

[20] M. Bauer, A. Horch, L. Xie, M. Jelali, and N. Thornhill. The current state of control loop performance monitoring – a survey of application in industry. *Journal of Process Control*, 38:1–10, 2016.

[21] M. Bauer and N.F. Thornhill. A practical method for identifying the propagation path of plant-wide disturbances. *Journal of Process Control*, 18(78):707–719, 2008.

[22] M. Bayazit and B. Őnőz. LL-moments for estimating low flow quantiles. *Hydrological Sciences Journal*, 47(5):707–720, 2002.

[23] I. Bazovsky. *Reliability Theory and Practice.* Dover Publications, Mineola, NY, 2004.

[24] T. Bedford and R. Cooke. *Probabilistic Risk Analysis: Foundations and Methods.* Cambridge University Press, Cambridge, UK, 2001.

[25] R.E. Bellman. *Dynamic Programming.* Princeton University Press, Princeton, NJ, 1957.

[26] I. Ben-Gal. Outlier detection. In O. Maimon and L. Rockach, editors, *Data Mining and Knowledge Discovery Handbook: A Complete Guide for Practitioners and Researchers*, pages 131–146. Springer, Boston, MA, 2005.

[27] I. Ben Nasr and F. Chebana. Homogeneity testing of multivariate hydrological records, using multivariate copula l-moments. *Advances in Water Resources*, 134:103449, 2019.

[28] J. Beran. *Statistics for Long-Memory Processes.* CRC Press, Boca Raton, FL, 1994.

[29] D. S. Bernstein. Feedback control: an invisible thread in the history of technology. *IEEE Control Systems Magazine*, 22(2):53–68, 2002.

[30] M.E. Biancolini. *Fast Radial Basis Functions for Engineering Applications.* Mathematics and Statistics. Springer, Cham, Switzerland, 2018.

[31] D. Bìlková. Trimmed L-moments: analogy of classical L-moments. *American Journal of Mathematics and Statistics*, 4(2):80–106, 2014.

[32] H.S. Black. Stabilized feedback amplifiers. *The Bell System Technical Journal*, pages 1–18, 1934.

[33] M. Bland. Estimating mean and standard deviation from the sample size, three quartiles, minimum, and maximum. *International Journal of Statistics in Medical Research*, 4(1):57–64, 2014.

[34] B. Bobee, L. Perreault, and F. Ashkar. Two kinds of moment ratio diagrams and their applications in hydrology. *Stochastic Hydrology and Hydraulics*, 7:41–65, 1993.

[35] H.W. Bode. Relations between attenuation and phase in feedback amplifier design. *The Bell System Technical Journal*, 19(3):421–454, 1940.

[36] S. Borak, A. Misiorek, and R. Weron. Models for heavy-tailed asset returns. In P. Cizek, W. K. Hărdle, and R. Weron, editors, *Statistical tools for finance and insurance*, pages 21–56. Springer, New York, NY, 2nd edition, 2011.

[37] J. Borneman. A study of estimates of sigma in small sample sizes. `https://www.isixsigma.com/sampling-data/a-study-of-estimates-of-sigma-in-small-sample-sizes/`. Accessed: 2023-11-23.

[38] S. Boulaaras, R. Jan, and VT Pham. Recent advancement of fractional calculus and its applications in physical systems. *The European Physical Journal Special Topics*, 232:2347–2350, 2023.

[39] S. L. Bressler and A. K. Seth. Wiener-granger causality: a well established methodology. *NeuroImage*, 58(2):323–329, 2011.

[40] M. D. Buhmann. *Radial Basis Functions: Theory and Implementations.* Cambridge Monographs on Applied and Computational Mathematics. Cambridge University Press, Cambridge, UK, 2003.

[41] W.B. Cannon. *The Wisdom of the Body.* W W Norton & Co Inc, New York, NY, 1932.

[42] A. Cerioli. Multivariate outlier detection with high-breakdown estimators. *Journal of the American Statistical Association*, 105(489):147–156, 2010.

[43] P. Chaber and P.D. Domański. Control assessment with moment ratio diagrams. In M. Pawelczyk, D. Bismor, S. Ogonowski, and J. Kacprzyk, editors, *Advanced, Contemporary Control*, pages 3–13. Springer Nature Switzerland, Switzerland, 2023.

[44] A. Chakraborty and C.P. Tsokos. A modern approach of survival analysis of patients with pancreatic cancer. *American Journal of Cancer Research*, 11(10):4725–4745, 2021.

[45] P.C. Chandrasekharan. *Robust Control of Linear Dynamical Systems.* Elsevier Science & Technology Books, London, UK, 1996.

[46] G. Chen. Generalized log-normal distributions with reliability application. *Computational Statistics & Data Analysis*, 19(3):309–319, 1995.

[47] L.J. Chen, J. Zhang, and L. Zhang. Entropy information based assessment of cascade control loops. In *Proceedings of the International MultiConference of Engineers and Computer Scientists, IMECS 2015*, 2015.

[48] Y. Chen, S. L. Bressler, and M. Ding. Dynamics on networks: assessing functional connectivity with granger causality. *Computational and Mathematical Organization Theory*, 15:329–350, 2009.

[49] M.A.A.S. Choudhury, S.L. Shah, and N.F. Thornhill. Diagnosis of poor control-loop performance using higher-order statistics. *Automatica*, 40(10):1719–1728, 2004.

[50] W. Clarke, C. Mohtadi, and P. S. Tuffs. Generalized predictive control – I. The basic algorithm. *Automatica*, 23(2):137–148, 1987.

[51] A. Clauset, C.R. Shalizi, and M.E. J. Newman. Power-law distributions in empirical data. *SIAM Review*, 51(4):661–703, 2009.

[52] R.G. Clegg. A practical guide to measuring the Hurst parameter. *International Journal of Simulation: Systems, Science & Technology*, 7(2):3–14, 2006.

[53] R. Cooke, D. Nieboer, and J. Misiewicz. *Fat-Tailed Distributions: Data, Diagnostics and Dependence.* Mathematical models and methods in reliability set. Wiley-ISTE, London, UK, Hoboken, NJ, 2014.

[54] D. Cousineau and S. Chartier. Outliers detection and treatment: a review. *International Journal of Psychological Research*, 3:58–67, 2010.

[55] D.R. Cox and D. Oakes. *Analysis of Survival Data.* Chapman and Hall, London, UK, 1984.

[56] CPC. *Univariate Controller Performance Assessment, Limited Trial Version.* University of Alberta, Computer Process Control Group, 2010.

[57] C.C. Craig. A new exposition and chart for the pearson system of frequency curves. *The Annals of Mathematical Statistics*, 7(1):16–28, 1936.

[58] H. Cramer. On the composition of elementary errors. *Scandinavian Actuarial Journal*, 1:13–74, 1928.

[59] T. Cui, Z. Wang, H. Gu, P. Qin, and J. Wang. Gamma distribution based predicting model for breast cancer drug response based on multi-layer feature selection. *Frontierts in Genetics*, 14:1095976, 2023.

[60] C.R. Cutler and B. L. Ramaker. Dynamic matrix control – a computer control algorithm. *Joint Automatic Control Conference*, 17:72, 1980.

[61] R. Cutler and B. Ramaker. Dynamic matrix control – a computer control algorithm. In *Proc. AIChE National Meeting, Houston, TX, US*, 1979.

[62] M. Czajkowski and M. Kretowski. The role of decision tree representation in regression problems – an evolutionary perspective. *Applied Soft Computing*, 48:458–475, 2016.

[63] L.R. da Silva, R.C.C. Flesch, and J.E. Normey-Rico. Analysis of anti-windup techniques in pid control of processes with measurement noise. *IFAC-PapersOnLine*, 51(4):948–953, 2018. 3rd IFAC Conference on Advances in Proportional-Integral-Derivative Control PID 2018.

[64] L.L. da Silva Costa and F. Ferraz do Nascimento. Regression in extremes using the four-parameter kappa distribution. *Communications in Statistics – Simulation and Computation*, 0(0):1–13, 2023.

[65] J. Danielsson, L.M. Ergun, L. de Haan, and Casper G. de Vries. Tail index estimation: quantile-driven threshold selection. Bank of Canada, Staff Working Paper 2019-28, 2019.

[66] P. Daoutidis, M. Zachar, and S.S. Jogwar. Sustainability and process control: a survey and perspective. *Journal of Process Control*, 44:184–206, 2016.

[67] L. Das, B. Srinivasan, and R. Rengaswamy. Data driven approach for performance assessment of linear and nonlinear Kalman filters. In *2014 American Control Conference*, pages 4127–4132, 2014.

[68] L. Das, B. Srinivasan, and R. Rengaswamy. Multivariate control loop performance assessment with Hurst exponent and mahalanobis distance. *IEEE Transactions on Control Systems Technology*, 24(3):1067–1074, 2016.

[69] R. Davidson and J.G. MacKinnon. *Estimation and Inference in Econometrics*. Oxford University Press, New York, NY, 1993.

[70] R. De Maesschalck, D. Jouan-Rimbaud, and D.L. Massart. The mahalanobis distance. *Chemometrics and Intelligent Laboratory Systems*, 50(1):1–18, 2000.

[71] D.Q.F. de Menezes, D.M. Prata, A.R. Secchi, and J.C. Pinto. A review on robust m-estimators for regression analysis. *Computers & Chemical Engineering*, 147:107254, 2021.

[72] L. Desborough and T.J. Harris. Performance assessment measures for univariate feedforward/feedback control. *The Canadian Journal of Chemical Engineering*, 71(4):605–616, 1993.

[73] D.A. Dickey and W. A. Fuller. Distribution of the estimators for autoregressive time series with a unit root. *Journal of the American Statistical Association*, 74(366):427–431, 1979.

[74] W. J. Dixon. Analysis of extreme values. *The Annals of Mathematical Statistics*, 21(4):488–506, 1950.

[75] P. D. Domański. Fractional order for multi-criteria control performance assessment. *IFAC-PapersOnLine*, 58(12):291–296, 2024. 12th IFAC Conference on Fractional Differentiation and its Applications, Bordeaux, France.

[76] P. D. Domański. Index ratio diagram – a new way to assess control performance. In *2024 European Control Conference (ECC)*, pages 2041–2046, Stockholm, Sweden, 2024.

[77] P.D. Domański. Non-Gaussian properties of the real industrial control error in SISO loops. In *Proc. of the 19th International Conference on System Theory, Control and Computing*, pages 877–882, 2015.

[78] P.D. Domański. Fractal measures in control performance assessment. In *Proc. of IEEE International Conference on Methods and Models in Automation and Robotics,Miedzyzdroje, Poland*, pages 448–453, 2016.

[79] P.D. Domański. Non-Gaussian and persistence measures for control loop quality assessment. *Chaos: An Interdisciplinary Journal of Nonlinear Science*, 26(4):043105, 2016.

[80] P.D. Domański. Multifractal properties of process control variables. *International Journal on Bifurcation and Chaos*, 27(6):1750094, 2017.

[81] P.D. Domański. Non-Gaussian statistical measures of control performance. *Control & Cybernetics*, 46(3):259–290, 2017.

[82] P.D. Domański. Statistical measures for proportional–integral–derivative control quality: simulations and industrial data. *Proceedings of the Institution of Mechanical Engineers, Part I: Journal of Systems and Control Engineering*, 232(4):428–441, 2018.

[83] P.D. Domański. Control quality assessment using fractal persistence measures. *ISA Transactions*, 90:226–234, 2019.

[84] P.D. Domański. Basic data-based measures. In *Control Performance Assessment: Theoretical Analyses and Industrial Practice*, pages 37–51. Springer International Publishing, Cham, Switzerland, 2020.

[85] P.D. Domański. Control feasibility study. In *Control Performance Assessment: Theoretical Analyses and Industrial Practice*, pages 359–367. Springer International Publishing, Cham, Switzerland, 2020.

[86] P.D. Domański. *Control Performance Assessment: Theoretical Analyses and Industrial Practice.* Springer International Publishing, Cham, Switzerland, 2020.

[87] P.D. Domański. Performance assessment of predictive control – a survey. *Algorithms*, 13(4):97, 2020.

[88] P.D. Domański. Statistical measures. In *Control Performance Assessment: Theoretical Analyses and Industrial Practice*, pages 53–74. Springer International Publishing, Cham, Switzerland, 2020.

[89] P.D. Domański. Statistical outlier labelling – a comparative study. In *Preprints of 7th International Conference on Control, Decision and Information Technologies, Prague, Czech Republic*, pages 439–444, 2020.

[90] P.D. Domański. Study on statistical outlier detection and labelling. *International Journal of Automation and Computing*, 17(6):788–811, 2020.

[91] P.D. Domański. Energy-aware multicriteria control performance assessment. *Energies*, 17(5):1173, 2024.

[92] P.D. Domański, YQ Chen, and M. Ławryńczuk. Outliers in control engineering – they exist, like it or not. In P.D. Domański, YQ Chen, and M. Ławryńczuk, editors, *Outliers in Control Engineering, Fractional Calculus Perspective*, pages 1–24. De Gruyter, Berlin, Boston, 2022.

[93] P.D. Domański, YQ Chen, and M. Ławryńczuk. Outliers in control engineering – they exist, like it or not. In P.D. Domański, YQ Chen, and M. Ławryńczuk, editors, *Outliers in Control Engineering: Fractional Calculus Perspective*, pages 1–24. De Gruyter, Berlin, Boston, 2022.

[94] P.D. Domański, YQ Chen, and M. Ławryńczuk, editors. *Outliers in Control Engineering: Fractional Calculus Perspective.* De Gruyter, Berlin, Boston, 2022.

[95] P.D. Domański, K. Dziuba, and R. Góra. PID control assessment using L-moment ratio diagrams. *Applied Sciences*, 14(8):3331, 2024.

[96] P.D. Domański and M. Gintrowski. Alternative approaches to the prediction of electricity prices. *International Journal of Energy Sector Management*, 11(1):3–27, 2017.

[97] P.D. Domański, S. Golonka, R. Jankowski, P. Kalbarczyk, and B. Moszowski. Control rehabilitation impact on production efficiency of ammonia synthesis installation. *Industrial & Engineering Chemistry Research*, 55(39):10366–10376, 2016.

[98] P.D. Domański, R. Jankowski, K. Dziuba, and R. Góra. Assessing control sustainability using l-moment ratio diagrams. *Electronics*, 12(11):2377, 2023.

[99] P.D. Domański and M. Ławryńczuk. Assessment of the GPC control quality using non-Gaussian statistical measures. *International Journal of Applied Mathematics & Computer Science*, 27(2):291–307, 2017.

[100] P.D. Domański and M. Ławryńczuk. Multi-criteria control performance assessment method for multivariate mpc. In *Proc. of 2020 American Control Conference, Denver, CO*, pages 1968–1973, 2020.

[101] P.D. Domański, M. Ławryńczuk, S. Golonka, B. Moszowski, and P. Matyja. Multi-criteria loop quality assessment: a large-scale industrial case study. In *Proc. of IEEE International Conference on Methods and Models in Automation and Robotics, Miedzyzdroje, Poland*, pages 99–104, 2019.

[102] P. Duan. Information theory-based approaches for causality analysis with industrial applications. *Thesis, University of Alberta*, 2014.

[103] K. Dziuba, R. Góra, P.D. Domański, and M. Ławryńczuk. Multicriteria ammonia plant assessment for the advanced process control implementation. *IEEE Access*, 8:207923–207937, 2020.

[104] E. Eftelioglu. Geometric median. In S. Shekhar, H. Xiong, and X. Zhou, editors, *Encyclopedia of GIS*, pages 701–704. Springer International Publishing, Cham, Switzerland, 2017.

[105] G. Elliott, T.J. Rothenberg, and J.H. Stock. Efficient tests for an autoregressive unit root. *Econometrica*, 64(4):813–836, 1996.

[106] W.R. Erdelen and J.G. Richardson. *Managing Complexity: Earth Systems and Strategies for the Future.* Routledge, Abingdon, UK, 2020.

[107] M. J. Falkowski and P.D. Domański. Causality analysis with different probabilistic distributions using transfer entropy. *Applied Sciences*, 13(10):5849, 2023.

[108] M.J. Falkowski. *Root Cause Analysis of control errors propagation in complex multi-loop systems.* PhD thesis, Faculty of Electronics and Information Sciences, Warsaw University of Technology, 2024.

[109] M.J. Falkowski, P.D. Domański, and E. Pawłuszewicz. Causality in control systems based on data-driven oscillation identification. *Applied Sciences*, 12(8):3784, 2022.

[110] J. Feder. *Fractals.* Springer, New York, NY, 2013.

[111] I. Fedotenkov. A review of more than one hundred pareto-tail index estimators. *Statistica*, 80(3):245–299, 2021.

[112] T.S. Ferguson. Maximum likelihood estimates of the parameters of the cauchy distribution for samples of size 3 and 4. *Journal of the American Statistical Association*, 73(361):211–213, 1978.

[113] S. Foss, D. Korshunov, and S. Zachary. *Heavy-Tailed and Long-Tailed Distributions*, pages 7–42. Springer, New York, NY, 2013.

[114] C.L.E. Franzke, T. Graves, N.W. Watkins, R.B. Gramacy, and C. Hughes. Robustness of estimators of long-range dependence and self-similarity under non-Gaussianity. *Philosophical Transactions of the Royal Society A*, 370(1962):1250–1267, 2012.

[115] X. Gabaix, D. Laibson, D. Li, H. Li, S. Resnick, and C. G. de Vries. The impact of competition on prices with numerous firms. *Journal of Economic Theory*, 165:1–24, 2016.

[116] J. Gao, R. Patwardhan, K. Akamatsu, Y. Hashimoto, G. Emoto, S.L. Shah, and B. Huang. Performance evaluation of two industrial mpc controllers. *Control Engineering Practice*, 11(12):1371–1387, 2003. 2002 IFAC World Congress.

[117] C. F. Gauss. *Theoria motus corporum coelestium in sectionibus conicis solem ambientium.* F. Perthes et I.H. Besser, Hamburg, Germany, 1805.

[118] M.G. Genton and A. Lucas. Comprehensive definitions of breakdown points for independent and dependent observations. *Journal of the Royal Statistical Society. Series B (Statistical Methodology)*, 65(1), 2003.

[119] J. Geweke and S. Porter-Hudak. The estimation and application of long memory time series models. *Journal of Time Series Analysis*, 4:221–238, 1983.

[120] M.E. Ghitany, E. Gemez-Deniz, and S. Nadarajah. A new generalization of the pareto distribution and its application to insurance data. *Journal of Risk and Financial Management*, 11(1):10, 2018.

[121] M. Giełczewski, M. Piniewski, and P.D. Domański. Mixed statistical and data mining analysis of river flow and catchment properties at regional scale. *Stochastic Environmental Research and Risk Assessment*, 36:2861–2882, 2022.

[122] P. J. Gisi. *Sustaining a Culture of Process Control and Continuous Improvement: The Roadmap for Efficiency and Operational Excellence*. Routledge, Taylor & Francis Group, 2018.

[123] G. C. Goodwin and K. S. Sin. *Adaptive Filtering Prediction and Control*. Dover Books on Electrical Engineering. Dover Publications, Mineola, NY, 2009.

[124] C.W.J. Granger. Investigating causal relations by econometric models and cross-spectral methods. *Econometrica*, 37(3):424–438, 1969.

[125] C.W.J. Granger and P. Newbold. Spurious regression in econometrics. *Journal of Econometrics*, 2(2):111–120, 1974.

[126] P. Grigolini, L. Palatella, and G. Raffaelli. Asymmetric anomalous diffusion: an efficient way to detect memory in time series. *Fractals*, 9(4):439–449, 2001.

[127] F. E. Grubbs. Sample criteria for testing outlying observations. *The Annals of Mathematical Statistics*, 21(1):27–58, 03 1950.

[128] Ł. Gruss, J. Pollert Jr., J. Pollert Sr., M. Wiatkowski, and S. Czaban. The application of new distribution in determining extreme hydrologic events such as floods. *Hydrology and Earth System Sciences Discussions*, 2020:1–31, 2020.

[129] G.J. Haarsma and M. Nikolaou. Multivariate controller performance monitoring: lessons from an application to a snack food process. In *AIChE Annual Meeting*, 1999.

[130] K. Haddad. Selection of the best fit probability distributions for temperature data and the use of L-moment ratio diagram method: a case study for NSW in Australia. *Theoretical and Applied Climatology*, 143:1261–1284, 2021.

[131] P. Hall. On some simple estimates of an exponent of regular variation. *Journal of the Royal Statistical Society: Series B (Methodological)*, 44(1):37–42, 1982.

[132] F.R. Hampel. *Contribution to the theory of robust estimation.* PhD thesis, University of California, Berkeley, 1968.

[133] F.R. Hampel, E.M. Ronchetti, P.J. Rousseeuw, and W.A. Stahel. *Robust Statistics: The Approach Based on Influence Functions.* John Wiley & Sons, New York, Chichetser, Brisbane, Toronto, Singapore, 1986. Wiley Series in Probability and Statistics.

[134] J. Hansen, S. Ahern, and A. Earnest. Evaluations of statistical methods for outlier detection when benchmarking in clinical registries: a systematic review. *BMJ Open*, 13(7), 2023.

[135] T.J. Harris. Assessment of closed loop performance. *Canadian Journal of Chemical Engineering*, 67:856–861, 1989.

[136] J. Haslett and A.E. Raftery. Space-time modelling with long-memory dependence: assessing ireland's wind power resource. *Journal of Applied Statistics*, 38:1–50, 1989.

[137] U. Hassler. Regression of spectral estimators with fractionally integrated time series. *Journal on Time Series Analysis*, 14(4):369–380, 1993.

[138] D.M. Hawkins. *Identification of Outliers.* Chapman and Hall, London, New York, 1980.

[139] H. L. Házen. Theory of servo-mechanisms. *Journal of the Franklin Institute*, 218:279–331, 1934.

[140] N.A. Heckert, J.J. Filliben, C.M. Croarkin, B. Hembree, W.F. Guthrie, P. Tobias., and J. Prinz. NIST/SEMATECH e-handbook

of statistical methods. NIST, US Department of Commerce, [08-February-2020], 2012.

[141] T. Hägglund. A control-loop performance monitor. *Control Engineering Practice*, 3(11):1543–1551, 1995.

[142] T. Hägglund. Automatic detection of sluggish control loops. *Control Engineering Practice*, 7(12):1505–1511, 1999.

[143] T. Higuchi. Approach to an irregular time series on the basis of the fractal theory. *Physica D: Nonlinear Phenomena*, 31(2):277–283, 1988.

[144] B.M. Hill. A simple general approach to inference about the tail of a distribution. *The Annals of Statistics,*, 3(5):1163–1174, 1975.

[145] J.E. Hirsch. An index to quantify an individual's scientific research output. *Proceedings of the National Academy of Sciences*, 102(46):16569–16572, 2005.

[146] T. K. Ho. Random decision forests. In *Proc. of 3rd International Conference on Document Analysis and Recognition*, volume 1, pages 278–282, 1995.

[147] P.W. Holland and R.E. Welsch. Robust regression using iteratively reweighted least-squares. *Communications in Statistics – Theory and Methods*, 6(9):813–827, 1977.

[148] J.D. Holmes and W.W. Moriarty. Application of the generalized pareto distribution to extreme value analysis in wind engineering. *Journal of Wind Engineering and Industrial Aerodynamics*, 83(1):1–10, 1999.

[149] J.R.M. Hosking. L-moments: analysis and estimation of distributions using linear combinations of order statistics. *Journal of the Royal Statistical Society. Series B (Methodological)*, 52(1):105–124, 1990.

[150] J.R.M. Hosking. Moments or L-moments? an example comparing two measures of distributional shape. *The American Statistician*, 46(3):186–189, 1992.

[151] J.R.M. Hosking. The four-parameter kappa distribution. *IBM Journal of Research and Development*, 38(3):251–258, 1994.

[152] J.R.M. Hosking. Some theory and practical uses of trimmed L-moments. *Journal of Statistical Planning and Inference*, 137(9):3024–3039, 2007.

[153] J.R.M. Hosking and James R. Wallis. *Regional Frequency Analysis: An Approach Based on L-Moments.* Cambridge University Press, Cambridge, UK, 1997.

[154] S.P. Hozo, B. Djulbegovic, and I. Hozo. Estimating the mean and variance from the median, range, and the size of a sample. *BMC Medical Research Methodology*, 5:13, 2005.

[155] B. Huang. *Multivariate Statistical Methods for Control Loop Performance Assesment.* PhD thesis, University of Alberta, Department of Chemicai and Material Engineering, Canada, 1997.

[156] N. E. Huang, Z. Shen, S. R. Long, M. C. Wu, H. H. Shih, Q. Zheng, N. C. Yen, C.C. Tung, and H. H. Liu. The empirical mode decomposition and the hilbert spectrum for nonlinear and non-stationary time series analysis. *Proceedings of the Royal Society of London. Series A: Mathematical, Physical and Engineering Sciences*, 454(1971):903–995, 1998.

[157] P.J. Huber and E.M. Ronchetti. *Robust Statistics, 2nd Edition.* John Wiley & Sons, Hoboken, NJ, 2009.

[158] M. Hubert and M. Debruyne. Minimum covariance determinant. *WIREs Computational Statistics*, 2(1):36–43, 2010.

[159] R. Huisman, K.G. Koedijk, C.J.M. Kool, and F. Palm. Tail-index estimates in small samples. *Journal of Business & Economic Statistics*, 19(2):208–216, 2001.

[160] H.E. Hurst, R.P. Black, and Y.M. Simaika. *Long-term storage: an experimental study.* Constable and Co Limited, London, UK, 1965.

[161] A. Hurwitz. On the conditions under which an equation has only roots with negative real parts. *Mathematische Annalen*, 46:273–284, 1895.

[162] Al.-H. Hussien and Al-R. Hamed. Ratio of products of mixture gamma variates with applications to wireless communications systems. *IET Communications*, 15(15):1963–1981, 2021.

[163] B. Iglewicz and D. C. Hoaglin. *How to detect and handle outliers.* ASQC Quality Press, Milwaukee, WI, 1993.

[164] J.O. Irvin. On a criterion for the rejection of outlying observation. *Biometrika*, 17:238–250, 1925.

[165] R. Isermann and M. Műnchhof. *Identification of Dynamic Systems: An Introduction with Applications.* Springer, Berlin, Heidelberg, Germany, 2011.

[166] M. Jafari-Mamaghani and J. Tyrcha. Transfer entropy expressions for a class of non-gaussian distributions. *Entropy*, 16(3):1743–1755, 2014.

[167] S. Jaffard. Multifractal formalism for functions. *SIAM Journal of Mathematical Analysis*, 28:944–970, 1997.

[168] S. Jaffard, B. Lashermes, and P. Abry. Wavelet leaders in multifractal analysis. In T. Qian, M. I. Vai, and Y. Xu, editors, *Wavelet analysis and applications*, pages 201–246. Birkhǻuser Verlag, Basel, Switzerland, 2007.

[169] S. Jain and N. Choudhary. Anomaly detection in image processing using artificial intelligence. In *2023 International Conference on Device Intelligence, Computing and Communication Technologies, (DICCT)*, pages 428–432, 2023.

[170] M. Jelali. An overview of control performance assessment technology and industrial applications. *Control Engineering Practice*, 14(5):441–466, 2006. Intelligent Control Systems and Signal Processing.

[171] M. Jelali. *Control Performance Management in Industrial Automation: Assessment, Diagnosis and Improvement of Control Loop Performance.* Springer, London, UK, 2013.

[172] A.F. Jenkinson. The frequency distribution of the annual maximum (or minimum) values of meteorological elements. *Quarterly Journal of the Royal Meteorological Society*, 81(348):158–171, 1955.

[173] H-D. J. Jeong, D. McNiclke, and K. Pawlikowski. Hurst parameter estimation techniques: a critical review. In *38th Annual ORSNZ Conference*, 2001.

[174] Y. Jia, J. Zhou, and D. Li. Performance assessment of cascade control loops with non-Gaussian disturbances. In *2018 Chinese Automation Congress*, pages 2451–2456, 2018.

[175] H. Jiang, R. Patwardhan, and S. L. Shah. Root cause diagnosis of plant-wide oscillations using the concept of adjacency matrix. *Journal of Process Control*, 19(8):1347–1354, 2009. Special Section on Hybrid Systems: Modeling, Simulation and Optimization.

[176] N.L. Johnson, S. Kotz, and N. Balakrishnan. *Continuous Univariate Distributions.* Number 2 in Wiley series in probability and mathematical statistics: Applied probability and statistics. John Wiley & Sons, New York, NY, 1995.

[177] R.A. Johnson and D.W. Wichern. *Applied Multivariate Statistical Analysis.* Prentice-Hall, Englewood Cliffs, NJ, 3 edition, 1992.

[178] R.E. Kalman. Contributions to the theory of optimal control. *Boletin de la Sociedad Matematica Mexicana*, 5:102–119, 1960.

[179] R.E. Kalman. A new approach to linear filtering and prediction problems. *Journal of Basic Engineering*, 82(1):35–45, 03 1960.

[180] R.E. Kalman. A new approach to linear filtering and prediction problems. *ASME A New Approach to Linear Filtering and Prediction Problems*, 82(1):35–45, 1960.

[181] J. Kaniuka, J. Ostrysz, M. Groszyk, K. Bieniek, S. Cyperski, and P.D. Domański. Study on cost estimation of the external fleet full truckload contracts. In *Proc. of the 20th International Conference on Informatics in Control, Automation and Robotics*, volume 2, pages 316–323, Rome, Italy, 2023.

[182] J. Kaniuka, J. Ostrysz, M. Groszyk, K. Bieniek, S. Cyperski, and P.D. Domański. Multicriteria machine learning model assessment – residuum analysis review. *Electronics*, 13(5):810, 2024.

[183] L.M. Kaplan and C.C. Jay Kuo. Fractal estimaton from noisy data via discrete fractional gaussian noise (DFGN) and the Haar basis. *IEEE Transactions on Signal Proccessing*, 41(12), 1993.

[184] T. Karagiannis, M. Faloutsos, and R. H. Riedi. Long-range dependence: now you see it, now you don't! *IEEE Global Telecommunications Conference GLOBECOM '02*, 3, 2002.

[185] S. Karra, M. N. Jelali, M. Karim, and A. Horch. Detection of oscillating control loops. In M. Jelali and B. Huang, editors, *Detection and Diagnosis of Stiction in Control Loops: State of the Art and Advanced Methods*, pages 61–100. Springer, London, UK, 2010.

[186] R.W. Katz, M. B. Parlange, and P. Naveau. Statistics of extremes in hydrology. *Advances in Water Resources*, 25(8):1287–1304, 2002.

[187] R.G. Kavasseri and R. Nagarajan. Evidence of crossover phenomena in wind-speed data. *IEEE Transactions on Circuits and Systems I: Regular Papers*, 51(11):2255–2262, 2004.

[188] R.G. Kavasseri and R. Nagarajan. A multifractal description of wind speed records. *Chaos, Solitons & Fractals*, 24(1):165–173, 2005.

[189] H. Kettani and J.A. Gubner. A novel approach to the estimation of the long-range dependence parameter. *IEEE Transactions on Circuits and Systems*, 53(6):463–467, 2006.

[190] L. Kirichenko, T. Radivilova, and Z. Deineko. Comparative analysis for estimating of the hurst exponent for stationary and nonstationary time series. *Journal, International Information Technologies & Knowledge*, 5:371–387, 2011.

[191] T.R. Kjeldsen, H. Ahn, and I. Prosdocimi. On the use of a four-parameter kappa distribution in regional frequency analysis. *Hydrological Sciences Journal*, 62(9):1354–1363, 2017.

[192] T.R. Kjeldsen and D.A. Jones. Prediction uncertainty in a median-based index flood method using L-moments. *Water Resources Research*, 42(7), 2006.

[193] L. B. Klebanov. Big outliers versus heavy tails: what to use? arXiv:1611.05410v1, Department of Probability and Mathematical Statistics, Charles University, Prague, Czech Republic, 2016.

[194] B-S. Ko and T.F. Edgar. Performance assessment of cascade control loops. *AIChE Journal*, 46(2):281–291, 2000.

[195] S. Kotz, T.J. Kozubowski, and K. Podgórski. Engineering sciences. In *The Laplace Distribution and Generalizations: A Revisit with Applications to Communications, Economics, Engineering, and Finance*, pages 277–287. Birkhäuser, Boston, MA, 2001.

[196] S. Kotz and S. Nadarajah. *Extreme Value Distributions. Theory and Applications.* Imperial College Press, London, UK, 2000.

[197] I. A. Koutrouvelis. Regression-type estimation of the parameters of stable laws. *Journal of the American Statistical Association*, 75(372):918–928, 1980.

[198] D. Koutsoyiannis. Climate change, the hurst phenomenon, and hydrological statistics. *Hydrological Science Journal*, 48(1):3–24, 2003.

[199] H. P. Kriegel, P. Kroger, and A. Zimek. Outlier detection techniques. Tutorial, 16th ACM SIGKDD Conference on Knowledge Discovery and Data Mining, Washington, DC, 2010.

[200] S. Y. Kung. *Kernel Methods and Machine Learning.* Cambridge University Press, Cambridge, UK, 2014.

[201] E.E. Kuruoglu. Density parameter estimation of skewed alpha-stable distributions. *IEEE Transactions on Signal Processing*, 49(10):2192–2201, 2001.

[202] D. Kwiatkowski, P.C.B. Phillips, P. Schmidt, and Shin Y. Testing the null hypothesis of stationarity against the alternative of a unit root. *Journal of Econometrics*, 54(1–3):159–178, 1992.

[203] J. C. Lagarias, J.A. Reeds, M. H. Wright, and P. E. Wright. Convergence properties of the nelder–mead simplex method in low dimensions. *SIAM Journal on Optimization*, 9(1):112–147, 1998.

[204] C.D. Lai, D. Murthy, and M. Xie. Weibull distributions and their applications. In H. Pham, editor, *Springer Handbook of Engineering Statistics*, pages 63–78. Springer, London, UK, 2006. Chapter 3.

[205] R. Landman, J. Kortela, Q. Sun, and S.L. Jamsa-Jounela. Fault propagation analysis of oscillations in control loops using data-driven causality and plant connectivity. *Computers and Chemical Engineering*, 71(4):446–456, 2014.

[206] X. Lang, N. ur Rehman, Y.Zhang, L. Xie, and H.Su. Median ensemble empirical mode decomposition. *Signal Processing*, 176:107686, 2020.

[207] K. H. Lee, B. Huang, and E. C. Tamayo. Sensitivity analysis for selective constraint and variability tuning in performance assessment of industrial MPC. *Control Engineering Practice*, 16(10):1195–1215, 2008.

[208] A.M. Legendre. *Nouvelles méthodes pour la determination des orbites des comètes.* Firmin Didot, Paris, France, 1805.

[209] D. Li, Q. Ge, P. Zhang, Y. Xing, Z. Yang, and W. Nai. Ridge regression with high order truncated gradient descent method. In *12th International Conference on Intelligent Human-Machine Systems and Cybernetics*, volume 1, pages 252–255, 2020.

[210] L. Li. Bayes estimation of parameter of laplace distribution under a new linex-based loss function. *International Journal of Data Science and Analysis*, 3(6):85–89, 2017.

[211] L. Li, Z. Li, Y. Zhang, and Y. Chen. A mixed-fractal traffic flow model whose hurst exponent appears crossover. In *2012 Fifth International Joint Conference on Computational Sciences and Optimization*, pages 443–447, 2012.

[212] M. Li, W. Zhao, and C. Cattani. Delay bound: fractal traffic passes through network servers. *Mathematical Problems in Engineering*, 2013, 2013.

[213] H. Lilliefors. On the Kolmogorov-Smirnov test for normality with mean and variance unknown. *Journal of the American Statistical Association*, 62:399–402, 1967.

[214] B. Lindner, L. Auret, and M. Bauer. Investigating the impact of perturbations in chemical processes on data-based causality analysis. part 1: Defining desired performance of causality analysis techniques. *IFAC-PapersOnLine*, 50(1):3269–3274, 2017. 20th IFAC World Congress.

[215] A. Lipi, K. Kumar, and S. Chakraborty. *Outliers: Detection and Analysis*. Mathematics Research Developments. Nova Science Publishers, Hauppauge, NY, 2022.

[216] C. Liu and D.B. Rubin. ML estimation of the t distribution using EM and its extensions, ECM and ECME. *Statistica Sinica*, 5:19–39, 1995.

[217] K. Liu, Y. Q. Chen, P.D. Domański, and X. Zhang. A novel method for control performance assessment with fractional order signal processing and its application to semiconductor manufacturing. *Algorithms*, 11(7):90, 2018.

[218] K. Liu, YQ Chen, and P. D. Domański. Control performance assessment of the disturbance with fractional order dynamics. In W. Lacarbonara, B. Balachandran, J. Ma, J. A. Tenreiro Machado, and G. Stepan, editors, *Nonlinear Dynamics and Control*, pages 255–264, 2020.

[219] L. Ljung. *System Identification: Theory for the User*. Prentice-Hall Information and System Sciences Series. Prentice-Hall, Upper Saddle River, NJ, 1987.

[220] R. Lopes and N. Betrouni. Fractal and multifractal analysis: A review. *Medical Image Analysis*, 13(4):634–649, 2009.

[221] F. Louzada, P. Ramos, and G. Perdoná. Different estimation procedures for the parameters of the extended exponential geometric distribution for medical data. *Computational and Mathematical Methods in Medicine*, 2016, 2016. Article ID 8727951.

[222] D. Luo, X. Wan, J. Liu, and T. Tong. Optimally estimating the sample mean from the sample size, median, mid-range, and/or mid-quartile range. *Statistical methods in medical research*, 27(6):1785–1805, 2018.

[223] A.M. Lyapunov. Problème général de la stabilité du mouvement. *Annales de la Faculté des sciences de Toulouse : Mathématiques*, 2(9):203–474, 1907. przedruk z rosyjskiego oryginału, *Comm. Soc. Math. Kharkow*, 1892.

[224] K. MacTavish and T.D. Barfoot. At all costs: A comparison of robust cost functions for camera correspondence outliers. In *12th Conference on Computer and Robot Vision*, pages 62–69, 2015.

[225] E.E. Maeda, J. Arevalo Torres, and C. Carmona-Moreno. Characterisation of global precipitation frequency through the L-moments approach. *Area*, 45(1):98–108, 2013.

[226] B. Majumder, S. Das, I. Pan, S. Saha, S. Das, and A. Gupta. Estimation, analysis and smoothing of self-similar network induced delays in feedback control of nuclear reactors. In *International Conference on Process Automation, Control and Computing*, 2011.

[227] B. B. Mandelbrot. *Les objets fractals: forme, hasard, et dimension.* Flammarion, Paris, France, 1975.

[228] B. B. Mandelbrot, A. Fisher, and L. Calvet. A multifractal model of asset returns. Technical Report 1164, Cowles Foundation Discussion Paper, 1997.

[229] B. B. Mandelbrot and R. L. Hudson. *The Misbehavior of Markets: A Fractal View of Financial Turbulence.* Basic Books, New York, NY, 2004.

[230] B.B. Mandelbrot. Possible refinements of the lognormal hypothesis concerning the distribution of energy dissipation in intermitent turbulence. In M. Rosenblatt and C. Van Atta, editors, *Statistical Models and Turbulence.* Springer, New York, NY, 1972.

[231] D. P. Mandic, N. Rehman, Z. Wu, and N. E. Huang. Empirical mode decomposition-based time-frequency analysis of multivariate signals: The power of adaptive data analysis. *IEEE Signal Processing Magazine*, 30(6):74–86, 2013.

[232] T. Matsuo, I. Tadakuma, and N. Thornhill. Diagnosis of a unit-wide disturbance caused by saturation in a manipulated variable. In *IEEE Advanced Process Control Applications for Industry Workshop, Vancouver, Canada*, volume 1, 2004.

[233] J. C. Maxwell. On governors. *Proceedings of the Royal Society*, 16:270–283, 1868.

[234] M. Mayadunne, G. Evans and B. Inder. An empirical investigation of shock persistence in economic time series. *Economic Record*, 71(2):145–156, 1995.

[235] J. H. McCulloch. Simple consistent estimators of stable distribution parameters. *Communications in Statistics – Simulation and Computation*, 15(4):1109–1136, 1986.

[236] K. G. Mehrotra, C. K. Mohan, and H. M. Huang. *Anomaly Detection Principles and Algorithms.* Terrorism, Security, and Computation. Springer, Cham, Switzerland., 2017.

[237] Q. W. Meng, F. Fang, and J. Z. Liu. Minimum-information-entropy-based control performance assessment. *Entropy*, 15(3):943–959, 2013.

[238] T. Miao and D.E. Seborg. Automatic detection of excessively oscillatory feedback control loops. In *Proceedings of the 1999 IEEE International Conference on Control Applications (Cat. No.99CH36328)*, volume 1, pages 359–364 vol. 1, 1999.

[239] Minitab. Estimate process variation with individuals data. `https://support.minitab.com/en-us/minitab/21/help-and-how-to/quality-and-process-improvement/control-charts/supporting-topics/estimating-variation/individuals-data/`. Accessed: 2023-11-23.

[240] N. Minorsky. Directional stability and automatically steered bodies. *Journal of American Society of Naval Engineering*, 34(2):284, 1922.

[241] G.S. Mudholkar and A.D. Hutson. LQ-moments: Analogs of L-moments. *Journal of Statistical Planning and Inference*, 71(1):191–208, 1998.

[242] J.F. Muzy, E. Bacry, and A. Arneodo. Wavelets and multifractal formalism for singular signals: Application to turbulence data. *Physical Review Letters*, 67(25):3515–3518, 1991.

[243] D. Pe na and V. J. Yohai. A review of outlier detection and robust estimation methods for high dimensional time series data. *Econometrics and Statistics*, 2023.

[244] O. Naifar and A.B. Makhlouf. *Fractional Order Systems – Control Theory and Applications.* Studies in Systems, Decision and Control. Springer, Cham, Switzerland, 2022.

[245] J. Nair, A. Wierman, and B. Zwart. *The Fundamentals of Heavy Tails: Properties, Emergence, and Estimation.* Cambridge Series in Statistical and Probabilistic Mathematics. Cambridge University Press, Cambridge, UK, 2022.

[246] N. M. Neykov, P. N. Neytchev, P. H. A. J. M. Van Gelder, and V. K. Todorov. Robust detection of discordant sites in regional frequency analysis. *Water Resources Research*, 43(6):W06417, 2007.

[247] Y. Nishikawa, N. Sannomiya, T. Ohta, and H. Tanaka. A method for auto-tuning of PID control parameters. *Automatica*, 20(3):321–332, 1984.

[248] A. Nissinen, S. Nuyan, and P. Virtanen. On-line performance monitoring in process analysis. In *Proceedings of Control Systems*, 2006.

[249] H. Nyquist. Regeneration theory. *The Bell System Technical Journal*, pages 126–147, 1932.

[250] M. Olteanu, F. Rossi, and F. Yger. Meta-survey on outlier and anomaly detection. *Neurocomputing*, 555:126634, 2023.

[251] A. Ordys, D. Uduehi, and M. A. Johnson. *Process Control Performance Assessment – From Theory to Implementation*. Springer-Verlag, London, UK, 2007.

[252] J.W. Osborne and A. Overbay. The power of outliers (and why researchers should always check for them). *Practical Assessment, Research, and Evaluation*, 9(6):1–8, 03 2004.

[253] S.K. Patakamuri, K. Muthiah, and V. Sridhar. Long-term homogeneity, trend, and change-point analysis of rainfall in the arid district of ananthapuramu, andhra pradesh state, india. *Water*, 12(1):211, 2020.

[254] M. Pawlak. Performance analysis of power boiler drum water level control systems. *Acta Energetica*, 29(4):81–89, 2016.

[255] J. Pearl. *Causality: Models, Reasoning and Inference*. Cambridge University Press, New York, NY, 2009.

[256] E. S. Pearson and C. Chandra Sekar. The efficiency of statistical tools and a criterion for the rejection of outlying observations. *Biometrika*, 28(3/4):308–320, 12 1936.

[257] K. Pearson. Contributions to the mathematical theory of evolution. – II. Skew variation in homogeneous material. *Philosophical Transactions of the Royal Society of London Series A*, 186:343–414, 1895.

[258] R.K. Pearson. *Mining Imperfect Data; Dealing with Contamination and Incomplete Records.* SIAM, Philadelphia, PA, 2005.

[259] M. Peel, Q. Wang, and T. Mcmahon. The utility L-moment ratio diagrams for selecting a regional probability distribution. *Hydrological Sciences Journal*, 46:147–155, 02 2001.

[260] B. Peirce. Criterion for the rejection of doubtful observations. *Astronomical Journal*, 2(45):161–163, 7 1852.

[261] C. K. Peng, S. V. Buldyrev, S. Havlin, M. Simons, H. E. Stanley, and A. L. Goldberger. Mosaic organization of DNA nucleotides. *Physical Review E*, 49(2):1685–1689, 1994.

[262] J.A. Perea. A brief history of persistence. *Morfismos*, 23(1):1–16, 2019.

[263] J. S. Perkiőmăki, T. H. Măkikallio, and H. V. Huikuri. Fractal and complexity measures of heart rate variability. *Clinical and experimental hypertension*, 27(2–3):149–158, 2005.

[264] E.E. Peters. *Chaos and Order in the Capital Markets: A New View of Cycles, Prices, and Market Volatility.* John Wiley & Sons, Hoboken, NJ, 1996. 2nd Edition.

[265] N. Pillay and P. Govender. A data driven approach to performance assessment of PID controllers for setpoint tracking. *Procedia Engineering*, 69:1130–1137, 2014.

[266] M. R. Pimentel and C. J. Munaro. Performance monitoring and retuning for cascaded control loops. In *Proc. of the 15th IEEE International Conference on Industry Applications*, pages 647–654, 2023.

[267] O. Podladchikova, B. Lefebvre, V. Krasnoselskikh, and V. Podladchikov. Classification of probability densities on the basis of Pearson's curves with application to coronal heating simulations. *Nonlinear Processes in Geophysics*, 10:323–333, 2003.

[268] H. Puchner and S. Selberherr. Simulation of ion implantation using the four-parameter kappa distribution function. In *4th International Conference on Solid-State and IC Technology*, pages 295–297, 1995.

[269] K. J. Åstrőm. Computer control of a paper machine – an application of linear stochastic control theory. *IBM Journal*, 11:389–405, 1967.

[270] K. J. Åstrőm. *Introduction to stochastic control theory*, volume 70 of *Mathematics in science and engineering*. Academic Press, London, UK, New York, NY, 1970.

[271] K. J. Åstrőm and L. Rundqwist. Integrator windup and how to avoid it. In *1989 American Control Conference*, pages 1693–1698, 1989.

[272] K.J. Åstrőm and T. Hăgglund. New tuning methods for PID controllers. In *Proceedings 3rd European Control Conference*, pages 2456–2462, 1995.

[273] K.J. Åstrőm and T. Hăgglund. Benchmark systems for pid control. In *IFAC Digital Control: Past, Present and Future of PlD Control*, pages 165–166, 2000.

[274] L. F. Recalde, R. Katebi, and H. Tauro. PID based control performance assessment for rolling mills: A multiscale PCA approach. In *2013 IEEE International Conference on Control Applications (CCA)*, pages 1075–1080, 2013.

[275] N. Rehman and D. P. Mandic. Multivariate empirical mode decomposition. *Proceedings of the Royal Society of London A: Mathematical, Physical and Engineering Sciences, The Royal Society*, 466(2117):1291–1302, 2009.

[276] W.J.J. Rey. *Introduction to Robust and Quasi-Robust Statistical Methods.* Springer-Verlag, Berlin, Heidelberg, New York, Tokyo, 1983.

[277] J. Richalet, A. Rault, J. L. Testud, and J. Papon. Model algorithmic control of industrial processes. *IFAC Proceedings Volumes*, 10(16):103–120, 1977. Preprints of the 5th IFAC/IFIP International Conference on Digital Computer Applications to Process Control, The Hague, The Netherlands, 14-17 June, 1977.

[278] J. Richalet, A. Rault, J.L. Testud, and J. Papon. Model predictive heuristic control: Applications to industrial processes. *Automatica*, 14(5):413–428, 1978.

[279] A. Robson and D. Reed. Statistical procedures for flood frequency estimation. In *Flood Estimation Handbook, 2nd edition*, volume 3 of the Flood Estimation Handbook. Centre for Ecology & Hydrology, Wallingford, UK, 1999.

[280] E. Ronchetti. The main contributions of robust statistics to statistical science and a new challenge. *METRON*, 79:127–135, 2021.

[281] F. Rosado. Outliers: The strength of minors. In A. Pacheco, R. Santos, M. R. Oliveira, and C. D. Paulino, editors, *New Advances in Statistical Modeling and Applications*, pages 17–27. Springer International Publishing, Cham, Switzerland, 2014.

[282] B. Rosner. Percentage points for a generalized esd many-outlier procedure. *Technometrics*, 25(2):165–172, 1983.

[283] S. M. Ross. *Introduction to Probability and Statistics for Engineers and Scientists.* Academic Press, London, UK, Boston, MA, 2014. Fifth Edition.

[284] P. J. Rousseeuw and K. Van Driessen. A fast algorithm for the minimum covariance determinant estimator. *Technometrics*, 41(3):212–223, 1999.

[285] P.J. Rousseeuw and C Croux. Alternatives to the median absolute deviation. *Journal of the American Statistical Association*, 88(424):1273–1283, 1993.

[286] P.J. Rousseeuw and M. Hubert. Anomaly detection by robust statistics. *WIREs Data Mining and Knowledge Discovery*, 8(2):e1236, 2018.

[287] P.J. Rousseeuw and A.M. Leroy. *Robust Regression and Outlier Detection.* John Wiley & Sons, New York, NY, 1987.

[288] P.J. Rousseeuw and S. Verboven. Robust estimation in very small samples. *Computational Statistics & Data Analysis*, 40(4):741–758, 2002.

[289] E.J. Routh. *A Treatise on the Stability of a Given State of Motion.* Macmillan & Co., London, UK, 1877.

[290] J. Royuela-del-Val, F. Simmross-Wattenberg, and C. Alberola-Lopez. libstable: fast, parallel, and high-precision computation

of α-stable distributions in R, C/C++, and MATLAB. *Journal of Statistical Software, Articles*, 78(1):1–25, 2017.

[291] D. B. Rubin. Estimating causal effects of treatments in randomized and nonrandomized studies. *Journal of Educational Psychology*, 66:688–701, 1974.

[292] D. Russel, J. Hanson, and E. Ott. Dimension of strange attractors. *Physical Review Letters*, 45(14):1175–1178, 1980.

[293] R. Sahu, M. Verma, and I. Ahmad. Regional frequency analysis using L-moment methodology – a review. In *Recent Trends in Civil Engineering*, pages 811–832. Springer, Singapore, 2021.

[294] T. I. Salsbury. A practical method for assessing the performance of control loops subject to random load changes. *Journal of Process Control*, 15(4):393–405, 2005.

[295] C. Scali, E. Marchetti, and A. Esposito. Effect of cascade tuning on control loop performance assessment. In *IFAC Conference on Advances in PID Control PID'12*, 2012.

[296] J. Schäfer and A. Cinar. Multivariable MPC system performance assessment, monitoring, and diagnosis. *Journal of Process Control*, 14(2):113–129, 2004.

[297] D. T. Schmitt and P. C. Ivanov. Fractal scale-invariant and nonlinear properties of cardiac dynamics remain stable with advanced age: a new mechanistic picture of cardiac control in healthy elderly. *American Journal of Physiology-Regulatory, Integrative and Comparative Physiology*, 293(5):R1923–R1937, 2007.

[298] T. Schreiber. Measuring information transfer. *Physical Review Letters*, 85(2):461–464, 2000.

[299] M. Schrőder. *Fractals, Chaos, Power Laws: Minutes from an Infinite Paradise.* W. H. Freeman and Company, New York, NY, 1991.

[300] E. Schulz, M. Speekenbrink, and A. Krause. A tutorial on Gaussian process regression: Modelling, exploring, and exploiting functions. *Journal of Mathematical Psychology*, 85:1–16, 2018.

[301] S. Schuster. Parameter estimation for the cauchy distribution. In *2012 19th International Conference on Systems, Signals and Image Processing*, pages 350–353, 2012.

[302] D. Seborg, T. F. Edgar, and D. Mellichamp. *Process dynamics & control.* John Wiley & Sons, New York, NY, 2006.

[303] D. E. Seborg, D.A. Mellichamp, T. F. Edgar, and F. J. Doyle. *Process dynamics and control.* John Wiley & Sons, Hoboken, NJ, 2010.

[304] E. Serrano and A. Figliola. Wavelet leaders: A new method to estimate the multifractal singularity spectra. *Physica A: Statistical Mechanics and its Applications*, 388(14):2793–2805, 2009.

[305] C. E. Shannon. A mathematical theory of communication. *The Bell System Technical Journal*, 27(3):379–423, 1948.

[306] S. S. Shapiro and M. B. Wilk. An analysis of variance test for normality (complete samples). *Biometrika*, 52(3–4):591–611, 1965.

[307] J. Sharpe and M. A. Juarez. Estimation of the pareto and related distributions – a reference-intrinsic approach. *Communications in Statistics – Theory and Methods*, 52(3):523–542, 2023.

[308] H. Sheng, Y.Q. Chen, and T.S. Qiu. *Fractional Processes and Fractional-Order Signal Processing.* Springer, London, UK, 2012.

[309] G. Shi, H.V. Atkinson, C.M. Sellars, and C.W. Anderson. Application of the generalized pareto distribution to the estimation of the size of the maximum inclusion in clean steels. *Acta Materialia*, 47(5):1455–1468, 1999.

[310] K. Shimizu and E.L. Crow. *Lognormal Distributions, Theory and Applications.* Routledge, New York, NY, 1988.

[311] F.G. Shinskey. How good are our controllers in absolute performance and robustness? *Measurement and Control*, 23(4):114–121, 1990.

[312] F.G. Shinskey. Process control: as taught vs as practiced. *Industrial and Engineering Chemistry Research*, 41:3745–3750, 2002.

[313] D. Shook. Special report: Process control-1: best practices improve control system performance. *Oil & Gas Journal*, 104(38), 2006.

[314] M.N.K. Sikder and F.A. Batarseh. Outlier detection using ai: a survey. In F.A. Batarseh and L.J. Freeman, editors, *AI Assurance*, pages 231–291. Academic Press, London, UK, New York, NY, 2023.

[315] G. P. Sillitto. Derivation of approximants to the inverse distribution function of a continuous univariate population from the order statistics of a sample. *Biometrika*, 56(3):641–650, 1969.

[316] T. Simková. Statistical inference based on L-moments. *Statistika : Statistics and Economy Journal*, 97:44–58, 03 2017.

[317] P. Singh, P. Verma, D. Perrotti, and K. K. Srivastava. *Environmental Sustainability and Economy.* Elsevier, Amsterdam, The Netherlands, 2021.

[318] S. Skogestad. Advanced control using decomposition and simple elements. *Annual Reviews in Control*, 56:100903, 2023.

[319] D. Sornette. Dragon-kings, black swans and the prediction of crises. *International Journal of Terraspace Science and Engineering*, 2(1):1–18, 2009.

[320] T. Spinner, B. Srinivasan, and R. Rengaswamy. Data-based automated diagnosis and iterative retuning of proportional-integral (PI) controllers. *Control Engineering Practice*, 29:23–41, 2014.

[321] B. Srinivasan, T. Spinner, and R. Rengaswamy. Control loop performance assessment using detrended fluctuation analysis (DFA). *Automatica*, 48(7):1359–1363, 2012.

[322] K.D. Starr, H. Petersen, and M. Bauer. Control loop performance monitoring – ABB's experience over two decades. *IFAC-PapersOnLine*, 49(7):526–532, 2016. 11th IFAC Symposium on Dynamics and Control of Process Systems Including Biosystems DYCOPS-CAB 2016.

[323] N.N. R. Ranga Suri, Narasimha Murty M, and G. Athithan. *Outlier Detection: Techniques and Applications. A Data Mining Perspective.* Intelligent Systems Reference Library. Springer Nature Switzerland AG, Cham, Switzerland, 2019.

[324] N.N. Taleb. *Fooled by Randomness: The Hidden Role of Chance in Life and in the Markets.* Penguin Books, New York, NY, 2001.

[325] N.N. Taleb. *Statistical Consequences of Fat Tails: Real World Preasymptotics, Epistemology, and Applications.* Technical Incerto Collection. STEM Academic Press, 2020.

[326] A.K. Tangirala. *Principles of System Identification: Theory and Practice (1st ed.).* CRC Press, Boca Raton, FL, 2015.

[327] M.S. Taqqu and V. Teverovsky. On estimating the intensity of long-range dependence in finite and infinite variance time series. In *A Practical Guide To Heavy Tails: Statistical Techniques and Applications*, pages 177–217, 1996.

[328] M.S. Taqqu, V. Teverovsky, and W. Willinger. Estimators for long-range dependence: an empirical study. *Fractals*, 03(04):785–798, 1995.

[329] M.S. Tavazoei. Notes on integral performance indices in fractional-order control systems. *Journal of Process Control*, 20(3):285–291, 2010.

[330] M. Templ, J. Gussenbauer, and P. Filzmoser. Evaluation of robust outlier detection methods for zero-inflated complex data. *Journal of Applied Statistics*, pages 1–24, 2019.

[331] V. Teverovsky and M.S. Taqqu. Testing for long-range dependence in the presence of shifting means or a slowly declining trend using a variance type estimator. *The Journal of Time Series Analysis*, 18:279–304, 1997.

[332] R. Thompson. A note on restricted maximum likelihood estimation with an alternative outlier model. *Journal of the Royal Statistical Society: Series B (Methodological)*, 47(1):53–55, 1985.

[333] N.F. Thornhill and T. Hägglund. Detection and diagnosis of oscillation in control loops. *Control Engineering Practice*, 5(10):1343–1354, 1997.

[334] S. Thudumu, P. Branch, J. Jin, and J. Singh. A comprehensive survey of anomaly detection techniques for high dimensional big data. *Journal of Big Data*, 7(42), 2020.

[335] G. L. Tietjen and R. H. Moore. Some grubbs-type statistics for the detection of several outliers. *Technometrics*, 14(3):583–597, 1972.

[336] D. Valério and J. S. da Costa. *An Introduction to Fractional Control.* Control, Robotics and Sensors. Institution of Engineering and Technology, Stevenage, UK, 2012.

[337] E. Vargo, R. Pasupathy, and L. Leemis. Moment-ratio diagrams for univariate distributions. *Journal of Quality Technology*, 42(3):1–11, 2010.

[338] R. Vincente, M. Wibral, M. Lindner, and G. Pipa. Transfer entropy – a model-free measure of effective connectivity for the neurosciences. *Journal of Computational Neuroscience*, 30:45–67, 2011.

[339] A. Visioli. Method for proportional-integral controller tuning assessment. *Industrial & Engineering Chemistry Research*, 45(8):2741–2747, 2006.

[340] R.M. Vogel and N.M. Fennessey. L moment diagrams should replace product moment diagrams. *Water Resources Research*, 29(6):1745–1752, 1993.

[341] S. von Manzoor. *New Techniques of Detection of Statistical Outliers.* LAP Lambert Academic Publishing, London, UK, 2013.

[342] B. Žmuk. Speeding problem detection in business surveys: benefits of statistical outlier detection methods. *Croatian Operational Research Review*, 8(1):33–59, 2017.

[343] H. Wainer. Robust statistics: A survey and some prescriptions. *Journal of Educational Statistics*, 1(4):285–312, 1976.

[344] X. Wan, W. Wang, J. Liu, and T. Tong. Optimally estimating the sample mean from the sample size, median, mid-range, and/or mid-quartile range. *BMC Medical Research Methodology*, 14:135, 2014.

[345] H. Wang, M. J. Bah, and M. Hammad. Progress in outlier detection techniques: A survey. *IEEE Access*, 7:107964–108000, 2019.

[346] Q. J. Wang. LH moments for statistical analysis of extreme events. *Water Resources Research*, 33(12):2841–2848, 1997.

[347] Q.J. Wang. Estimation of the GEV distribution from censored samples by method of partial probability weighted moments. *Journal of Hydrology*, 120(1):103–114, 1990.

[348] W. Wei and H. Zhuo. Research of performance assessment and monitoring for multivariate model predictive control system. In *2009 4th International Conference on Computer Science Education*, pages 509–514, 2009.

[349] E. Weiszfeld. Sur le point pour lequel la somme des distances de n points donnes est minimum. *Tohoku Mathematical Journal*, pages 355–386, 1937.

[350] R. Weron. Estimating long range dependence: finite sample properties and confidence intervals. *Physica A*, 312:285–299, 2002.

[351] D.L. Whaley. *The Interquartile Range: Theory and Estimation.* PhD thesis, Faculty of the Department of Mathematics, East Tennessee State University, 8 2005. Electronic Theses and Dissertations. Paper 1030.

[352] H.P. Whitaker, J. Yamron, and A. Kezer. *Design of Model Reference Adaptive Control Systems for Aircraft.* Report Massachusetts Institute of Technology Instrumentation Laboratory. M.I.T. Instrumentation Laboratory, Cambridge, MA, 1958.

[353] N. Wiener. The theory of prediction. In E.F. Beckenbach, editor, *Modern mathematics for engineers*, pages 3269–3274. McGraw-Hill, New York, NY, 1956.

[354] C. J. Wild and G. A. F. Seber. *Nonlinear Regression.* Wiley Series in Probability and Statistics. John Wiley & Sons, Hoboken, NJ, 2003.

[355] J. Wolberg. Kernel regression. In *Data Analysis Using the Method of Least Squares: Extracting the Most Information from Experiments*, pages 203–238. Springer, Berlin, Heidelberg, Germany, 2006.

[356] Y. Wu and J. Li. Hurst parameter estimation method based on haar wavelet and maximum likelihood estimation. *Journal of Huazhong Normal University*, 52(6):763–775, 2013.

[357] Z. Wu and N. E. Huang. Ensemble empirical mode decomposition: a noise-assisted data analysis method. *Advances in adaptive data analysis*, 1(01):1–41, 2009.

[358] C. Xia, J. Howell, and N. F. Thornhill. Detecting and isolating multiple plant-wide oscillations via spectral independent component analysis. *Automatica*, 41(12):2067–2075, 2005.

[359] H. Yamazaki and R. Lueck. Why oceanic dissipation rates are not lognormal? *Journal of Physical Oceanography*, 20(12):1907–1918, 1990.

[360] F. Yang, T. Chen, S.L. Shah, and P. Duan. *Capturing Connectivity and Causality in Complex Industrial Processes.* Springer Briefs in Applied Sciences and Technology. Springer, Cham, Heidelberg, New York, Dordrecht, London, 2014.

[361] F. Yang and D. Xiao. Progress in root cause and fault propagation analysis of large-scale industrial processes. *Journal of Control Science and Engineering*, 2012(1):478373, 2012.

[362] M. Yağci and Y. Arkun. An integrated application of control performance assessment and root cause analysis in refinery control loops. *IFAC-PapersOnLine*, 53(2):11650–11655, 2020. 21st IFAC World Congress.

[363] H. You, J. Zhou, H. Zhu, and D. Li. Performance assessment based on minimum entropy of feedback control loops. In *2017 6th Data Driven Control and Learning Systems (DDCLS)*, pages 593–598, 2017.

[364] J. Yu and S. J. Qin. Statistical mimo controller performance monitoring. part i: Data-driven covariance benchmark. *Journal of Process Control*, 18(3):277–296, 2008.

[365] Z. Yu and J. Wang. Performance assessment of static lead-lag feedforward controllers for disturbance rejection in PID control loops. *ISA Transactions*, 64:67–76, 2016.

[366] Z. Yu, J. Wang, B. Huang, J. Li, and Z. Bi. Design and performance assessment of setpoint feedforward controllers to break tradeoffs in univariate control loops. *IFAC Proceedings Volumes*, 47(3):5740–5745, 2014. 19th IFAC World Congress.

[367] H. Yue and H. Wang. Minimum entropy control of closed-loop tracking errors for dynamic stochastic systems. *IEEE Transactions on Automatic Control*, 48(1):118–122, 2003.

[368] J. Zhang, M. Jiang, and J. Chen. Minimum entropy-based performance assessment of feedback control loops subjected to non-Gaussian disturbances. *Journal of Process Control*, 24(11):1660–1670, 2015.

[369] J. Zhang, L. Zhang, J. Chen, J. Xu, and K. Li. Performance assessment of cascade control loops with non-Gaussian disturbances using entropy information. *Chemical Engineering Research & Design*, 104:68–80, 2015.

[370] Z. Zhang. Performance assessment for the feedforward-cascade drum water level control system. *International Journal of Control and Automation*, 9(12):96–98, 2016.

[371] Y.M. Zhao, W.F. Xie, and X.W. Tu. Performance-based parameter tuning method of model-driven PID control systems. *ISA Transactions*, 51(3):393–399, 2012.

[372] B. Zheng. *Analysis and auto-tuning of supply air temperature PI control in hot water heating systems.* PhD thesis, Dissertation of University of Nebraska, 2007.

[373] L. Zhong. Defect distribution model validation and effective process control. In *Proceedings of the SPIE*, volume 5041, pages 5041: 41–8, 2003.

[374] J. G. Ziegler and N. B. Nichols. Optimum settings for automatic controllers. *Transactions of the American Society of Mechanical Engineers*, 64:759–768, 1942.

[375] A. Zimek and P. Filzmoser. There and back again: Outlier detection between statistical reasoning and data mining algorithms. *Wiley Interdisciplinary Reviews: Data Mining and Knowledge Discovery*, 8(6), 11 2018.

[376] K. Życzkowski. Rényi extrapolation of shannon entropy. *Open Systems & Information Dynamics*, 10:297–310, 01 2003.

Index

Printed in the United States
by Baker & Taylor Publisher Services